H. Sizun

Radio Wave Propagation for Telecommunication Applications

H. Sizun

Radio Wave Propagation for Telecommunication Applications

With 161 Figures

 Springer

Hervé Sizun
France Télécom
Recherche et Développement
avenue des usines 6
90007 Belfort
France

Translator: Pierre de Fornel.
Originally published in French by Springer-Verlag, Paris 2003

Cataloging-in-Publication Data applied for
Bibliographic information published by Die Deutsche Bibliothek
Die Deutsche Bibliothek lists this publication in the Deutsche Nationalbibliografie;
detailed bibliographic data is available in the Internet at <http://dnb.de>

ISBN 3-540-40758-8 Springer Berlin Heidelberg New York

Springer is a part of Springer Science+Business Media

springeronline.com

© Springer-Verlag Berlin Heidelberg 2005
Printed in Germany

Typesetting: Digital data supplied by translator
Cover-Design: Design & Production, Heidelberg
Printed on acid-free paper   62/3020 hu 5 4 3 2 1 0

Foreword

The first radio links, wireless telegraphy, were established at the beginnings of the twentieth century by Marconi, who drew upon the theory developed by Maxwell and upon the experimental researches conducted by Hertz. In France, such renown scientists, mathematicians, physicists and experimenters as Poincaré, Blondel and the Général Ferrié played a crucial role in the development of radiocommunications, more particularly through theoretical and experimental researches which contributed to a better understanding of the different propagation media. Following the Second World War, the researchers and engineers of the newly created *Centre National d'Etudes des Télécommunications* (CNET), among whom may be mentioned Jean Voge, François du Castel, André Spizzichino or Lucien Boithias, made decisive contributions to the understanding of the propagation of radio waves, in particular in the context of their application to telecommunications. Although the CNET has now become *France Telecom Recherche & Développement,* the present book is in keeping with this approach, which has been going on for more than half a century. By providing the reader with some of the most recent researches in this field, Hervé Sizun offers here an essential complement to the work by Lucien Boithias *Radiowave Propagation*, first published in 1983 in the *Collection Technique et Scientifique des Télécommunications,* and published in an English version in 1987 by McGraw-Hill.

As a field of research, the propagation of radio waves remains intensively investigated, and the researches conducted in this context have led to a large number of results with applications not only in the field of radiocommunications, but also in such other fields as remote detection. While Maxwell's equations have of course remained unchanged, the ever increasing telecommunication needs require to develop a modelling of the propagation of radio waves under an ever wider range of conditions as regards such parameters as the environment, the frequency range or the bandwidth.

A first reason explaining the renewal of activity in the field of radio propagation is the notion of mobility. Telecommunication services should indeed be provided for the largest and most varied types of environments, at home, in workplaces and also increasingly during displacements. Although this can be to some extent achieved through the use of the routing capabilities of fixed networks, radio nevertheless possesses a considerable intrinsic advantage in this context. However, since non line-of-sight situations tend more than often to become the rule, this very mobility induces a number of problems in terms of propagation. While the physical characteristics of the atmosphere still exert their influence, these characteristics nonetheless tend to become secondary compared to the influence of the configuration of human land-use.

A second reason which may be offered for the renewal of interest in the propagation of radio waves is the tendency towards the use of higher frequency ranges and correlatively towards the increase of the bandwidth of the transmitted signals. During the first half of the twentieth century, the waves used in radio links had a wavelength larger than one meter. In terms of bandwidth, such waves simply cannot offer the capacities of several tens and hundreds of megahertz, nor their equivalent information rates, that present day telecommunication needs call for. As a result, this has led to the implementation of decimetre, centimetre and millimetre-length waves.

The relation between the size of obstacles and the wavelength is a decisive parameter when constructing approximations of physical laws which are to be applied to the propagation of radio waves. It is therefore apparent that significant differences in the modelling of propagation phenomena will arise at this point. Finally, high rate digital transmissions are particularly sensitive to the frequency selectivity of the propagation channel and this broadband characterisation is essential for a better evaluation of the effects induced by propagation distortions on the quality of the numerical signal. The major topics in the study of propagation thus undergo evolutions which are driven by the changes in telecommunication needs. These evolutions are of course taken into account in the present book. For instance, the relative importance of propagation studies devoted to ionospheric links, to interurban links or to forward-scatter microwave links has somewhat declined in the last recent years compared to propagation studies devoted to mobiles in and outside buildings or to short-distance radio links operating in the millimetre band.

These developments are well reflected in this book, where the interactions between the theoretical understanding of the physical environment, the construction of approximations of electromagnetic laws adequate to specific conditions of propagation and experimental contributions are clearly shown. Special emphasis is also laid in this book on significant results which are relevant for the evaluation of the influence of propagation on numerical communications and on their quality.

The author has had the relatively rare opportunity of successively working over the course of his career on several of the main topics which are addressed in this book: first on the ionosphere, and later on point-to-point microwave links and mobile radio links. This has allowed him in this book to bring together different fields of study which too often are considered apart from one another. Further, the author was part in the conception and the realisation of several series of measurements whose results, after contributing to the enrichment of the experimental databases of the *International Telecommunication Union,* are presented here. This experience allows him to stand back and fully appreciate the actual significance of the different parameters involved in the propagation of radio waves. The practical character of the results presented in this book should also be emphasised, as well as the variety of models which are described here. As might be indeed stressed here, there does not, and will not, exist any universal model that could be applied to all the multiple problems that radiocommunications present us with.

From a didactic point of view, this book is organised into four main parts, consisting each of one or two chapters where the most significant results are presented, and of different appendices exploring in more depth certain crucial points. Chapters 2 and 3 are thus devoted to the characteristics of the physical environment, the atmosphere, which is involved in all the situations studied in this book, and to a presentation of the main results concerning the mechanisms of propagation, refraction, reflection or diffraction. These two chapters are supplemented by Appendices B, E, K and O, where the properties of hydrometeors and the calculation methods are examined in more detail. The second part of the book is devoted to the propagation of radio waves in ionised media: refractions and reflections in the ionosphere are indeed still being used for certain long-distance links, As a complement, Appendices A and E present a description of the solar activity and of the disturbances induced by this activity. The third part of the book, consisting of Chapters 5 and 6, supplemented by Appendices D, I and J, is devoted to an analysis of horizontal and oblique fixed links, primarily in direct line-of-sight propagation. Finally, the last part of the book addresses the characterisation of radio mobile propagation: the different applicable types of models are thus successively described in Chapter 7, while Appendices L, M and N address in more detail such subjects as geographical databases, the experimental measurement of the directions of arrival as well as some specific problems associated with their modelling.

By way of conclusion, this book which, as should be the case for any good book devoted to the propagation of radio waves, harmoniously combines a theoretical, a physical and an experimental approaches to this subject, should provide a large amount of practical information for designers of radiocommunication systems who more than often tend to view the propagation of radio waves as a source of multiple and varied distortions rather than as a fascinating field of studies.

Jean-Claude Bic

To my parents
To my wife Marie-Thérèse,
To my children Jean-Michel, Ronan and Guénaelle.

Acknowledgements

I would like first to express my most sincere gratitude to the Profs. Jean Mevel and Louis Bertel from the Radioelectricity Laboratory of the University of Rennes, who introduced me to research, and to Mr. Jacques Papet-Lépine, who welcomed me at France Telecom Research & Development, formerly known as the *Centre National d'Etudes des Télécommunications*, within the Ionospheric and Radioelectric Measurement Department.

I would also link to thank those under which I had the opportunity of conducting my researches, both in the field of ionospheric propagation (HF) in Lannion: Rudi Hanbaba, Patrick Lassudrie-Duchesne, and in the field of higher frequencies (VHF, UHF, SHF, EHF and even optical), located in Belfort: Monique Juy, Jean-Claude Bic, Jean Vuichard for their confidence, and who provided me with their encouragements, their assistance and their most helpful advice.

My deepest gratitude go to all of those whose researches have contributed to the work presented in this book. May I mention here more particularly:

My colleagues in Lannion : Roland Fleury, Yvon Le Roux, Jacky Meynard, Jean-Pierre Le Pape, etc.

My colleagues in Belfort: Jean-François Bourdeilles, Lionel Chaigneaud, Sabine Durieux, Valérie Guillet, Bertrand Guisnet, Philippe Laspougeas, Erwan Le Fur, Benoît Linot, Nadine Malhouroux, Sandrine Mourniac, Isabelle Siaud, etc. Laurent Blanchard, Stéphane Morucci and Olivier Veyrunes for their researches on long-distance propagation (mechanisms, field measurements and jamming), on vegetation attenuation and the influence of hydrometeors on the propagation of millimetre waves respectively.

My colleagues in Paris: Armand Lévy, Jean-Marc Raibaud, Arturo Ortega-Molina

The members of the *Centre d'Etude de l'Environnement Terrestre et Planétaire* (CETP), Jacques Lavergnat, Peter Gole, Monique Dechambre, for their researches on the cartography of the refraction index and the study of the influence of vegetation on propagation.

I apologise to all those who I certainly forget to mention here.

I am also indebted to M. Louis Martin, researcher at FTR&D, for his great experience and for his sensible advice on propagation issues.

I would like to address my congratulations to the members of the 'Radio Measurement and Experimentation' Laboratory at FTR&D for the realisation and the operation of the different radio links reported in this book in the best conditions that could have been hoped for: Jean-Yves Thiriet, Christophe Pradal, Philippe

Leclerc, Laurent Cartier, Gilles Cucherousset, Christian Degoulet, Isabelle Eimer and Dominique Tscheiller.

I would also like to thank Jean-Claude Imbeaux, in charge of the 'Antennas and Propagation' Department at FTR&D, and Frédérique de Fornel, from the University of Dijon, for their insightful remarks which contributed to the success of this book.

Finally, I would like to thank Pierre-Noël Favennec for his constant interest in this book, and for supervising its publication.

Contents

1 Introduction to the Propagation of Radio Waves

1.1 Introduction

One of the prerequisites for the development of telecommunication services is the understanding of the propagation of the waves, either acoustic, electromagnetic, radio or light waves, which are used for the transmission of information.

In this work, we shall limit ourselves to the study of radio waves: this term apply to the electromagnetic waves used in radio communications. Their frequency spectrum is very broad, and is divided into the following frequency bands : ELF waves (f < 3 kHz), VLF (3-30 kHz), LF waves (30-300 kHz), MF waves (300-3000 kHz), HF (3-30 MHz), VHF waves (30-300 MHz), UHF waves (300-3000 MHz), SHF waves (3-30 GHz), EHF waves (30-300 GHz) and sub-EHF waves (300-3000 GHz).

1.1.1 Propagation Mechanisms

Radio waves propagate in space according to several different physical mechanisms: free-space propagation or line-of-sight propagation, reflection, transmission, diffraction, scattering and wave guiding.

In free space a wave propagates without encountering any obstacle. The surface of the wave is the set of all points reached at a certain time after the moment of emission of the wave within a homogeneous medium. The attenuation in free space results from the scattering of energy which occurs as the wave propagates away from the transmitter. Free-space attenuation is a function of the distance and the frequency. The excess attenuation compared to free-space attenuation is defined as the difference between the path loss and free-space attenuation (atmospheric absorption, hydrometeor attenuation, building penetration loss, vegetation attenuation, attenuation due to diffraction, etc).

Reflection is the phenomenon whereby vibrations or waves are reflected at a surface according to Snell-Descartes law. This phenomenon occurs when a propagating wave impinges upon a surface with large dimensions compared to the wavelength. A distinction is commonly drawn between specular reflection, occurring in the presence of a perfectly plane, homogeneous surface, and diffuse reflection, which takes place in the presence of a rough surface, i.e. a surface presenting irregularities. The reflection coefficient is defined as the ratio between the received energy flux and the incident energy flux.

The phenomenon of transmission is the process whereby vibrations or waves propagate through a medium, for instance vacuum, the air or an obstacle, without a change of frequency according to Snell-Descartes law. Different types of transmission are usually distinguished. In regular transmission, the wave propagates through an object without diffusion. In diffuse transmission a phenomenon of diffusion occurs at a macroscopic scale independently of the refraction laws: the incident wave, while being transmitted, is scattered over a range of different angles. At last, mixed transmission is a partly regular and partly diffuse transmission. The transmission coefficient is defined as the ratio between the transmitted energy flux and the incident energy flux.

The building penetration loss is defined as the power attenuation that an electromagnetic wave undergoes as it propagates from outside a building towards one or several places inside this building. This parameter is determined from the comparison between the external field and the field present in different parts of the building where the receiver is located.

The phenomenon of diffraction occurs when waves impinge upon an obstacle or an aperture with large dimensions compared to the wavelength. This phenomenon is one of the most important factors in the propagation of radio waves, and results in disturbances affecting the propagation of these waves, for instance the bending of the path around obstacles or beam divergences.

Scattering is the phenomenon whereby the energy of an electromagnetic wave is distributed in a propagation medium along several directions after meeting a rough surface or heterogeneities with small dimensions compared to the wavelength.

The emitted energy can be channelled along a given direction using a waveguide. The propagation is achieved in this case by successive reflections of the waves off the surfaces of the waveguide. Certain environments, for instance canyon streets, corridors, or tunnels behave like waveguides with respect to the propagation of radio waves.

Interferences result from the superposition of oscillations or waves of same nature and equal frequency. These interferences can be either constructive when the different paths arrive in phase, leading to a signal reinforcement, or destructive, causing in this case a fading of the signal. It might be further noted that the mobile itself moves inside this figure of interferences, so that it propagates successively through luminous and dark regions (interference fringes), which results in a fading of the signal.

After a wave has been emitted, a wave may follow different paths between the emitter and the receiver. Depending on the nature of the obstacles that the waves encounter during their propagation, they are submitted to different phenomena of reflection, for instance at walls or at atmospheric or ionospheric layers, as well as different refraction, transmission, scattering or guiding phenomena. This results in a multitude of elementary paths. Each such path is characterised at receiver level by an attenuation, a delay and a specific phase difference. This mode of propagation is referred to as a multipath propagation. The different waves propagated along such multiple paths interfere at the reception.

For a radio transmission system, the propagation channel is defined as the physical ratio between signal $e(t)$ at the modulator output and signal $s(t)$ at the demodulator input. This concept takes into account the microwave channel as well as the emission and reception antennas. The propagation channel is generally described through its time-dependent impulse response $h(t, \tau)$, where τ is the delay and t represents the time dependence (and accordingly, since the vehicle is moving, the space dependence). The impulse response is a function of two variables, and expresses the three characteristics of the channel: its attenuation, its variability (t) and its selectivity (τ). The dual variables by τ and t Fourier transforms respectively are the frequency and the Doppler speed

1.1.2 Propagation Environment

The propagation environment is the geographical environment considered for the description of the propagation of waves between a transmitter and a receiver. This environment is generally described from the physical parameters of the medium, like the pressure, the temperature, the humidity or the refractive index and from geographical databases containing data concerning the topography, the vegetation and land use, the street axes and buildings. Geographical databases are constructed and maintained through a complex process combining satellite and aerial photographs or the maps of buildings with complex digitalisation processes. Depending on the physical base station antenna and on its geographical coverage area, these databases allow the definition of four different types of cell with respect to the propagation of radio waves: macrocell, small cell, microcell and picocell. The characteristics of each of these cells are dependent on the location, on the power and on the height of the base station antenna height as well as on the geographical environment.

The largest cell is the macrocell, with an activity radius of the order of several ten thousand kilometres. The environment of cells of this type is generally rural or mountainous, and the base station antenna is located at an elevated point: the typical height of the base station is 15 metres on a mast and 20 metres on top of a building. The geographical coverage area is predominantly rural and induces for a number of paths important delays (up to 30 μs). Further, due the limited number of diffusers and the distance between them, no significant fast fading occurs.

With the increase in users, in urban areas for the most part, the dimensions of the cells had to be decreased in order to reduce the reuse distance of the allocated frequencies. The most current urban cell is the small cell. Its coverage area has a radius lower than a few kilometres and the base station antenna is located above roof level, i.e. from 3 to 10 metres above ground level. The maximum duration of the impulse response is 10 μs.

In very dense urban areas, small cells are replaced by microcells with an activity radius of a few hundreds metres. The antennas are located below roof level, and the waves are guided by the streets. The maximum duration of the impulse response is 2 µs.

Picocells, with a radius equal to a few tens of metres, correspond to communications occurring in the building where the base station antenna is located. The maximum duration of the impulse response is 1 µs.

1.1.3 Antennas

Antennas are devices used either for the emission or for the reception of radio waves. An emitting antenna is a device supplied by an electric power generator at a certain frequency and radiating radio waves in space. These waves are generated through the emission of a variable current along the emitting antenna. A receiving antenna is a device whose function is to transmit to a receiver the effects of the radio waves emitted by a distant source. The interaction between an antenna and an electromagnetic wave produces on the antenna a variable current identical to the current that would have been necessary for this antenna to emit the wave.

The shapes and dimensions of the emitting and receiving antennas depend on their intended use as well as on the frequency. Among the different forms of antennas we may for instance mention linear, helical, reflector, loop, horn and patch antennas. The main characteristics of antennas are their radiation pattern, the power gain, the directivity, the beamwidth, the aperture, the polarisation, the current distribution along the antennas, their effective height and their impedance.

1.1.4. Selectivity

In the presence of significant time differences between the multiple paths, the transfer function is no longer constant over the entire width of the spectrum: in these conditions, the path loss is dependent on the frequency, and accordingly the propagation channel shall be described as being frequency selective. Different selectivity parameters are deduced from the average power delays profile: among these parameters we may mentioned here the mean delay, the root-mean square delay spread, the delay interval, the delay window and the correlation bandwidth.

The correlation bandwidth is defined as the frequency at which the autocorrelation function of the transfer function, i.e. the Fourier transform of the power of the impulse response, intersects with a given threshold value, which may be equal to 50 or 90 percent compared to the peak value.

In narrow band communication, the used frequency band is lower than the correlation bandwidth. A signal in narrow band is therefore characterised by a nearly constant amplitude within this frequency band. The propagation channel cannot be studied save through the consideration of the attenuation. In order to compensate for the possible increase of attenuation, one generally resorts in analog to the use of a power margin and in digital to a frequency hopping.

In broadband communication, the used frequency band is higher than the correlation bandwidth. The presence of multiple paths leads to the temporal spreading of the received signals, revealed by presence of power peaks in the impulse response, and to a major fading in frequency domain. The different spectral components of the emitted signal are not affected in the same way over the used frequency band. This phenomenon, associated with the temporal spreading of the signal, results in the appearance of inter-symbol interferences due to the superposition of the delayed preceding symbols on the last emitted symbol. The possibility that such inter-symbol interferences may occur imposes a higher limit to the bit data rate. This limit can be improved if an equaliser is used.

1.1.5 Propagation Modelling

The propagation of radio waves is described through the modelling of the different physical mechanisms (free-space attenuation, atmospheric attenuation, vegetation and hydrometeor attenuation, attenuation by diffraction, building penetration loss, etc). This modelling is necessary for the conception of telecommunication systems and, once they have been designed, for their actual field deployment.

In the first case propagation models are implemented in software in order to simulate the transmission chain: this process allows to identify and reproduce the relevant characteristics of the propagation channel and to evaluate systems in terms of quality and error rate. These models are based on the consideration of the impulse response and its evolution in space and time, and rely on generic or typical environments rather than on geographical databases.

In the second case propagation models are implemented in engineering tools for the prediction different parameters useful for the field deployment of systems, for the study of the radio coverage (selection of the emission sites, frequency allocation, powers evaluation, antenna gains, polarisation) and for the definition of the interferences occurring between distant transmitters.

The analysis of propagation has its place in the study of the different types of links: ionospheric links, fixed links, point-to-point or microwave links, Earth-satellite links and mobile radio links. These different types of links will be successively considered in the course of this book.

1.2 Overview of the Book

This book devoted to the propagation of radio waves is aimed at complementing the excellent book written by L. Boithias *Radio Wave Propagation*. It results from researches conducted at France Telecom R&D by different researchers on the one hand on and from an extensive bibliographic compilation of studies in this field on the other hand. This book is organised in seven chapters and fifteen appendices.

The first and present chapter is an introduction to the propagation of radio waves, to the frequency spectrum, the different uses of models, the evolution of ideas and technologies which has led to the development of the present-day systems (radio relay systems, ionospheric links, transmission by geostationary or medium or low earth orbit satellites, mobile radio communication, etc). It defines the different notions associated with the propagation of radio waves: free-space attenuation, the different propagation mechanisms (reflection, transmission, diffraction, scattering, guiding), interferences, multipath propagation, propagation environment, coverage cells, narrow and broadband propagations, correlation bandwidth, antennas, etc.

Chapter 2 provides first a description of the structure and composition of the Earth's atmosphere. The different parameters used for its characterisation are then introduced, before considering the main weather phenomena which occur within the atmosphere.

The Earth's atmosphere is structurally divided into two main regions: the homosphere and the heterosphere. The homosphere is the layer of the atmosphere extending from the surface to the altitude of approximately 90 kilometres, while the homosphere is the region of the atmosphere extending beyond this altitude. The two main components of the homosphere, nitrogen and oxygen, are present in this layer in constant proportions. In contrast, light gases, such as nitrogen, hydrogen and helium are prevailing in the heterosphere.

The homosphere is itself subdivided into three layers differentiated by their temperature gradient with respect to the altitude: the troposphere, extending from the surface to the altitude of approximately 15 kilometres, the stratosphere, between 15 and 45 kilometres, and the mesosphere, from 45 to 80 kilometres.

The heterosphere is likewise subdivided into two layers: the thermosphere extending at altitudes between 80 and 1000 kilometres, and the exosphere, extending beyond 1000 kilometres.

The ionosphere located within thermosphere is characterised by the presence of ions and electrons resulting from an ionisation of the different atmospheric constituents by the solar radiation.

The constituents of the atmosphere vary with the altitude. They are classified into major atmospheric constituents (N_2, O_2, Ar, CO_2, He, Kr, CH_4, H_2), minor atmospheric constituents (water vapour primarily) and aerosols, which are fine particles suspended in the atmosphere.

The Earth's atmosphere can be characterised through different atmospheric parameters, such as the pressure, the temperature, the humidity, the dew-point, the water vapour partial pressure, the saturation vapour pressure and the water vapour density. Each of these parameters will be defined within this chapter.

Several different weather phenomena take place in the atmosphere. After defining the various processes involved in these phenomena, like evaporation, condensation, solidification, fusion, superfusion or reverse sublimation, such weather phenomena as the wind, turbulence, advection, subsidence, meteors, fog, mist, precipitations, clouds and auroras are addressed.

Chapter 3 is entirely devoted to electromagnetic waves. These waves are the propagation mode of electromagnetic disturbances characterised by a simultaneous variation of an electric field and a magnetic field. Electromagnetic waves are transverse waves propagating at the speed of light in vacuum. Their spectrum or frequency range of these waves is very broad, and they include radio waves, used for radio communications and where the wavelength λ ranges from a few tenths of millimetres to a few tens of thousands kilometres, infrared rays (0.8 $\mu m < \lambda < 300$ μm), visible rays (0.4 $\mu m < \lambda < 0.8 \mu m$), ultraviolet rays (0.001 $\mu m < \lambda < 0.4$ μm), X-rays (0.1 Å $< \lambda < 100$ Å) and gamma rays ($\lambda < 0.1$Å).

The first section of this chapter is devoted to the fundamental properties of electromagnetic waves and approaches the following topics : the electromagnetic parameters, the electric and magnetic fields, the electric and magnetic induction, Maxwell's equations, the propagation velocity of a wave, the wavelength, the frequency, the characteristic impedance of the propagation medium, the Poynting vector, the refractive index, polarisation, cross-polarisation, depolarisation, the cross-polarisation discrimination or decoupling ratio and the cross-polarisation isolation.

Different mechanisms of propagation, which depend on the environment where the wave propagates, are considered in the second section: reflection, either specular or diffuse, transmission, diffraction, scattering and guiding, as well as the models associated with these phenomena.

The third section of this chapter defines the different parameters used for the study of propagation both in narrowband, where the signal has a nearly constant amplitude in the used frequency band, and in broadband, where due to the presence of multiple paths the signal has no longer a constant amplitude but is affected by major fading effects in the frequency domain. The following topics are approached in this section : the different paths that a wave may follow between an emitter and a receiver (line-of-sight, reflected, transmitted, diffracted and scattered paths), the study of Fresnel ellipsoids for the characterisation of radio visibility, the main characteristics of the signal (attenuation, variability, selectivity), the different representations of the channel (time - delay representation, delay - Doppler shift representation, Doppler shift - frequency representation, temporal attenuation representation) and the broadband representation of the radio channel (average delay profile, average delay, delay spread, delay interval, delay window, correlation bandwidth).

The next four chapters consider the propagation of radio waves for different types of links: ionospheric links, terrestrial fixed links, Earth-satellite links and radio mobile links.

Chapter 4 is devoted to the propagation of waves over long distances by ionospheric refraction and reflection in the high frequency range (3-30 MHz).

The study of ionospheric refraction is developed on the basis of the magneto-ionic theory developed by Appelton and Hartree: the equations for the refractive index and the polarisation ratio are presented in this context. Approximations are then introduced for different conditions of propagation, in quasi longitudinal propagation with respect to the magnetic field, in quasi transverse propagation and in the general case. A definition of the different parameters involved in the propagation of radio waves in the ionosphere, like the absorption, the phase velocity, the group velocity, the ordinary and extraordinary critical frequencies or the real and virtual reflection heights, is provided.

When a radio wave penetrates inside the ionosphere, the presence in this medium of electrified particles combined with the influence of the magnetic field, causes different modifications affecting the essential characteristics of the wave, its trajectory, its frequency and the absorption it is submitted to. The problems associated with the calculation of trajectories are also considered in this chapter.

Finally, since the conditions of propagation in the ionosphere are greatly variable in time and space, it is essential, in order to maintain an ionospheric link under satisfactory conditions, to develop forecasts predicting the usable frequency band. Different methods of forecast have therefore been developed depending on the duration chosen for the forecast. Forecasts are classified as short-term, medium-term or long-term forecasts depending on whether they are established over twenty-four hours, over a week or over more than a month. These different types of forecasts will be described in this chapter.

Although the last decades have seen a remarkable development of satellite transmissions, ionospheric links still play an important part in radio communications, where they find privileged fields of application in maritime communications and broadcasting services.

Chapter 5 is devoted to the propagation mechanisms involved in point-to-point fixed links, and more specifically in terrestrial fixed links, also known as radio relay systems. A definition of the different radio atmospheric parameters, such as the refractive index, the modified refractive index, the standard atmosphere for refraction or the variability of the refractive index is first provided, before considering such topics as the phenomenon of refraction, the trajectory of radio waves, the radius of curvature of the paths and the effects of the variations of the refractivity index in the subrefraction and superrefraction cases.

Experimental results of interest for the study of the refractive index, of the refractivity gradient and of the cumulative distribution of the refractivity gradient are then described. A modelling of the cumulative distribution of the refractivity gradient, based on the evaluation of the median and on the percentage of time at which the gradient is lower or higher than median, is also proposed.

The main propagation mechanisms are described in the course of this chapter: line-of-sight propagation, duct propagation, reflection at elevated layers, diffraction and tropospheric scatter. The fluctuations of the scattered field, the scatter geometry and the models for the path loss due to tropospheric scatter are more particularly examined in this context.

The results of a forward-scatter terrestrial link experiment are then presented: these results concern such phenomena as tropospheric scatter, tropospheric radio duct, reflection at elevated layers, spherical diffraction, spherical superrefraction and the dynamic and statistical characteristics of the path loss. A comparison with a forward-scatter maritime link is also undertaken.

The influence of rain on a horizontal 800 metre link at four different frequencies (30, 50, 60 and 94 GHz) is also examined: after describing the experimental setup, theoretical and statistics results are presented concerning the precipitation rates, the path loss, the frequency scaling, the dynamic characteristics of rain attenuation and the distributions for the rain intensity and for the attenuation due to rain.

Chapter 6 addresses the propagation of radio waves between the Earth and a satellite. In this context, the atmospheric paths and the influence of the ground can be neglected: the study of the propagation of radio waves between the Earth and a satellite is therefore reduced, besides free-space attenuation, to the study of phenomena related to the refractive indices inside the troposphere and the ionosphere, to the absorption due to atmospheric gases, in particular oxygen and water vapour, and to the attenuation induced by hydrometeors like clouds, rain, fog, snow or ice.

The following topics are considered in the course of this chapter: free-space attenuation, the different phenomena associated with the refractive indices in the troposphere and in the ionosphere, in the case of either an absorbing or a non-absorbing medium and either in the presence or the absence of a magnetic field, refraction, delay and propagation time distortion, directions of arrival and scintillations. The different types of attenuation affecting the propagation of radio waves in this context are also examined: atmospheric attenuation, attenuation due to hydrometeors like clouds, fog or rain, attenuation due to cross-polarisation, building penetration loss, attenuation due to vegetation.

Chapter 7 is devoted to mobile radio links. Compared with the three previously mentioned links (ionospheric, terrestrial and Earth-satellite links), mobile radio links are based on the concept of a non line-of-sight propagation between the transmitter, i.e. the base station and a mobile receiver. The propagation of radio waves is generally achieved through a variety of propagation mechanisms: reflection, for instance at mountainsides or at walls, diffraction at edges, either horizontal (roofs) or vertical (corners of buildings), scattering by vegetation or guiding in street canyons. This results in the existence of a multitude of elementary paths at the reception, each such path being characterised by an attenuation, a delay and a phase difference leading to constructive or destructive interferences.

In this chapter, the modelling of the propagation in different environments (rural, suburban and urban) will be more particularly emphasised: the different types of theoretical, empirical, statistical and semi-empirical models are presented, as well as the different uses of these models. The following models are then considered: macrocell models used in rural or mountainous environments, microcell models, small cell models, ray launching models, building penetration loss models, indoor propagation models, broadband models, for instance path models and representation models (deterministic propagation models, either with or without frequency hopping), ray tracing models and geometrical models.

The use of simulation software for broadband models is also approached in this chapter. These software tools can be used for proceeding to the evaluation of the quality of service (QoS) of a digital transmission where the distortions induced by propagation play an essential part.

In addition to the chapters described above, this book includes fifteen appendices aimed at addressing in more depth certain specific topics.

Appendix A provides a detailed description of the Sun and of the solar activity. The atmosphere of the Sun is divided into three main layers: the photosphere, which delimits its visible contour, the chromosphere and the corona. The sunspots present in the photosphere are associated with strong magnetic fields and therefore play an important part in solar activity and in the disturbances affecting the relations between the Earth and the Sun. The solar wind plays an important part in the configuration of the magnetosphere, i.e. the region of the circumterrestrial space subjected to the influence of the Earth's magnetic field. The solar activity is commonly characterised either by the Wolf number or by the radio flux at the 10.7 centimetre wavelength (2800 MHz).

The microphysical properties of hydrometeors, like rain, drizzle, snow, hail and fog, are addressed in Appendix B. For each of these hydrometeors, different models are presented for the determination of some of their characteristics, like their density, form, size or fall speed.

Appendix C is entirely devoted to the frequency spectrum and describes successively the different frequency bands are: for each band, the atmospheric and terrestrial influence, as well as the system considerations and the associated services are indicated (Hall 1989). The frequency allocation for UMTS by the IMT-2000 is then given for the different geographical areas (ITU, Europe, United States and Japan), before listing the frequency bands used in satellite communications. At last, additional information concerning the P, L, S, X, K, Q, V and W bands is provided.

The phenomenon of cross-polarisation induced by the atmosphere will be considered in Appendix D. Indeed, while orthogonal polarisations are used in order to increase the line capacity of a given link without increasing the bandwidth, the presence for instance of asymmetrical raindrops or ice crystals in the atmosphere where the waves propagate causes a part of the energy emitted with a given polarisation to become orthogonally polarised, thereby causing interferences between the two communication channels. Different parameters, like the cross-polarisation discrimination and the cross-polarisation isolation are considered for the characterisation of this phenomenon. Special attention is given to the models which have been developed for the determination of the cross-polarisation discrimination due to rain, to snow and in clear atmosphere respectively.

A description of Fresnel equations used in the evaluation of the reflection and transmission coefficients, either simple or multiple, is provided in Appendix E for different types of polarisations: horizontal, vertical and unspecified polarisations.

Appendix F is aimed at complementing Chapter 4 on the propagation of electromagnetic waves in the ionosphere. Different types of disturbances observed within the Earth's atmosphere after a solar flare are described in this appendix. These disturbances may be either radio electric like sudden ionospheric disturbances or polar cap absorption events, geomagnetic like magnetic storms, ionospheric like ionospheric storms or atmospheric like polar auroras.

Appendix G is devoted to the sounding methods of the ionosphere used for determining the characteristics of the ionospheric propagation medium, like for instance the critical frequencies and the heights of the different layers, the Doppler frequencies, the diffusion function or the angles of arrival. These methods include bottomside vertical or oblique soundings of the ionosphere, topside vertical soundings of the ionosphere, backscatter soundings and incoherent scatter soundings. A description of riometers as well as of low frequency and very low frequency receivers is also presented in this appendix.

Appendix H provides a brief description of the terrestrial magnetic field. The different magnetic indices K, K_p, A_p, A_a, Dst and AE used for characterising the terrestrial magnetic field are then defined.

The attenuation due to rain is one of the most important factors to consider for the design of telecommunication systems, and more particularly for the design of satellite telecommunication systems. This subject is considered in Appendix I, and different statistical models of rain attenuation are presented.

Different models that have been developed in order to determine the attenuation due to vegetation are then described in Appendix J. These models include the exponential decay model, the modified exponential decay model, the Rice model, the ITU-R model, the Al Nuaimi Hammoudeh model and the Stephens model, as well as radiative transfer models like the MIMICS model or the Karam-Fung model. The results of two experiments conducted in the UHF band are presented. In the first experiment, a 2270 metre fixed link with a 160 metre length inside the vegetation, was considered in order to investigate the daily and seasonal influence of vegetation depending on the meteorological parameters. In the second experiment, mobile links were set up in a wooded area along a 52 kilometre long route in order to investigate the effects of the penetration distance inside the foliage, depending on the period of the day and the season, and for vegetation depths ranging from a few meters and 6 kilometers.

Appendix K is devoted to a survey of different diffraction models, either for the diffraction by a single or multiples knife edges, or for the diffraction by a rounded obstacle. The Millington method, the Vogler method, the Epstein-Peterson method, the Shibuya method, the Deygout method and the Giovanelli method for the calculation of the attenuation due to diffraction by multiple knife edges are introduced. The Wait method and the ITU-R method for the diffraction by a rounded obstacle are then described.

The measurements of the field strength, of the impulse response and of the directions of arrival for mobile radio links are addressed in Appendix L. These measurements are necessary for the development, the optimisation and the validation of propagation models, which are then implemented in engineering tools in order to define, design and set up communication systems. The main features of the propagation channel sounder AMERICC defined and developed at France Telecom R&D are then briefly described. This channel sounder can be used in particular for obtaining a precise characterisation of the radio propagation channel over a bandwidth ranging from 0 to 250 MHz around a carrier frequency fixed between 1.9 and 60 GHz.

The experimental determination and the mathematical modelling of directions of arrival are more particularly addressed in Appendix M. Different determination methods of the angles of arrival are presented in this appendix: linear methods like Fourier analysis or phase reconstruction, as well as non-linear or high-resolution methods, like the MUSIC method or the method based on the estimate of the maximum probability.

Appendix N is devoted more specifically to geographical databases. These databases allow the description of propagation environments. They generally include data relating to the topography, the vegetation and the land use, the streets axes and the buildings.

At last, different methods for the determination of the electromagnetic field after its interaction with a structure, a half-plane or a dihedron for instance, either metallic or dielectric, are presented in Appendix O. These methods include the geometric optical method, the geometrical theory of diffraction, the uniform theory of diffraction, the finite difference in time domain method, the moment method and parabolic equation methods.

2 The Earth's Atmosphere

2.1 Structure

The atmosphere of the Earth is the gaseous envelope surrounding it and interdependent with the different movements of the Earth. The atmosphere may be regarded as a series of concentric layers delimiting different regions, the two main such regions being the homosphere, which extends from the surface of the Earth up to approximately 80 or 90 kilometres, and the heterosphere, extending beyond these altitudes. The two main components of the homosphere, nitrogen and oxygen, are present in this layer in constant proportions, whereas light gases, such as nitrogen, hydrogen and helium are prevailing in the heterosphere.

2.1.1 The Homosphere

Three principal atmospheric layers are defined in the homosphere: the troposphere, the stratosphere and the mesosphere. As can be seen in Fig. 2.1, these layers are differentiated by their temperature gradient with respect to the altitude.

The Troposphere

The troposphere is the lowest layer of the atmosphere, and is characterised by the regular decrease of temperature as altitude rises, at the average rate of - 5 to - 6° C per kilometre. This is where most meteorological phenomena, including the formation of clouds, develop. The troposphere itself is subdivided into two layers: the turbulent layer and the free atmosphere. The turbulent layer extends from the surface up to the attitude of 1500 metres over plains, and to higher altitudes over high-relief areas. A number of important interactions, either mechanical or thermal, between the terrestrial surface and the atmosphere are taking place in the turbulent layer. Beyond this region, the effects of the temperature at the surface become negligible.

The altitude of the upper limit of the troposphere varies between 8 kilometres high at the poles and 18 kilometres high at the equator depending on the geographical latitude as well as on seasons and meteorological conditions. In this region, known as the tropopause, temperature varies from 190 K at the equator and

220 K at the poles. The tropopause slows down the ascending convection movements and constitutes an upper limit for clouds, with the exception however of cumulonimbus.

The Stratosphere

The region extending above the tropopause, known as the stratosphere, is a layer where the temperature rises with increasing temperature, first slowly up to 20 kilometres, then at a slower rate until it reaches a maximum value approaching 290 K at approximately the altitude of 50 kilometres. The region where temperature is the highest is referred to as the stratopause.

The increase in temperature observed in the stratosphere is caused by the absorption by this layer of a part of the ultraviolet radiation emitted by the Sun. The stratosphere forms therefore a regulating filter which creates the conditions for life to exist at the surface of the Earth.

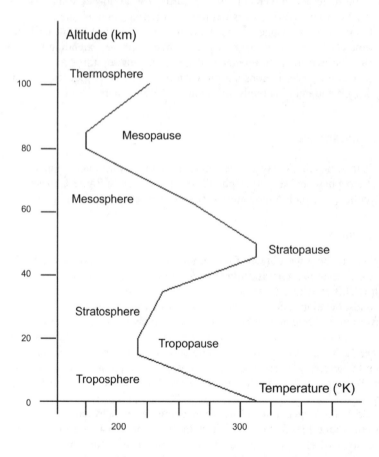

Fig. 2.1. The atmospheric temperature profile as a function of the altitude

The Mesosphere

The layer of the atmosphere lying above the stratopause is known as the mesosphere. In this region, the infrared emission by carbon dioxide, a component present in the atmosphere in very small quantities, is enough to cause a reduction of the temperature down to a minimum value between 150° and 210° K. The region where this minimum temperature occurs is called the mesopause, and extends at altitudes between 80 and 90 kilometres.

2.1.2 The Heterosphere

Two main atmospheric layers are defined in the heterosphere: the thermosphere and the exosphere.

The Thermosphere

The thermosphere is the layer of the atmosphere extending above the mesosphere. It is a region of increasing temperature with altitude. Here temperature rises with increasing altitude until the thermopause is reached at the altitude of approximately 1000 kilometres. At these altitudes temperature varies from 500 K during the night in periods of minimal solar activity to 1 750 K during the day in periods of maximal solar activity. For more detail the reader is referred to Appendix A devoted to solar activity.

The Exosphere

The outermost region of the heterosphere, known as the exosphere, extends at altitudes higher than 1000 kilometres. At distances equal to several times the terrestrial radius, the exosphere tends to merge with the end of the solar corona. Due to the reduced gravitational attraction exerted at these altitudes by the Earth, molecules may leak out of the atmosphere into outer space.

2.1.3 The Ionosphere

The existence of ionised regions within the thermosphere can be experimentally demonstrated. The term ionosphere will be employed to describe these regions of the upper atmosphere, where charges, either positive or negative, are present in quantities large enough to influence the trajectory of radio waves. The existence of these charges results from the ionisation by the solar rays of the components of the overall neutral atmosphere, and therefore from a photoelectric effect. This phenomenon was experimentally demonstrated around 1925 by different physicists, like Appleton, Barnett, Breit, Tuve or Marconi, in the wake of experiments conducted on the propagation of radio waves. The ionosphere extends over a few hundreds kilometres above the mesosphere and is divided by convention into three sublayers, the D, E and F layers. Fig. 2.2 provides a schematic representation of the structure of the ionosphere.

The D layer

The D layer extends at heights ranging approximately from 75 to 95 kilometres. In this region, ionisation is the main cause of the absorption of the high frequency radio waves reflected off higher layers. The electronic concentration in the D layer displays an important diurnal variation and a marked seasonal variation: it passes by a maximum (10^8 to 10^9 electrons/m^3) shortly after the local solar midday and its values of night are low. It reaches its maximum in summer, while in winter abnormally high electronic concentrations, due to the modification of the neutral atmosphere composition, can be observed.

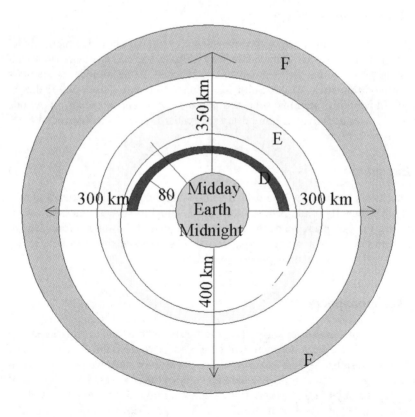

Fig. 2.2. Schematic representation of the ionosphere

The E layer

The E region, located between 95 and 150 kilometres approximately, consists of an E normal layer (E layer) and a sporadic E layer E (Es layer): the normal E layer is a regular layer, with an electronic concentration strongly dependent on the distance to the solar zenith and on the solar activity. It presents a daily maximum around midday and a seasonal maximum in summer.

The Es layer is an irregular, parcelled out layer which is only to a very limited extent directly influenced by the ionising radiation emitted by the Sun. This layer might be compared to plasma clouds in suspension within the E layer.

The F layer

The F layer extends at altitudes above approximately 150 kilometres. This region is sometimes subdivided in two layers, referred to as the F1 and F2 layers respectively. The distinction between these two layers no longer holds during the night.

The F1 layer is the region extending at altitudes ranging from 150 to 210 kilometres and presents a regular stratification at moderate latitudes. The maximum electronic concentration in this region is approximately equal to 2.10^{11} electrons/m^3.

The F2 layer is the highest and most ionised reflective layer. The electronic concentration in this layer, which varies between 5.10^{11} electrons/m^3 during the night and 20.10^{11} electrons/m^3 during the day, is heavily influenced by neutral winds, by diffusion and by different other dynamic effects.

The value of the electronic density in each of these layers depends primarily on the solar activity which, as is well known, follows cycles. The ionosphere is not a stable ionised medium, and it presents considerable variations that may attenuate and even possibly enhance the transmission of a signal propagating at a high frequency between the transmitter and the receiver. Disturbances may arise suddenly with high magnitudes, leading to the interruption of the communications in a given frequency band. These disturbances can be of very different nature. Among disturbances more specifically associated to solar activity, we may mention here sudden ionospheric disturbances, polar cap absorptions and ionospheric storms. In order to predict the conditions of propagation at high frequencies and thereby limiting the effects induced by such disturbances, forecasting services have been set up in a number of countries. The forecasts provided by theses services allow, for a given link, to plan and select antennas as well as adequate frequencies and exploitation schedules.

In the last decades, experiments concerned with the artificial modifications of the ionosphere have been conducted by geophysicists. These experiments are directed at:

- the modification of some parameter of the ionospheric plasma in order to obtain, through the measurement of this parameter, information on the local properties of the medium and their evolution,
- the excitation of plasma instabilities,

– the generation of electromagnetic waves in the range of the great wavelengths (between 10 and 10 000 kilometres).

A variety of different techniques are used in the context of these experiments. The most important techniques and their associated applications are the following ones:

– the injection of a large quantity of some chemically active gas or of a plasma cloud in the ionosphere. This technique allows to study the dispersion of the injected gases, the corresponding photochemistry and the plasma instabilities associated with high plasma density gradients,
– the emission of radiations at very low frequencies, either from the ground or from space, in order to excite the instabilities contained within the magnetospheric plasma. This technique is used for generating hydromagnetic emissions, for precipitating particles from the radiation belts into the ionosphere and for studying the coupling and the energy transfers occurring between different regions of the system formed by the magnetosphere and the ionosphere,
– the injection in the ionosphere of a beam of charged particles, in order to modify the characteristics of the ionosphere, create artificial auroras and investigate the interactions between the particles thus injected and the ambient plasma,
– the heating of the ionosphere using high-power transmitters, in order to modify some properties of the ionosphere and generate instabilities in the ionospheric plasma as well as electromagnetic waves in the range of large wavelengths.

Considered as a whole, the Earth's atmosphere presents a general circulation driven by the rotation of the Earth and by the differences on a global scale of temperature due to solar radiations. A number of phenomena of very different nature occur within the atmosphere: these phenomena may be meteorological, optical (reflection, refraction, etc), acoustical (propagation of sound waves), chemical, electric or magnetic. The atmosphere is also the propagation medium of several types of oscillatory or quasi-oscillatory and/or turbulent natural or artificial phenomena.

These phenomena results from the different forms of energy that may be present in the atmosphere and from the existence of physical mechanisms allowing energy transfers from one form to another (Cohn 1972). Example of such mechanisms are longitudinal compression waves, more commonly known as acoustic waves or sound waves, which result from a double energy transfer between the pressure and the speed through Newton's law of motion. Electromagnetic waves are generated through a double energy transfer between the electric and magnetic fields $\vec{E}$ and $\vec{H}$ as described by Maxwell's equations (see Tables 2.1 and 2.2).

Table 2.1. Forms of energy and waves generated by energy transfers between two forms of energy

Forms of energy		Generated Waves
Electric energy	Magnetic field energy	Electromagnetic waves
Electric energy	Kinetic energy	Electrostatic waves
Magnetic energy	Kinetic energy	Alfven waves
Potential energy	Kinetic energy	Sound waves

Table 2.2. Forms of energy and waves generated by energy transfers between three forms of energy

Accumulation mode of energy			Generated waves
Electric energy	Magnetic field energy	Kinetic energy	Ionic electromagnetic waves
Electric energy	Kinetic energy	Potential energy	Electro-hydrodynamic waves
Magnetic energy	Kinetic energy	Potential energy	Magneto-hydrodynamic waves

Although the troposphere occupies but a negligible part of the volume of the atmosphere, it contains the four fifths of the total mass of atoms and particles present in the atmosphere. It might also be noted that the elements which are at the origin of most atmospheric propagation phenomena, for instance transmission, thermal emission and atmospheric scattering, are for the larger part present below the stratopause. These phenomena originate at the level of the molecules, clouds, fogs, mists and aerosols composing the atmospheric medium.

In atmospheric media, and more particularly in the troposphere, the propagation of electromagnetic waves is strongly influenced by:

- the gas composition of the atmosphere,
- the presence of aerosols, that is, small particles of variable size (ranging from 0.01 to approximately 100 µm) in suspension in the air,
- hydrometeors such as rain, snow, hail,
- lithometeors such as dust, smoke or sand.

Further, due to the energy transfers occurring at the interface between the air and the ground (relief, microrelief, vegetation), the gradient of the refractive index of atmospheric media is also influenced by variations in the spatial and temporal meteorological parameters, for instance in the pressure, the temperature, the humidity or the visibility.

If these modifications arise at a small scale, the corresponding effects are turbulent phenomena. If they take place at a larger scale, the corresponding effects are atmospheric refraction phenomena of electromagnetic nature, leading for instance to focusing or defocusing phenomena.

2.2 Atmospheric Composition

The constituents of the atmosphere vary with altitude: gases become increasingly light and increasingly rare as altitude rises. They are generally classified into three different categories: components with fixed density or majority components, components with variable density or minority components and aerosols (Cojan 1990).

2.2.1 Major Atmospheric Constituents

The major constituents of the atmosphere have a quasi-uniform distribution until altitudes ranging from 15 to 20 kilometres. Among these constituents, the most important are the nitrogen (N_2): 78.095 percent of the total volume, oxygen (O_2): 20.93 percent, argon (Ar): 0.93 percent and carbon dioxide (CO_2): 0.03 percent.

2.2.2 Minor Atmospheric Constituents

To the major constituents of the atmosphere must be added constituents present in very small quantity like neon (Ne), helium (He), krypton (Kr), methane (CH_4) or hydrogen (H_2). The concentration in minor constituents depends on the geographical location as regards the latitude or the altitude, on the environment (continental or maritime) and on the weather conditions.

Water vapour is the main such constituent of the atmosphere. The concentration in water vapour of the atmosphere varies strongly in time and space, and depends on climatic and geographical parameters: while it may reach 2 percent of the composition of the air at the sea level in maritime environments, its presence is negligible beyond the tropopause, i.e. at altitudes higher than 20 kilometres.

Water in the atmosphere appears either in solid form, for instance snowflakes or ice crystals, in liquid form, for instance clouds, rain or fog, or in vapour form. These three phase forms induce different absorption or scattering effects on the propagation of radio waves depending on the frequency range (millimetre waves, optical waves).

The vapour content is determined from the atmospheric humidity and can be defined in three different ways:

– the absolute humidity, expressed in $g.m^{-3}$, determines the mass of water vapour per unit air volume,
– the relative humidity, expressed in percentage, can be defined as the ratio between the absolute humidity and the maximum quantity of vapour that could be contained in the air at the same temperature and at the same pressure,
– the number of millimetres of precipitable water w_0 per unit distance, usually per kilometre.

Another major variable component is ozone which is found in variable proportions depending on the season, the altitude and the latitude. Ozone is present in large quantities in the stratosphere, where it forms a layer absorbing the ultraviolet radiation emitted by the Sun. The presence of ozone in the atmosphere results from the dissociation of oxygen by solar radiation. The absorption by ozone of ultraviolet solar radiation also accounts for the higher temperatures in certain regions

2.2.3 Aerosols

Aerosols are extremely fine particles suspended in the atmosphere with a very low fall speed caused by gravity. Their size generally lies between 10^{-2} and 100 μm. Aerosols may be either liquid or solid, in the form for instance of microscopic dust or of salt crystals in maritime environments.

The presence of aerosols may induce severe disturbances on the propagation of optical and infrared waves, since their dimensions are very close to the wavelength at these frequencies. It is not the same in the range for instance of centimetre and millimetre waves, where the wavelength is much higher than the size of aerosols.

Aerosols influence the conditions of atmospheric transmission and scattering due to their geometrical sizes, their structures and their chemical nature. Aerosols are present in the troposphere, and they may also be found in the stratosphere in the form of dust of volcanic origin.

A variety of different of particles and materials can be observed in the vicinity of the terrestrial surface. Due to gravity, large-sized particles are gradually eliminated as the altitude increases. The nature of these particles primarily depends on the environment, for example whether it is a rural, urban or maritime environment. In rural media, aerosols are more particularly formed by dust lifted by the winds whereas in coastal regions they consist for the most part of salt water and liquid crystal droplets.

The size distribution of aerosols varies only to a limited extent between the vicinity of the surface and the upper troposphere. The difference rather lies in their concentration and in their average radius: being heavier, aerosols of larger dimensions, i.e. of radius larger than 0.2 μm tend to be mainly distributed near the surface. Large aerosols may nevertheless be present at much higher altitudes, due to the existence of convection phenomena and turbulent flows.

Depending on their density, aerosols are classified either as mists, with visibility higher than 1000 metres and consisting in the origin of microscopic dust (0.5 μm), or as fogs, with visibility lower than 1000 metres, which form when the mist particles turn into droplets or ice crystals. Clouds have essentially the same structure as fogs, the difference between fogs and clouds being only a difference in their relative altitudes.

2.3 Atmospheric Parameters

The Earth's atmosphere is characterised by a number of different parameters: the pressure, the temperature, the humidity, the direction and speed of the winds, precipitations, evaporation, radiation, sunshine duration, horizontal visibility, electronic density, etc.

2.3.1 Atmospheric Pressure

The atmospheric pressure is defined as the pressure, or force per unit area, exerted by the atmosphere on a surface by virtue of its weight. It is equivalent to the weight of a vertical column of air extending above a surface of unit area to the outer limit of the atmosphere.

Atmospheric pressure can be measured either using a direct reading barometer or a recording barometer called a barograph. In temperate latitudes, the values of the atmospheric pressure fluctuate between 950 and 1050 hPa. The atmospheric pressure decreases with increasing altitude.

2.3.2 Temperature

Variations in the atmospheric conditions can be subjectively described in terms of relative concepts of cold or heat. The objective quantity corresponding to these concepts is the temperature, which is measured precisely with the use of thermometers. Temperature is measured either in degrees C (t) or in K (T). The temperature T (K) is connected to the temperature t (°C) by the following formula:

$$T(K) = t(°C) + 273.15 \qquad (2.1)$$

The propagation of heat occurs through a variety of physical processes: conduction, advection or horizontal movement, convection or ascending vertical movement, subsidence or downward vertical movement, turbulence and radiation.

2.3.3 Relative Humidity

The relative humidity of the air, expressed as a percentage H, is defined as the ratio between the actual quantity of vapour contained in the air and the maximum quantity of vapour that the air could contain at the same temperature. When the relative humidity of the air reaches 100 percent, air is said to be saturated: water vapour turns into liquid state and condenses in the form of fine droplets. This phenomenon occurs for instance in fogs and clouds.

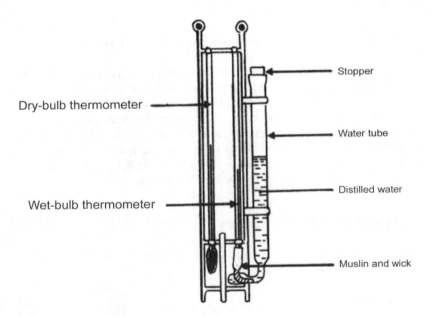

Fig. 2.3. Representation of a psychrometer

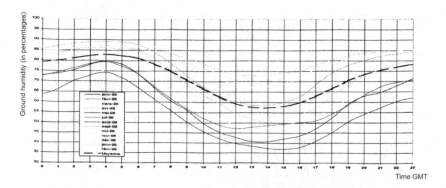

Fig. 2.4. Variation of the average daily humidity observed at ground level from January 1998 to February 1999 in Sélestat

The air humidity is commonly measured with a psychrometer or with a hygrometer. A psychrometer is a device consisting of two thermometers, as represented in Fig. 2.3:

- the first one, called the dry-bulb thermometer, measures the temperature t_s of the air.
- the second one, called the wet-bulb thermometer, measures the so-called wet-bulb temperature of the air. This thermometer is covered with a muslin sleeve which is kept constantly moist with distilled and clean water. Due to the latent heat of evaporation, the temperature of this thermometer decreases, and since evaporation depends on the humidity of the air, the drier the air is, the more the temperature decreases. The recorded temperature t_m is lower or equal to the temperature t_s of the air. The relative humidity of the air can therefore be determined from the difference between these two temperatures.

2.3.4 Water Vapour Partial Pressure

Several different methods can be found in the literature for the evaluation of this parameter, expressed in hPa.

ITU-R Relation

The equation suggested by the ITU-R defines the water vapour partial pressure as a function of the temperature t of the air expressed in degrees C (ITU-R P.453):

$$e_s = a \exp\left(\frac{bt}{t+c}\right) \tag{2.2}$$

The values of the coefficients a, b and c for liquid water are $a = 6.1121$, $b = 17.502$ and $c = 240.97$ respectively. These values are valid for temperatures between -20° and +50°C with a precision of +/- 0.20 percent.

Moupfouma Relation

The water vapour partial pressure can be expressed as a function of the wet temperature t_m as described in the following equation (Moupfouma 1985):

$$e_s = 1.33322\left(4.583 + 0.33329t_m + 0,01047t_m^2 + 2.264\times10^{-4}t_m^3 + 9.081\times10^{-7}t_m^4 + 3.975\times10^{-8}t_m^5\right) \quad (2.3)$$

Goff-Grath Relation

The water vapour partial pressure is expressed here by the following equation:

$$Log(e_s) = 23.8319 - \frac{2948.964}{T} - 5.028\log_{10}(T) - 29810.16e^{-aT} + 25.2193e^{\frac{-b}{T}} \quad (2.4)$$

where $a = 0.0699382$ and $b = 2999.924$.

2.3.5 Dew Point

This term refers to the temperature to which a given parcel of air must be cooled at constant pressure and constant water vapour content in order for saturation to occur, i.e. for the relative humidity to reach 100 percent.

2.3.6 Saturation Vapour Pressure

The saturation vapour pressure is the vapour pressure that the air would have if it were saturated. Different methods exist for the evaluation of this parameter, usually expressed in hPa. For instance, the saturation vapour pressure can be determined from the relative humidity H, expressed in percentages, and from the water vapour partial pressure, expressed in hPa, by using the following equation (ITU-R P.453):

$$e = \frac{He_s}{100} \quad (2.5)$$

A different expression for the saturation vapour pressure is the Assaman relation, which provides a definition of this parameter from the atmospheric pressure

P, from the water partial pressure and from the dry (t_s) and wet (t_m) temperatures of the air (Moupfouma 1985):

$$e = e_s - 0.00066 \times P \times (t_s - t_m) \times (1 + 0.00115 \times t_m)$$ (2.6)

2.3.7 Water Vapour Density

The water vapour density ρ of the air, also referred to as the absolute humidity and expressed in g/m^3, is a function of the temperature T in K and of the water vapour pressure expressed in hPa. The following equation yields it (ITU-R P.453):

$$\rho = \frac{216.7 \times e}{T}$$ (2.7)

2.4 Weather Phenomena

2.4.1 Sunshine Duration

The daily duration of sunshine during the day is commonly measured with a heliograph.

2.4.2 Solar Radiation

Solar radiation is the electromagnetic radiation emitted by the Sun. This energy increases the temperature at the surface of the Earth, which causes a heating of the atmosphere. The solar radiation is expressed in W/m^2 and is measured using a pyranometer, which is a device consisting of a detector in the form of a photodiode with an output in volts proportional to the received radiation. The sensitivity of this device is generally a cosine function of the incidence angle of the received radiation.

2.4.3 Evaporation

Evaporation is the process whereby a liquid changes into vapour. The troposphere contains water vapour, i.e. water present in gas state in variable proportions, and which results from two different evaporation processes: the physical evaporation occurring above the oceans, the seas, the lakes, the rivers or from wet grounds, and the physiological evaporation and transpiration of the vegetable cover. Evaporation depends on the temperature and the pressure. It might be noted here that the amount of energy necessary to the evaporation of water causes a decreases of the temperature, whereas the reverse process of condensation results in a release of energy accompanied by an increase of the temperature.

2.4.4 Condensation

The phenomenon of condensation is the process whereby water change phase state from a vapour to a liquid. It results either from an increase of the water vapour, due for instance to evaporation and causing a decrease of the value of the dew point, or from a cooling of the air: in this case, the relative humidity of the air increases, and if the temperature of the air further decreases beyond the dew point, a condensation of the excess water vapour will occur.

Condensation due to the increase of water vapour leads to the formation of fogs or clouds of limited extent.

Condensation due to the cooling of the air may be occur either by contact with a colder surface or by the mechanical uplift of the air. Two main types of cooling by surface contact exist: radiation cooling, resulting in the formation of dew, or of hoar frost if the temperature is below the dew point, and advection cooling, when moist air flows over colder surfaces.

Cooling by mechanical uplift occurs in three different forms: cooling at fronts, i.e. at the boundary of two air masses with different temperatures and humidity, cooling by orographic lift, where air is lifted by hills or mountains, and cooling by convection, associated with ascending air currents. Cooling by convection is the most important phenomenon, resulting in the formation of great cloud formations, and especially those associated with fronts.

Condensation is made possible by the presence in the atmosphere of condensation nucleus. Condensation nucleus are fine particles, either liquid or solid, in the form either of sea salt crystals, of mineral particles like dust, sand or smoke, or of electrical particles, upon which water condenses to form cloud droplets

2.4.5 Solidification and Melting

Solidification is the change of a substance from the liquid to the solid state. At temperatures lower than 0 °C, water is solidified in the form of snow or ice. The reverse phase transition of a liquid to a solid is called melting.

2.4.6 Superfusion

Water droplets in the atmosphere often remain in liquid state at temperatures significantly lower than 0°C, down to temperatures of several tens of degrees. A liquid in this state is said to be undercooled or superfused. This phenomenon can be observed more particularly in fog and clouds where water droplets superfused at temperatures down to - 40°C can be observed.

2.4.7 Reverse Sublimation

Water vapour may crystallise without any phase transition to a liquid state. Hoar frost is an example of such a phenomenon: it consists in deposit by reverse sublimation of ice crystals onto vegetation and other surface objects and occurs when the air is too cold for being saturated, while the temperature of the object falls to the dew point

2.4.8 Wind

Wind is the horizontal movement of air and results from the distribution of pressures over the Earth's surface. It is measured with an anemometer which delivers an instantaneous speed: the average speed of the wind can then be determined using an integrator. The speed of the wind can be expressed either in metres per second (m/s), in kilometres per hour (km/h) or in knots (Kt). A knot is a unit of speed equal to the velocity at which a one-minute meridian arc is travelled in one hour. The direction of the wind is expressed in degrees with respect to the geographical North and is indicated by a wind vane. The vertical component of the wind is referred to as turbulence.

2.4.9 Turbulence

Turbulence is an irregular motion of the air resulting from the formation of vertical currents. Turbulence is present in the atmosphere in the form of whirlwinds with variable dimensions, ranging from the molecular to the hemisphere scale. The rotation axes of whirlwinds are unspecified and can be vertical, horizontal or oblique. Different types of turbulence are to be distinguished: clear air turbulence, mechanical turbulence, convective turbulence, orographic turbulence, upslope turbulence, etc. An air mass is said to be stable if it tends to go back to its initial level after a weak disturbance. In the reverse case, if it tends to move away from its initial level, it is described as unstable.

Clear air turbulence. This term refers to the different forms of turbulence caused by the erratic wind shears occurring in cloudless air at altitudes ranging from 7000 to 12000 metres in the vicinity of convective clouds.

Mechanical Turbulence. This type of turbulence results from frictions between the air and the ground. This phenomenon generally occurs in the lower one kilometre of the atmosphere, referred to as the boundary layer. The degree of turbulence increases both with the roughness of the ground and with the speed of the wind.

Convective turbulence. This type of turbulence develops from the ground inside an unstable thermal structure. This phenomenon appears in the form of quasi-vertical columns with diameters varying from a few tens of metres to a few kilometres.

Upslope turbulence. This type of turbulence results from a combination of strong vertical currents in the air and of friction turbulence due to roughness of the slopes and mountainsides.

Orographic turbulence. This type of turbulence is similar to mechanical turbulence but is caused by hills or mountains: the higher the obstacle is, the higher will be the vertical extent of the turbulence. The degree of turbulence depends on the shape of the obstacle, on the speed of the wind and on the stability of the air mass.

2.4.10 Advection

The phenomenon of advection is the horizontal flow of air masses with different properties. This process allows humidity and water transfers between the air and the ground or the sea surface, thereby modifying the structure and composition of the lower layers of the atmosphere. An occurrence of such a phenomenon in continental regions is in winter when oceanic air advances over the ground previously cooled due to the stagnation of air masses. This phenomenon is described as an advection of maritime air masses. The difference in the temperatures of the maritime air and the temperatures of the colder continental regions results in fogs or in fine rains.

2.4.11 Subsidence

Subsidence is the downward vertical motion of air masses, and is the opposite of atmospheric convection. This type of motion occurs when an air mass previously heated in the vicinity of the surface rises in altitude and subsides. This causes a compression of the air mass, leading to an increase of the temperature, while the colder air below this air mass diverges horizontally. The sky clears while patches of blue sky appear here and there through the clouds.

The causes for subsidence may be thermal, as in the case of anticyclones formed by radiation cooling on snow. They are however more frequently dynamic and due to such phenomena as the divergence of different air masses, the downward movements of cold air masses or the differential speed between two superimposed layers.

2.4.12 Nebulosity

Nebulosity is defined as the fraction of the sky covered with clouds at a given time.

2.4.13 Meteors

Atmospheric meteors are phenomena observed in the atmosphere or at the surface of the Earth. They may consist either of a suspension, a precipitation, a deposit of solid, liquid or aqueous particles, or appear in the form of phenomena of optical or electric nature. They are classified into different categories: hydrometeors, lithometeors, photo-meteors and electro-meteors.

Hydrometeors are meteors consisting of either liquid or solid water particles resulting from the condensation or the sublimation of water vapour. Different varieties of hydrometeors exist: while some of them are liquid droplets in suspension in the atmosphere, like for instance clouds, fogs or ice fog, other hydrometeors are precipitations of either liquid or solid particles, like rain, drizzle, snow or hail, or on the contrary aggregations of particles raised by the wind, like drifting snow, blowing snow or spray. At last, hydrometers may appear in the form of a deposit of particles forming directly at the surface of the ground contact without having been precipitated: this is the case with such phenomena as white dew, hoar frost, rime or glaze. For more detail concerning the form, size and fall speed of hydrometeors, the reader is referred to Appendix B where the microphysical properties of hydrometeors are more extensively examined

Lithometeors are either particles in suspension in the air, like dry mist, sand fog, or smoke or particles, for instance sand or dust, blown by the wind and integrated into the atmosphere.

Photometeors are atmospheric optical phenomena produced by the reflection, refraction, diffraction or interference of light emitted by the Moon or the Sun. Examples of photometeors are halo phenomena, coronae, irisations, glories, rainbows, Bishop's rings, mirages, shimmers, scintillations, green flashes and twilight colours.

Electro-meteors are phenomena resulting from the interaction of electric charges with the atmosphere. Examples of such phenomena are storms, lightning, thunder, St. Elmo's fires and polar auroras.

2.4.14 Fog and Mist

These phenomena are due to the presence of fine water droplets (with diameter lower than 100 µm) in the atmospheric layer in contact with the ground. These droplets form when the moist air is cooled below its dew point: the air becomes saturated and the water vapour contained in the air condenses in the form of fine water droplets. The formation of clouds is based on the same principles: clouds differ from fogs only to the extent that the base of a fog is at the Earth's surface while clouds are above the surface.

According to the international definition, a fog occurs when the horizontal visibility is reduced below one kilometre and when humidity is close or equal to 100 per cent. The visibility is determined by the maximum distance beyond which a prominent object can no longer be seen and identified by unaided, normal eyes, and is measured using either a transmissometer or a diffusiometer. The reduction in the visibility depends on the nature of the fog, on the volume concentration and on the size distribution of the droplets. Fog appears in the form of a white uniform film. It can move in the form of benches with variable sizes, more or less spaced and more or less mobile. While its formation may be very fast, its dissipation is sometimes slow.

A light fog where the visibility is higher than one kilometre is usually called a mist. Mist generally occurs in very hot and moist and hot days: the air seems to be less transparent, although visibility is reduced to a lesser degree than in the case of a fog.

A dry fog is a type of light fog which instead of being composed of water droplets is formed by dust particles.

Fogs are classified according to the physical process whereby water vapour is condensed: radiation fog, advection fog, upslope fog, evaporation fog and mixing fog:

– radiation fog is a type of fog generated by the radiative cooling of an air mass during the night. It forms when the surface releases the heat that it has accumulated during the day and becomes colder : the air is in contact with this surface is cooled below the dew point, causing the condensation of water vapour, which results in the formation of a ground level cloud. This type of fog occurs more particularly in valleys.

– advection fog is generated when the warm, moist air flows over a colder surface: the air in contact with the surface is cooled below its dew point, causing the condensation of water vapour. Sea fog in coastal areas is a form of advection fog, formed when warmer sea air flows over colder land. This type of fog appears more particularly in spring when southern displacements of warm, moist air masses move over snow covered regions.

– upslope fog is a type of fog formed when moist air is lifted by the westward side of hills, mountain slopes or elevated plains: as it ascends up the slope, the air expands and is cooled below its dew points.

– evaporation fog is a fog due to an increase of the water vapour contained in a cold air mass. It forms more particularly in autumn and in winter when the cold air flows over a relatively warmer surface for instance a lake or a pond: this causes an evaporation of water, resulting in an increase of the value of the dew point. This type of fog generally appears in the form of smoking columns.

– mixing fog results from the cooling of a warmer air mass by mixing with a colder air mass. Different conditions must however be fulfilled in order that such a fog be generated: an important difference in temperature and a high degree of mixing of the two air masses and a high humidity. This type of fog is generally not very dense and appears mostly in the form of mist.

2.4.15 Precipitations

As condensation intensifies, the diameter of the droplets from which clouds are formed increases, either by coalescence, i.e. by agglomeration, or by the absorption of the steam around them. When their fall speed increases, precipitation occurs, either in the form of drizzle, if the diameter of the droplets lies between 0.1 and 0.5 millimetres, or in the form of rain, if the droplets are of larger dimensions. If the temperature falls beyond 0° C, hydrometeors are present in solid form (snow and hailstones). For further information on the characteristics of precipitations, the reader is referred to Appendix B devoted more specifically to the microphysical properties of hydrometeors.

2.4.16 Clouds

Clouds are aggregates of extremely small water droplets or of ice crystals in suspension in the atmosphere after the water vapour present in the atmosphere has condensed, due either to an increase of water vapour or to the cooling of an air mass. The condensation of water vapour is made possible by the presence in the atmosphere of condensation nucleus, which may be either fine solid particles, like sea salt crystals, mineral particles, like dust, sand or smoke, or electric particles. The water droplets thus formed have a very low fall speed due to their low density and to the resistance of the air, and remain suspended in the atmosphere under the influence of the ascending air currents.

While the particles thus formed are of small size and lower than 100 μm, their concentration may reach high levels, up to a few hundreds particles per cm^3.

The 0° C isotherm defines the limit between liquid and solid phases. The melting layer of precipitation is the layer beyond the 0° C isotherm where snowflakes melt and turn into raindrops. The altitude of the melting layer determines the rain height below which a liquid precipitation can be observed.

The existence of clouds strongly depends on the climate of the region. Their presence is characterised by a nebulosity index indicating the fraction of covered sky and expressed in tenths of percentage.

Clouds are classified by appearance or by altitude of occurrence. High-level clouds form above 6000 metres and are primarily composed of ice crystals. Cirrus, cirrostratus and cirrocumulus clouds are typical examples of such clouds. Mid-level clouds typically occur at altitudes ranging from 2000 to 6000 metres, and are primarily of water droplets, although they may also be contain ice crystals if the temperatures is low enough. Typical mid-level clouds include altostratus and altocumulus clouds. Low-level clouds are found at altitudes below 2000 meters and are mostly composed of water droplets. Typical low-level clouds include stratus, stratocumulus and nimbostratus clouds.

2.4.17 Auroras and Auroral Activity

Auroras are the illumination of the sky occurring at night in high magnetic latitudes. They are classified as auroras borealis and auroras australis depending on whether they appear in the northern or southern hemisphere. This phenomenon occurs through the appearance in the sky of a coloured blue-green arc with pink or red edges and with a height of several hundreds of kilometres. Auroras may be either short, with a duration of only a few minutes, or last all night.

The maximum luminosity of most visible auroras is at altitudes extending between 100 and 130 kilometres. These phenomena are generated by the precipitation through auroral horns of particles, for the most part electrons, into the Earth's atmosphere. This precipitation of particles is caused by the increase of magnetospheric hot plasma which appears after a solar flare and by the resulting flow of plasma which occurs, due to influence the convection electric field, from the geomagnetic tail towards the surface of the Earth. The convection electric field is itself directed overall from dawn to dusk and is generated by the interactions occurring between the solar wind and the terrestrial magnetic field. This field acts as a generator in the electric circuit formed by the currents which flow in the magnetosphere, in the ionosphere and along the terrestrial magnetic field lines. The precipitation of particles through auroral horns characterises the auroral activity and occurs inside the auroral oval.

The auroral activity and the extension of the auroral oval are monitored using polar orbit satellites: data are available at different websites (NOAA 2003).

References

Arden AL, Kerker M (1951) Scattering of electromagnetic waves from two concentric spheres. J. Appl. Phys. vol 22: 1242-1246

Bringi VN, Seliga TA (1977) Scattering from axisymetric dielectrics or perfect conductors imbedded in an axisymetric dielectric. IEEE Trans. A.P. vol AP-25 4: 575-580

Cohn GI (1972) Magneto-hydrodynamic wave phenomena in sea water. AGARD Conference Proceeding on Electromagnetics of the Sea 77 pp 25-1 - 25-20

Cojan Y, Fontanella JC (1990) Propagation du rayonnement dans l'atmosphère. E 4030 Technique de l'ingénieur Traité Electronique

Douglas RH (1963) Hail size distributions of Alberta hail samples. Mc Gill Univ., Montreal, Stormy Wea. Gp. Sci. Rep. MW-36: 55-71

Douglas RH (1964) Hail size distributions. World Conference on Radiometeorology 11th Weather Radar, Boulder, Colorado, pp. 146-149

Gunn KLS, Marshall JS (1958) The distribution with size of aggregate snowflakes. J. Meteorol. vol 15: 452-461

ITU-R P.453 (1995) The radio refractive index: its formula and refractivity data. Rec. ITU-R P.453-1

Masson BJ (1971) The physics of clouds. Second edition, Clarendon Press, Oxford

Moupfouma F, Martin L (1985) L'acquisition de données de propagation et de radiométéo-rologie par le canal du système de collecte par satellite Argos. NT/LAB/MER/201 CNET

NOAA   (2003)   Auroral   oval   from   the   NOAA   POES   satellite. www.sel.noaa.gov/pmap/index.html

Oguchi T (1983) Electromagnetic wave propagation and scattering in rain and other hydrometeors. Proc. IEEE vol 71 9

Pruppacher HR, Beard KV (1970) A wind tunnel investigation of the internal circulation and shape of water drops falling at terminal velocity in air. Quart. J. R. Met. Soc. vol 96: 247-256

Pruppacher HR, Pitter (1971) A semi-empirical determination of the shape of clouds and rain drops. J. Atmos. Sci. 28: 86-94

Renaudin M (1991) Météorologie. Cepadues Editions

Smith PL, Musil DJ, Webber SF, Spahn JF, Johnson GN, Sand WR (1976) Raindrop and hailstone size distributions inside hailstones, Proc. Cloud Phys. Conf., Amer. Meteor. Soc. Boston, Mass., Boulder, Colorado, pp. 252-257

Stock (1993) La météo de A à Z, La météorologie nationale, Paris

Willis JT, Browning KA, Atlas D (1964) Radar observations of ice spheres in free fall. J. Atmos. Sci. 21 103: 348-420

3 Electromagnetic Waves and Propagation Characteristics

3.1 Basic Properties of Electromagnetic Waves

3.1.1 Electromagnetic Parameters characterising Wave Propagation

The propagation of an electromagnetic wave is described through different parameters: the electric and magnetic fields $\vec{E}$ and $\vec{H}$, the electric flux density $\vec{D}$ and the magnetic induction $\vec{B}$ (Jouguet 1978). Only vectors $\vec{E}$ and $\vec{B}$ generate effects allowing the determination of the electromagnetic field. The vectors $\vec{D}$ and $\vec{B}$ are linked to vectors $\vec{E}$ and $\vec{H}$ through the following linear relations:

$$\vec{D} = \varepsilon * \vec{E} \qquad (3.1)$$

$$\vec{B} = \mu * \vec{H} . \qquad (3.2)$$

The ε and μ coefficients depend on the nature of the medium where the electromagnetic wave propagates. In the case of homogeneous, isotropic media, ε and μ are constants, whereas inside vacuum, these coefficients, respectively referred to as the permittivity and magnetic permeability of the medium, assume the following values:

$$\varepsilon_0 = 10^{-9} / 36 \, \pi = 8.842 * 10^{-12} \; \textit{(Farad per metre)} \qquad (3.3)$$

$$\mu_0 = 4 \, \pi * 10^{-7} \; \textit{(Henry per metre).} \qquad (3.4)$$

For any material medium, these coefficients can be deduced from the previous values using the following two equations:

$$\varepsilon = \varepsilon_0 * \varepsilon_r; \tag{3.5}$$

$$\mu = \mu_0 * \mu_r. \tag{3.6}$$

where ε_R and μ_R are the permittivity and the relative magnetic permeability of the medium respectively.

The refractive index of a given medium can be characterised from its permittivity and permeability using the following equation:

$$n = \sqrt{\varepsilon\mu} \tag{3.7}$$

The displacement of the particles with density ρ induces a current characterised by its density, represented by a vector $\vec{J}$. The displacement of electric charges described by this vector can be caused simply by the effect of the electric field $\vec{E}$: in this case $\vec{J}$ is proportional to $\vec{E}$ and, for a homogeneous isotropic medium, the following equation applies:

$$\vec{J} = \sigma * \vec{E} \tag{3.8}$$

where the σ-coefficient, measured in Mhos per metre or Siemens per metre (Sm^{-1}), is the electric conductivity of the medium.

As an aside, it might be pointed out that a displacement of charged particles can also be induced through the action of an external cause entirely independent from the electromagnetic field. An example would be the action exerted by a generator with its own energy source. In this case the index 0 is affected to the vector $\vec{J}$, thereby leading to the following general equation:

$$\vec{J} = \vec{J}_0 + \sigma * \vec{E} \tag{3.9}$$

3.1.2 Maxwell's Equations

The relations between the five parameters ($\vec{E}$, $\vec{B}$, $\vec{D}$, $\vec{H}$ and $\vec{J}$) defining the electromagnetic state of a given medium are expressed through Maxwell's equations:

$$\mathrm{rot}\,\vec{H} = \vec{J} + \partial\vec{D}/\partial T \tag{3.10}$$

$$\mathrm{rot}\,\vec{E} = -\partial\vec{B}/\partial T \tag{3.11}$$

$$\mathrm{div}\,\vec{D} = \rho \qquad (3.12)$$

$$\mathrm{div_B}\,\vec{B} = 0 \qquad (3.13)$$

where ρ is the density of the free charged particles.

These equations were originally put forth by Maxwell, in order to account for experimental results obtained by Ampère and Faraday. The resolution in free space of these equations leads, for each vector $\vec{E}$, $\vec{B}$, $\vec{D}$ and $\vec{H}$ to Helmholtz wave equation. Thus, for the $\vec{E}$ field, we are presented with the following equation:

$$\left(\Delta - \varepsilon\mu\,\frac{\partial^2}{\partial t^2}\right)\vec{E} = 0 \qquad (3.14)$$

where Δ is the Laplacian operator grad(div) - rot(rot). Restricting ourselves to the sinusoidal mode, the solution can be written down as follows:

$$E = E_0 \exp[\,j(\omega t - kz)] \qquad (3.15)$$

where:

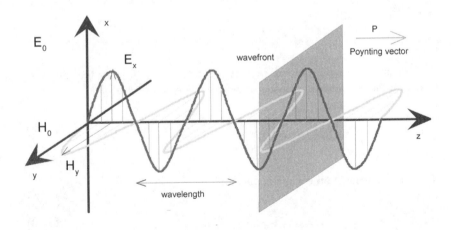

Fig. 3.1. Schematic representation of the propagation of an electromagnetic wave

- E_0 is the field amplitude,
- $\omega = 2\pi f$ is the angular frequency,
- t is the elapsed time,
- k is the wave number,
- z is the distance travelled along the z-axis.

The wave number represents the rate of variation with respect to distance of the field phase (the phase of the wave has a kz radian variation over a distance of z metres).

At the boundary surface of two media with different dielectric characteristics, the tangential components of the electric and magnetic fields are continuous. The same applies to the normal components of the electric and magnetic flux densities.

3.1.3 Propagation Velocity (Phase Speed)

It results from Helmholtz equations that the oscillations with constant phase of the $\vec{E}$ vector propagate in space in the form of a wave with a propagation factor γ expressed by the following equation:

$$\gamma^2 = \varepsilon\mu \frac{\partial^2}{\partial t^2} \tag{3.16}$$

while γ^2 is expressed in the case of a non-absorbing medium by the equation:

$$\gamma^2 = \frac{1}{v^2} \frac{\partial^2}{\partial t^2} \tag{3.17}$$

A comparison between the two previous relations readily yields the following equation for the propagation velocity of electromagnetic waves:

$$v = \frac{1}{\sqrt{\varepsilon\mu}} \tag{3.18}$$

The following equation is valid in vacuum:

$$\frac{1}{\sqrt{\varepsilon_0\mu_0}} = c = 3.10^8 \, m/s \tag{3.19}$$

where c is the speed of light.

3.1.4 Wavelength and/or Frequency

The resolution of Helmholtz equations in the case of a plane electromagnetic wave indicates that the electric field $\vec{E}$ and magnetic field $\vec{H}$ vectors are perpendicular to each other and oscillate in phase at each point in time and space. Waves with such properties are referred to as transverse electromagnetic waves (TEM).

The relation between the wavelength and the frequency is expressed by the equation:

$$\lambda = \frac{v}{f} = \frac{2\pi}{\omega\sqrt{\varepsilon\mu}} = vT \qquad (3.20)$$

where f is the wave frequency expressed in Hertz, ω is the pulsation and T is the period. The latter represents the distance over which the variation of the wave phase is equal to 2π.

The wavelength and/or frequency characterise an electromagnetic wave, along with its propagation and its applicability. Different frequency ranges are to be distinguished here: ELF (f < 3 kHz), VLF (3 kHz <f < 30 kHz), LF (30 kHz <f <300 kHz), MF (300 kHz <f <3 MHz), HF (3 MHz <f <30 MHz), VHF (30 MHz <f <300 MHz), UHF (300 MHz <f <3 GHz), SHF (3 GHz <f <30 GHz), EHF (30 GHz <f <300 GHz), …infrared (3 THz <f <430 THz) or luminous waves (430 THz <f <860 THz).

For more detail concerning terrestrial influence, systems considerations and services associated with each frequency ranges, the reader is referred to Appendix C devoted to the frequency spectrum.

3.1.5 Characteristic Impedance of the Propagation Medium

The ratio of the amplitude of the electric field to the amplitude of the magnetic field is given by the following equation:

$$E_x/H_y = (\mu/\varepsilon)^{1/2} = Zc \qquad (3.21)$$

As can be demonstrated, this ratio, referred to as the characteristic impedance of the propagation medium, has the same physical dimensions as an electric resistance. The characteristic impedance assumes in vacuum the following value:

$$Z_0 = (\mu/\varepsilon)^{1/2} = 377 \text{ ohms} \qquad (3.22)$$

It may be pointed out here that the characteristic impedance does not depend exclusively on the propagation medium, but also on the nature of the wave, for example whether it is a TEM wave *(transverse electromagnetic wave)*, a spherical wave or a guided wave. For this reason, the Z_0 parameter is more properly termed the wave impedance.

3.1.6 Poynting Vector

The Poynting vector describes the amplitude and direction of the power flow transported by the wave per square metre of surface parallel to the (x, y) plane, that is, the power density. The Poynting vector is measured in watts per square metre (W/m^2), and its instantaneous value is given by the following equation:

$$P = E \times H \tag{3.23}$$

The average value of the Poynting vector over a given period, which represents the real power transported by the wave, is more generally used, and is determined by the equation:

$$P_{moyen} = \frac{E_0 . H_0}{2} \tag{3.24}$$

3.1.7 Refractive Index

The equation for the refractive index is as follows:

$$n = \frac{1}{\sqrt{\varepsilon_r \mu_r}} \tag{3.25}$$

From this equation, the propagation velocity turns out to be equal to:

$$v = \frac{1}{\sqrt{\varepsilon_0 \varepsilon_r \mu_0 \mu_r}} = \frac{1}{\sqrt{\varepsilon_0 \mu_0}} \times \frac{1}{\sqrt{\varepsilon_r \mu_r}} = \frac{c}{n} \tag{3.26}$$

In an absorbing medium, the refractive index may be written in the following complex form:

$$n = n' - in'', \tag{3.27}$$

while the field of the wave becomes:

$$E = E_0 \exp[j(\omega t - nk_0 z)] \tag{3.28}$$

$$E = E_0 \exp[j(\omega t - (n' - in'')k_0 z)] \tag{3.29}$$

$$E = E_0 \exp(-n'' z)\exp[j(\omega t - n' k_0 z)] \tag{3.30}$$

The exp$(-n''z)$ term stands here for the attenuation that the wave undergoes as it travels inside the propagation medium. The imaginary component of the refractive index represents the wave attenuation coefficient (α in m^{-1}), and is a function both of the wave frequency and of the permittivity, permeability and conductivity (Sm^{-1} or (Ωm)$^{-1}$) of the medium. Propagation media are characterised by the dielectric parameters such as their permittivity, their permeability and their conductivity.

The electromagnetic field decreases exponentially as a function of the distance. The distance at which the field decreases by an e ratio (i.e. is divided by e) is called the penetration depth or skin depth (δ in m). The following equation defines the penetration depth:

$$\delta = \frac{1}{\alpha} \tag{3.31}$$

3.1.8 Wave Polarisation

An electromagnetic wave has two active orthogonal components, its electric and magnetic field vectors. As can be seen in Fig. 3.1, these two vectors are in phase both in time and space.

The polarisation plane is defined by the direction of propagation along the z-axis and the direction of the electric field along the x-axis. Fig. 3.1 represents a vertically polarised wave: polarisation can be seen to occur in the same direction than the direction of emission of the wave. The use of a vertical dipole antenna allows the emission of such vertically polarised waves. Conversely, if the electric field is horizontal, the waves will be referred to as transverse electromagnetic waves (TEM waves) with horizontal polarisation.

Maxwell's equations being linear, several transverse electromagnetic waves can be combined together in the following way:

$$E_x = E_1 * e^{j(\omega t - \beta z)} \tag{3.32}$$

$$E_y = E_2 * e^{j(\omega t - \beta z + \Psi)} \tag{3.33}$$

where $\beta = \omega / v = 2\pi f / c\lambda$, and ψ is the phase difference of the second wave with respect to the first wave.

Within a given wave plane (for example the plane $z = 0$), the extremity of the resulting electric field $\vec{E} = \vec{E_x} + \vec{E_y}$ will describe an ellipse. The equation for this ellipse can be written in a parametric form by taking the real components of the two following equations and by assuming that $z = 0$:

$$E_x = E_1 * \cos\omega t \tag{3.34}$$

$$E_y = E_2 * \cos(t\ \omega + \psi) \tag{3.34}$$

Waves of this type will be referred to as elliptically polarised plane waves.

In the case where $E_1 = E_2$ and $\psi = \pi/2$, the ellipse turns into a circle, while the wave becomes circularly polarised. Depending on the direction along which $\vec{E}$ rotates, the polarisation can be either left-handed (levogyrous) or right-handed circular (dextrogyrous). These two polarisation cases are represented in Fig. 3.2.

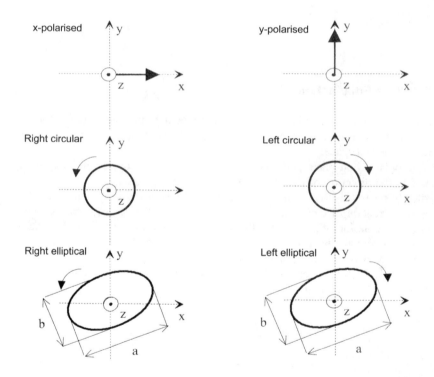

Fig. 3.2. Different polarisation states for a plane wave propagated along the z direction

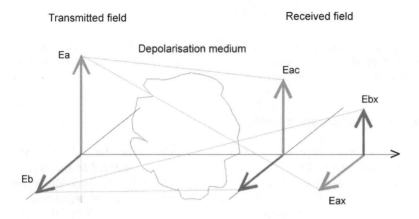

Fig. 3.3. Schematic representation of the phenomenon of cross-polarisation

3.1.9 Depolarisation

Depolarisation is a phenomenon whereby all or part of a radio wave emitted with a given polarisation has no longer any determined polarisation after propagation.

3.1.10 Cross-Polarisation

The phenomenon of cross-polarisation is the appearance, during the propagation, of a polarisation component, known as the cross-polar component, orthogonal to the initial polarisation or polar component. This phenomenon can originate in rain or in hydrometeors in troposphere, in multi-path propagation, in tropospheric or in ionospheric scintillation.

3.1.11 Cross-Polarisation Discrimination or Decoupling Ratio

For an electromagnetic wave emitted with a given polarisation, the cross-polarisation discrimination is defined as the ratio between the power received at the reception point with its initial polarisation and the power received with an orthogonal polarisation. The cross-polarisation discrimination thus expresses the degree to which a wave emitted with a given polarisation is found in orthogonal polarisation after propagation. The value of this parameter, expressed in decibels, depends on the nature of the propagation medium, as can be seen in Fig. 3.3.

$$XPD = 20\log (Eac/Eax) \tag{3.36}$$

The cross-polarisation discrimination is intimately related to the copolar attenuation of the wave. Several models can be found in the literature (see Appendix D).

3.1.12 Cross-Polarisation Isolation

Let us consider two components with orthogonal polarisations *Ea* and *Eb* emitted within the same plane. The cross-polarisation isolation is defined as the ratio of the copolar power *Eac* (or *Ebc*) to the cross-polar power *Ebx* (or *Eax*). These two ratios play an important part in the design of systems. It might be noted that the *Eac/Ebx* ratio value is not necessarily equal to that of the *Ebc/Eax* ratio. The cross-polarisation isolation expresses the degree of interference at the reception of two signals simultaneously transmitted in orthogonal polarisation. It depends on the nature of the propagation medium and is expressed in decibels by the following equation:

$$XPI = 20 \log(Eac/Ebx) \tag{3.37}$$

For more detail on the various cross-polarisation isolation models existing in literature, the reader is referred to Appendix D devoted to the cross-polarisation due to the atmosphere.

3.2 Propagation Mechanisms

Within an homogeneous medium, waves propagate in the form of rectilinear rays. The phase varies regularly and progressively as the wave propagates along its path, and follows Snell-Descartes laws.

3.2.1 Reflection

The phenomenon of reflection occurs when a wave comes against a surface with large dimensions and presenting only small irregularities compared to the wavelength. The relations between the reflected and incident fields are described by Fresnel equations, which are addressed at more length in Appendix E. Specular and diffuse reflections are to be distinguished here.

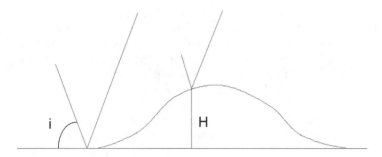

Fig. 3.4. Phase difference created by a surface irregularity with height H

Specular reflection

The phenomenon of specular reflection occurs at all frequencies in the presence of a perfectly plane and homogeneous surface, as for instance with obstacles like the ground, the façades of buildings and more generally plane surfaces. The path loss induced by reflections of this nature can be derived from Fresnel equations and depends on the dielectric characteristics of the reflective surface: its conductivity σ, its permittivity ε or relative permittivity $\varepsilon_r = 36\pi\varepsilon10^9$, and its relative complex permittivity $\varepsilon'_r = \varepsilon_r - j.60.\sigma.\lambda$. Values for these parameters are available in the literature: for example at the 60 GHz frequency, the following values were found: $\varepsilon'_r = 4$ for walls and $\varepsilon'_r = 3$ for the ground. Concerning the characteristics with respect to the frequency of the different types of grounds (very dry, fairly dry or wet) or water (sea water, pure water, ice or fresh water), the reader is referred to Recommendation UIT-R P.527. It should be noted that the values of the dielectric characteristics display a great variability depending on the nature of the materials and that the measurement of these parameters turns sometimes to be quite delicate.

Diffuse Reflection

The reflection off rough surfaces, that is, surfaces presenting height irregularities at various points, leads to a type of reflection known as diffuse reflection. If an incident wave strikes such a surface, it will no longer be reflected in a single direction but on the contrary it will be diffused in several different directions. In order to determine if a given reflection is diffuse or specular, use is generally made of the Rayleigh criterion. This leads to considering the height h of the surface irregularities and the incidence angle i.

Given two waves reflected off the surface under study, the irregularity with height H induces a path difference $\delta = 2H \sin(i)$ between these two waves, and consequently a phase difference $\Delta\Phi = 4\pi\,H\,\sin(i)/\lambda$. This situation is represented in Fig. 3.4. If the height H is small enough, the two waves are in phase, which leads us back to the previous case of specular reflection. If it is not, the surface is to be regarded as rough. The Rayleigh criterion states that reflection is diffuse if $\Delta\Phi > \pi/2$, i.e. when $H > \lambda/8\sin(i)$. The roughness of a given surface therefore depends on the frequency, on the incidence angle and on the height of the irregularities.

An example of a phenomenon of diffuse reflection will be described hereafter, the case of a simple reflection off the façade of a building in a street. The transmitter and receiver are separated from each other by a distance D and both are at a distance W from the façade (Fig. 3.5).

The specular reflection angle I can be determined according to the laws of geometrical optics:

$$\sin(i) = \frac{W}{\sqrt{W^2 + D^2/4}} \tag{3.38}$$

The Rayleigh roughness criterion is then written in the form:

$$\text{if } H > \frac{\lambda\sqrt{W^2 + D^2/4}}{8W}\text{, then reflection is diffuse.} \tag{3.39}$$

Fig. 3.6 presents graphically results obtained for a distance W equal to five metres. As can be seen in this figure, the critical height is a function of the distance D and a decreasing function of the frequency. In the case of surfaces with a low roughness, of the order of a centimetre, this curve shows that the surface diffusion needs to be taken into account as soon as the frequency becomes higher than 10 GHz.

Two models are generally used for the representation of the phenomenon of reflection: the Lambert model and the Phong model.

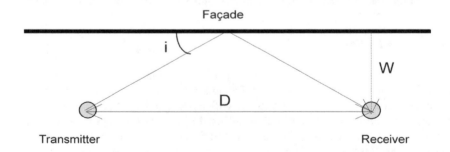

Fig. 3.5. Geometry of a reflection off a façade

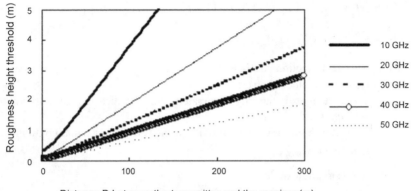

Distance D between the transmitter and the receiver (m)

Fig. 3.6. Roughness height beyond which reflection is to be regarded as diffuse according to the Rayleigh criterion

Lambert Model. Some surfaces are very irregular and reflect radiation in all directions, independently of the direction of the incident radiation. These surfaces are referred to as diffuse and can be represented using the Lambert model. This model, very simple and easily implemented in computation software, is described by the following equation:

$$R(\theta_0) = \rho R_i \frac{1}{\pi} \cos(\theta_0) \qquad (3.40)$$

where ρ is the surface reflection coefficient, while R_i represents the incident power, and θ_0 is the observation angle.

Phong Model. The reflection pattern is adequately represented with the Lambert model for a number of different rough surfaces. An exception however is the region of space in the vicinity of specular reflection: here the diagram presents an important component which is not allowed for by the Lambert model. In the Phong model, the reflection diagram is considered as the sum of two components, the diffuse and specular components. The percentages associated with each of these components depend on the characteristics of the surface and are a parameter of the model. The diffuse component is as in the Lambert model, while the specular component is introduced as a function of the incidence angle θ_i and the observation angle or reflection angle θ_0. The following equation describes this model:

$$R(\theta_i, \theta_0) = \rho \frac{R_i}{\pi} \left[r_d \cos(\theta_0) + (1 - r_d) \cos^m (\theta_0 - \theta_i) \right]$$ (3.41)

where:

- ρ is the surface reflection coefficient,
- R_i represents the incident power,
- r_d is the percentage of intensity associated with the diffused rays. Its values range between 0 and 1,
- m is a parameter describing the directivity of the reflected specular component,
- θ_i is the incidence angle,
- θ_0 is the observation angle.

As can be observed, the Lambert model may be obtained from the Phong model by assuming that the percentage r_d is equal to one. The Phong model, depending on the observation and incidence angles, is more complex than the Lambert model, and its implementation leads thus to longer computation times. The reflection pattern presents a principal lobe centred in the direction of the specular reflection.

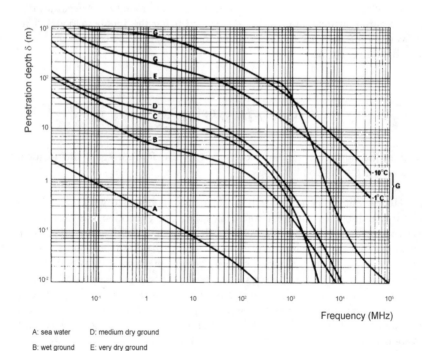

A: sea water   D: medium dry ground

B: wet ground   E: very dry ground

C: fresh water   F: ice (fresh water)

Fig. 3.7. The penetration depth as a function of the frequency (ITU-R P.527)

3.2.2 Transmission

Transmission is the phenomenon whereby an electromagnetic wave travels through an obstacle. The transmitted waves propagating through buildings or vegetation play in general a limited part. The transmitted field is connected to the incident field through Fresnel equations. The penetration depth of the radio electrical energy δ is defined as the distance at which the intensity of the wave is attenuated down to a ratio of $1/e$ (or equivalently to 37 per cent) of its value at the surface. Fig. 3.7 represents the dependence of the penetration depth on the frequency for different types of ground and water (ITU-R P.527). As can be observed, at frequencies higher than 10 GHz, the penetration depth almost assumes a null value. This is the case for all the media presented in Fig. 3.7, with the exception however of ice, where the penetration depth assumes values ranging from a few centimetres to a few tens centimetres.

For further information concerning reflection and transmission coefficients, the reader is referred to Appendix E devoted to Fresnel equations.

3.2.3 Diffraction

A path is said to be diffracted when the wave has travelled round such obstacles as for instance buildings, hills, vegetation, roofs, roads or other structures. The phenomenon of diffraction occurs when a wave meets an edge with large dimensions compared to the wavelength, and it is one of the most important factors involved in the propagation of radio waves (Boithias 1983). The *Geometrical Theory of Diffraction* (GTD) allows a representation of this phenomenon in the form of rays (McNamara 1990). This theory shows that, save in free space, the additional attenuation due to diffraction has a frequency response of the form $10*\log_{10}(f)$ in the case of the diffraction by an edge. Diffraction has a strong influence in frequency bands not higher than a few GHz. At higher frequencies, especially at frequencies higher than 15 GHz, the attenuation due to diffraction may reach a considerable level compared to the attenuation due simply to reflection.

In order to evaluate the attenuation caused by diffraction in frequency bands of interest for us, let us consider here an edge with null thickness and with a height h, either positive or negative. Let d be the total distance, d_1 and d_2 the distances from the edge to the transmitter and the receiver respectively, and h the height of the obstacle extending above the straight line joining the two ends of the path. Let us now define a variable called v in the following way:

$$v = h\sqrt{\frac{2}{\lambda}\left(\frac{1}{d_1}+\frac{1}{d_2}\right)}$$

(3.42)

The field E present at the reception point will then be expressed in amplitude and phase from the free space field E_0 through the following equation:

$$\frac{E}{E_0} = \frac{1}{1+j} \int_v^\infty e^{j\pi \frac{t^2}{2}} dt \qquad (3.43)$$

where E_0 is the field that would exist in the absence of the edge.

The corresponding power ratio is then written as follows:

$$\frac{P}{P_0} = \frac{1}{2}\left[\left(\frac{1}{2} - \xi(v)\right)^2 + \left(\frac{1}{2} - \eta(v)\right)^2\right] \qquad (3.44)$$

where $\xi(v)$ and $\mu(v)$ are the Fresnel integrals defined by the following equations:

$$\xi(v) = \int_0^v \cos\frac{\pi t^2}{2} dt \qquad (3.45)$$

$$\mu(v) = \int_0^v \sin\frac{\pi t^2}{2} dt \qquad (3.46)$$

If v is negative, i.e. if the top of the edge is below the straight line joining the transmitter to the receiver, then P/P_0 oscillates towards the level of free space. If v is positive, P/P_0 decreases regularly as the obstruction of the edge increases. If v is null, the transmitter and the receiver are aligned with the top of the edge and the attenuation is equal to 6 dB.

Approximate expressions are available for the obstruction cases (Boithias 1987):

$$10\log\frac{P}{P_0} = -13 - 20\log v \quad \text{(more specifically valid for } v > 1.5) \qquad (3.47)$$

$$10\log\frac{P}{P_0} = -6.9 + 20\log\left[\sqrt{(v-0.1)^2)+1} - v + 0.1\right] \qquad (3.48)$$

The second equation is more specifically valid for $v > -0.7$. It can be employed in the vicinity of $v = 0$.

In order to determine orders of magnitude, let us proceed here to an evaluation of the first approximated expression at a distance of one kilometre, for two heights equal to 1 and 10 metres respectively and at four different frequencies: 20, 40, 60 and 100 GHz. The results are summarised in Table 3.1.

Table 3.1. Values for the attenuation due to diffraction

	Diffraction attenuation [dB]	
	$D = 1$ km, $H = 1$ m	$D = 1$ km, $H = 10$ m
Frequency = 20 GHz	10	30.3
Frequency = 40 GHz	14	33.3
Frequency = 60 GHz	16	35
Frequency = 100 GHz	18	37

As can be seen in this table, at these frequencies, this type of propagation leads to attenuation levels that would be prejudicial to the link balance. This is the reason why in general the possibility of using diffracted paths in the deployment of millimetre wave links is not even considered. Consequently, millimetre wave links are generally regarded as being in line-of-sight. Therefore, it will be assumed that there are no obstacles present within the first Fresnel ellipsoid. Experimental measurements (Hammoudeh 1997; Thomas 1985) and numerical simulations (Hammoudeh 1997) have revealed that this phenomenon is negligible compared to specular reflection.

3.2.4 Diffusion

Although most of the time obstacles are of large size with respect to the wavelength ($\lambda < 2$ cm), their surface must still often be regarded as rough : this is the case with such obstacles as the balconies, the frontage windows, the urban furniture or the vegetation. The phenomenon of diffusion appears when an obstacle presents a number of irregularities of wavelength size.

3.2.5 Guiding

Certain environments, for instance street canyons, corridors and tunnels, behave like waveguides as regards to the propagation of radio waves following upon multiple successive reflections off walls. This is an application of ray theory together with the theory of propagation modes, especially when the wavelength is very small compared to the transverse section of the tunnel.

3.3 Main Physical Phenomena and Frequency Dependence

3.3.1 Path Propagation

An important feature of radio mobile propagation is its multipath character. The notion of multipath propagation also appears in the study of ionospheric propagation by reflection off the different layers. In mobile radio propagation, signals often follow paths with different lengths, propagate with different propagation times and arrive at the reception point with different phases. These paths can be classified as follows:

- direct paths, where the wave propagates along the straight line joining the two ends of the path, i.e. the emission and the reception points,
- transmitted paths : these paths generally play but a limited part, since propagation along the direct paths is often prevented by such obstacles as the vegetation or buildings,
- reflected paths, where the waves arrive at the receiver after reflecting off such obstacles as the façades of buildings, the ground or mountain sides. The waves associated to these paths contain enough energy for being usable.
- diffracted paths, corresponding to the possibility for waves of travelling round obstacles by diffraction (clumps of vegetation, buildings, edges of roofs, hills, etc.),
- scattered paths: this phenomenon appears more particularly when the wave strikes either a rough surface, like for instance the vegetation, the ground or the sea, or heterogeneities with small dimensions compared to the wavelength. The very notion of a smooth surface is intimately connected with the frequency.

These various paths, with different amplitude and phase, interfere at the time when they arrive unto the receiver. These interferences can be either constructive when the different paths arrive in phase, leading to a signal reinforcement, or destructive, causing in this case a fading of the signal. It might be further noted that the mobile itself moves inside this figure of interferences, so that it propagates successively through luminous and dark regions (interference fringes), which results in a fading of the signal.

The presence of the multiple paths and the displacements of the receiver imply that three fundamental properties of the propagation channel are to be examined: its attenuation, its variability and its selectivity in frequency.

Whereas in an analogical context of communication, the concept of attenuation is adequate for the study of the propagation channel, it is no longer the case in a numerical context of communication. In this case, the fading associated with the variability and the selectivity causes a qualitative deterioration of the communication which is independent from the attenuation.

3.3.2 Fresnel Ellipsoids

A study of the propagation of electromagnetic waves between a transmitter E and a receiver R, leads to a subdivision of the propagation space into a family of ellipsoids, referred to under the name of Fresnel ellipsoids, and schematically represented in Fig. 3.8. These ellipsoids have their foci located at the emitter E and the receiver R, so that at any point M of any such ellipsoid, the following equation holds true:

$$EM + MR = ER + n/2 \, \lambda \qquad (3.49)$$

where n is an integer number characterising the ellipsoid under consideration (for example, $n = 1$ characterises the first Fresnel ellipsoid, etc.) and λ is the wavelength.

It is in general assumed that the propagation occur in line-of-sight (direct visibility), i.e. with negligible diffraction phenomena in the absence of obstacles inside the first Fresnel ellipsoid. The first Fresnel ellipsoid delimits the region of space where almost all the energy is present.

The radius of the nth ellipsoid at a point of the propagation located at distances d_1 from E and d_2 from R is given by the following relation (where $d_1+d_2=d$, provided that d_1 and d_2 are large compared to the radius of the ellipsoid):

$$R_n = \sqrt{\frac{n\lambda d_1 d_2}{d_1 + d_2}} \, . \qquad (3.50)$$

The radius of the first ellipsoid is therefore equal to:

$$R = \sqrt{\frac{\lambda d_1 d_2}{d_1 + d_2}} \qquad (3.51)$$

while the maximum value at the middle of this path is:

$$r_{max} = \frac{1}{2} \sqrt{\lambda d} \, . \qquad (3.52)$$

For instance, with $d = 1$ km,

 – $f = 900$ MHz, $\lambda = 33$ cm and $r_{max} = 9$ m,
 – $f = 1800$ MHz, $\lambda = 16.5$ cm, and $r_{max} = 6.4$ m.

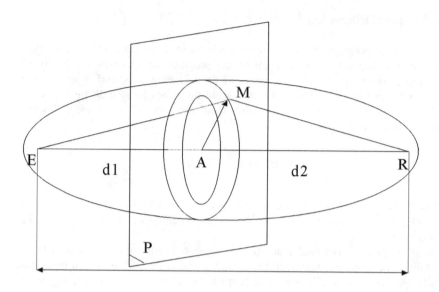

Fig. 3.8. Schematic representation of the Fresnel ellipsoids

3.3.3 Free-Space Attenuation

Different models of attenuation have been defined in the literature (see the ITU-R.1994 P.525). A distinction is generally drawn between the path loss in free-space and the attenuation in excess compared to the latter.

Free-space attenuation is defined as the transmission loss caused by the dispersion of the energy of the wave that would occurs were the antennas to be replaced by isotropic radiators placed inside a perfectly dielectric, homogeneous, isotropic and unlimited environment, the distance between the antennas remaining unchanged.

The equation for free-space attenuation is:

$$A_0 = \left(\frac{4\pi d}{\lambda} \right)^2 .$$
(3.53)

This equation can be rewritten in logarithmic form, and becomes:

$$A_0 \left(dB \right) = 32.4 + 20 \log_{10} \left(f \right) + 20 \log_{10} \left(d \right)$$
(3.54)

where d is the distance in kilometres between the transmitter and the receiver, λ is the wavelength in kilometres and f is the frequency in MHz.

As an example, for a distance d equal to one kilometre, the following values were found:

- $f = 900$ MHz, $A_0 = 91.4$ dB,
- $f = 1800$ MHz, $A_0 = 97.4$ dB,

The method described above supposes that only one of the radio paths has its first Fresnel ellipsoid cleared.

Free-space attenuation has an expression in d^2 (energy conservation), which means that a 6 dB loss occurs each time the distance is doubled.

The frequency response (f^2) depends on the configuration of the transmission, for example whether it is a point-to-point propagation or a mobile radio propagation. These two configurations will be successively addressed here.

Point-to-Point Links

Assuming that both the transmitting and receiving antennas are pointed, the reasoning proceeds from the consideration of the volume of the antennas, i.e. their equivalent surface. The received power Pr, expressed as a function of the emitted power Pe, is equal to:

$$Pr = Pe * A_e * A_r * f^2 \tag{3.55}$$

where A_e and A_r are the equivalent surfaces of the transmitting and receiving antennas respectively. In this case a 6 dB gain is realised each time the frequency is doubled.

Mobile Radio Links

In a mobile radio context of communication, the base station is not aware of the actual location of the mobile. The size of the receiving antennas decreases with increasing frequency. The antennas used here are omnidirectional and therefore they are not pointed in any specific direction. The received power is the product of the gains of the transmitting (G_e) and receiving (G_r) antennas by the emitted power divided by f^2:

$$Pr = Pt * G_e * G_r / f^2 \tag{3.56}$$

A 6 dB loss occurs each time the frequency is doubled.

3.3.4 Variability

The mobile radio environment is most fluctuating: the passage of vehicles and people, the wind in trees, the opening of doors, to name but a few examples, cause fluctuations of the radio paths and generate fast variations of the observed signal. These phenomena, combined with the movements of vehicles, create a temporal and a spatial variability inside the propagation channel.

The variations of the signal have a random character. A statistical analysis allows evaluating the effect of the multi-path on the transmission of mobile radio systems. For instance, both the received field law and the fading average time are essential parameters for the determination of the dimensions of transmission systems and of the devices protecting them against fading (interlacing and diversity techniques).

A modelling of the fast variations of the signal has been proposed and shall be described here (Clarke 1968). Assuming that the mobile moves inside an interference figure generated by the superposition of a high number of plane waves with random and independent amplitudes, phases and directions, it can be demonstrated by applying the central limit theorem that the received field (i.e. the complex envelope) is a Gaussian variable. The envelope of the narrow band signal, representing the received power, follows in that case a Rayleigh law. Correspondingly the term Rayleigh fading will be used here.

The following equation yields the probability density of a random variable R which follows a Rayleigh law:

$$f_R(x) = \frac{x}{\sigma^2} \exp\left(-\frac{x^2}{2\sigma^2}\right), \ x \geq 0 \qquad (3.57)$$

$$f_R(x) = 0, \ x < 0 \qquad (3.58)$$

The Rayleigh fading is characterised by the parameter σ, where σ^2 represents the average power of the signal. This phenomenon is observed in the presence of multiple paths in a configuration of blocked links where the contributions of all the paths are equivalent.

If a dominant path exists (for example in line-of-sight situations or in open environments like suburbs or rural environments), the envelope of the narrow band signal, representing the received power, will follow a Rice law with a probability density given by the equation:

$$f_R(x) = \frac{x}{\sigma^2} \exp\left(-\frac{x^2+r^2}{2\sigma^2}\right) I_0\left(xr/\sigma^2\right), \ x \geq 0 \qquad (3.59)$$

$$f_R(x) = 0, \ x < 0 \qquad (3.60)$$

where I_0 is a modified Bessel function of the first kind and zeroth order.

The Rice fading is characterised by the two parameters R and σ. This phenomenon can be observed if one of the paths is dominating over the others. The Rice law corresponds to a Rayleigh-type law in the case when $R = 0$ (i.e. in the absence of a line-of-sight path) and allows to identify a direct path and its preponderance. This type of fading is also characterised by a K coefficient, referred to as the Rice parameter, and defined by the equation:

$$K = 10 \log\left(\frac{r^2}{2\sigma^2}\right) \qquad (3.61)$$

The K parameter represents the relation existing between the power of the direct path (line-of-sight propagation) and the power contribution of the secondary paths which follow a Rayleigh distribution. The larger the K parameter, the more important the direct path is in comparison with the multiple paths and the less loaded the link is. Conversely, if the preponderance of the direct path is weak, ($K \leq -5$ dB), the Rice distribution can be identified to a Rayleigh distribution (CCIR 1990). The value for K is dependent upon the nature of the environment (urban dense, suburban, rural, etc).

The envelope of the mobile radio signal can be characterised through different laws, as for instance the Weibull and Nakagami laws (Braun 1991). These laws are less frequently used than the Rayleigh and Rice laws since the latter, being generated from Gaussian variables, are more readily implemented in computation software. For more information on the subject of the different laws used for the characterisation of the distribution of the rapid variations of mobile radio signals, the reader is referred to Recommendation ITU-R P.1057 (ITU-R 1994).

In order to determine the variation law followed by the signal, statistical tests, for instance the Kolmogorov-Smirnov test (Barbot 1992), needs to be applied to the rapid variations caused by the interferences occurring between the received waves.

Provided that the statistical characteristics of the signal (probability density function, cumulative distribution function) are available, it becomes possible to determine the different parameters relevant for operating a radio system. Among these parameters the following two may be mentioned:

- the probability for the signal to fall below a certain threshold. When the power of the signal is lower than the noise threshold allowed by the receiver (interferences, noise of thermal or industrial origin, etc) the signal is obscured by the noise. In this case the receiver will not be able to correctly interpret the transmitted data.
- the statistical fading duration. During fading the transmission of data is interrupted. The slower the mobile moves, the longer the cut-offs periods are.

Some orders of magnitude at the 900 MHz frequency range are reported in Table 2.1 (Guisnet 1998). As can be seen, fading values higher than 10 dB remain very frequent: such values still appear in 10 percent of the cases. For a 4 km/h pedestrian walking speed, the average duration of the cut-offs is 40 ms; it falls to 4 ms for a vehicle moving at a 40 km/h speed. Hence, whereas in the first case the communication is attenuated or interrupted during relatively long periods, the communication with a mobile at high speed presents a quasi-uniform time distribution of errors.

Table 3.2. Statistical characterisation of major fading

		Average duration of fading [ms]	
Level of fading compared to the average field	P (percentage that x < threshold)	f = 900 MHz v = 4 km/h	F = 900 MHz v =40 km/h
-5 dB	27	80	8
-10 dB	10	40	4
-20 dB	1	12	1
-30 dB	0.1	4	0.4

The variability resulting from the movement of the mobile can also be characterised through the Doppler spectrum obtained from the power spectral density function $T(f_0, t)$. A line with a pure frequency at the emission broadens in proportion with the speed of the mobile. The Doppler spectrum is distributed over the frequency interval $[f_0 - f_d, f_0 + f_d]$ where f_0 is the emitted frequency and f_d is the maximum Doppler frequency as given by the equation:

$$f_d = f_0 \frac{v}{c} \qquad (3.62)$$

where v is the speed of the mobile and c is the velocity of the electromagnetic wave.

The classical Doppler spectrum is given by the following equation:

$$S(f) = \frac{1}{\sqrt[2]{1 - \left(\dfrac{f}{f_d}\right)^2}} \qquad (3.63)$$

where f belongs to the interval $[-f_d, +f_d]$.

The Doppler spectra deduced from experimental measurements (Gollreiter 1994; Codit 1994; Dersch 1994) reveal a clear asymmetry in the spatial frequency domain. This asymmetry points to the existence of privileged directions of arrival, especially in line-of-light situations.

The reverse of the width of the Doppler spectrum, or Doppler dispersion, is referred to as the coherence time. It is defined as the time during which function $T(f_0, t)$ can be regarded as being almost constant, i.e. the time during which a mobile receiver would not detect the variability of the channel.

Techniques used for the prevention of the major fading of signals include data interlacing techniques and diversity (micro-diversity, frequency) techniques (Fechtel 1993; Rappaport 1996; Visoz 1998).

3.3.5 Frequency Selectivity

In the presence of significant differences between the delays of the multiple paths, the transfer function is no longer constant over the entire width of the spectrum, and the path loss depends now on the frequency. In these conditions, the channel will be said to be frequency selective.

For the purposes of evaluating the performances of a complete transmission chain, maintaining the quality of transmission of the numerical signals and designing new systems, it is therefore essential to develop a broadband modelling of the propagation channel.

Since the multiple paths are linearly superimposed, the channel is generally represented in the form of a temporal variable linear filter of the wide sense stationary uncorrelated scattering (WSSUS) type, with no correlation between the different diffusers (Bello 1963). The echoes reaching the receiver have their origins in a set of non-correlated sources: two echoes with different propagation delays are not correlated. To demonstrate that this condition is fulfilled is extremely difficult; in actual practice it is generally assumed that it applies to distances equal to a few wavelengths.

The radio channel will be represented by its time-dependent impulse response h (t, τ), where τ is the delay and t represents the time dependence (and accordingly, since the vehicle is moving, the space dependence). The impulse response is a function of two variables, and expresses the three characteristics of the channel: its attenuation, its variability (t) and its selectivity (τ).

The dual variables by τ and t Fourier transforms respectively are the frequency and the Doppler speed (Parsons 1992, Katterbach 1995).

Radio Mobile Channel Representation

Four different possible representations of the radio mobile propagation channel exist. The relations between these representations can be graphically represented as follows, where F and F^{-1} respectively represent the direct and opposite Fourier transforms (Bello 1963):

Time - Delay Representation. In this model, the output signal $y(t)$ is described as a function of the input signal $x(t)$, in the form of a convolution equation with a time-dependent core:

$$y(t) = \int_0^\infty h(t,\tau)x(t-\tau)d\tau \qquad (3.64)$$

where $h(t,\tau)$ is the impulse response at time t to a radio impulse $x(t)$ emitted at time $t - \tau$. The propagation channel is completely characterised by the impulse response, which can be used for identifying the different echoes from their delays of propagation.

Delay - Doppler Shift Representation. From a perspective concerned with the physical analysis of propagation paths, the delay - Doppler shift representation $S(\tau,v)$ offers a number of advantages. This representation allows for instance to follow the evolution of the different propagation paths during the displacement of a mobile moving at constant speed. Further, in the case of a stable propagation path, the evolution of the Doppler shift shall provide information pertaining for the determination of the angle of arrival. The function $S(\tau,v)$ is defined by the following equation:

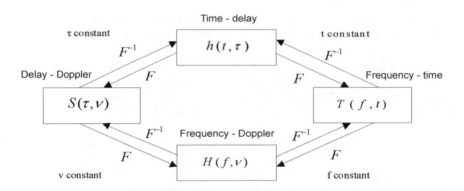

Fig. 3.9. Schematic representation of the different Fourier transforms of the impulse response

$$y(t) = \int\limits_{-\infty}^{+\infty} \int\limits_{-\infty}^{+\infty} x(t-\tau)S(\tau,v)e^{2\pi j\tau t}dvd\tau \tag{3.65}$$

with the relation:

$$h(t,\tau) = \int\limits_{-\infty}^{+\infty} S(\tau,v)e^{2\pi jvt}dv \tag{3.66}$$

Doppler Frequency Shift - Frequency Representation. The $H(f,v)$ function, dual of the $h(t,\tau)$ function, can be used in order to represent the Doppler frequency shift as a function of frequency. This function is defined as follows:

$$Y(f) = \int\limits_{-\infty}^{+\infty} X(f-v)H(f,v)dv \tag{3.67}$$

where $X(f)$ and $Y(f)$ are the Fourier transforms of $x(t)$ and $y(t)$ respectively.

Not only does this function allow directly identifying each frequency shift, it also lends itself to the characterisation of the frequency selectivity of the channel.

Time - Attenuation Representation. The transfer function $T(f, t)$ can be used for studying the time-dependence of the various effects induced by the multiple paths (temporal attenuation and, if the mobile is moving, the resulting spatial attenuation) and for characterising the narrow-band propagation channel. The following equation defines the transfer function:

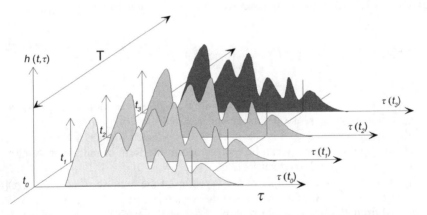

Fig. 3.10. Schematic representation of the temporal evolution of the impulse response of the propagation channel

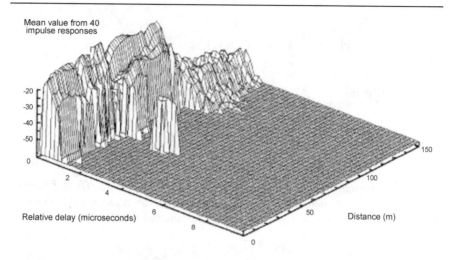

Fig. 3.11. Evolution of the impulse response at a street corner street in a microcell environment (Paris, 900 MHz, France Telecom R&D sounder)

Broadband Representation of the Mobile Radio Channel

The selectivity of the propagation channel is generally qualified in terms of parameters deduced from the average power profile of the impulse response. Among these parameters the most generally used are the average delay, the rms. delay spread, the delay interval, the delay window and the correlation bandwidth of the channel (Failly 1989).

Average Delay Profile. The average power delay profile $P(\tau)$ of the impulse response is defined from the impulse response $h(t, \tau)$ by the following equation:

$$P(\tau) = \frac{1}{T} \int\limits_0^T \left| h(t,\tau) \right|^2 dt \qquad (3.68)$$

The average delay profile corresponds to an average over a certain time interval during which the wide sense stationary uncorrelated scattering property is actually fulfilled. As shown in Fig. 3.9, the interval T is selected in such a way that the measured impulse responses can be represented in the form of random stationary and ergodic processes (Lavergnat 1997).

Fig. 3.10 presents an example of the variation of the impulse response $|h(t, \tau)|$ in the microcell context of a mobile turning at a street corner. The received power is reduced from 30 dB and the form of the response is significantly changed.

Average delay. The average delay is the average value of the delays weighted by their power. It is given by the first order moment of the impulse response:

$$\tau_m(t) = \frac{1}{P_m} \int_{\tau_{LOS}}^{\tau_3} (\tau - \tau_{LOS}) P(t,\tau) d\tau \tag{3.69}$$

where τ_{LOS} is the delay in line-of-sight, τ_3 is the instant when $P(\tau)$ exceeds the cut-off threshold for the last time and P_m is the total energy of the impulse response, as defined by the following equation:

$$P_m = \int_{\tau_0}^{\tau_3} P(\tau) d\tau \tag{3.70}$$

where $P(\tau)$ is the impulse response power density, τ is the excess delay and τ_0 is the instant when $P(\tau)$ exceeds the cut-off threshold for the first time.

Delay spread. The delay spread or standard deviation of the delays weighted by their power is given by the square root of the second central moment of the impulse response:

$$Delay-spread(t) = \sqrt{\frac{1}{P_m} \int_{\tau_0}^{\tau_3} \tau^2 P(\tau) d\tau - \left[\frac{1}{P_m} \int_{\tau_0}^{\tau_3} \tau P(\tau) d\tau \right]^2} \tag{3.71}$$

The delay dispersion emphasises the risk of the occurrence of inter-symbol interferences as well as the disturbance phenomena that strong, remote echoes may induce.

Delay interval. The delay interval at X dB is defined as the time interval between instant τ_0 when the amplitude of the impulse response first exceeds a given threshold and instant τ_3 when for the last time the amplitude goes under this threshold.

Delay Window. The y percent delay window measures the time interval $\tau_2-\tau_1$ containing y percent of the total energy of the impulse response. Times τ_1 and τ_2 are as defined by the following equation (see Fig. 3.12):

$$\int_{\tau_1}^{\tau_2} P(\tau) d\tau = \frac{y}{100} \int_{\tau_0}^{\tau_3} P(\tau) d\tau = \frac{y}{100} P_m \tag{3.72}$$

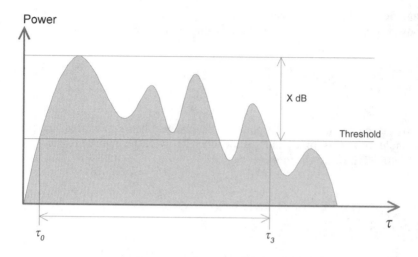

Fig. 3.12. Example of an average delay profile, emphasising the delay interval at X dB

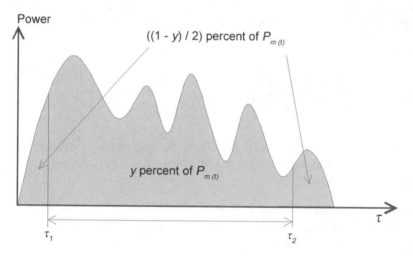

Fig. 3.13. Example of a power delay profile, emphasising the delay window containing y percent of the total energy of the impulse response

Correlation Bandwidth. The definition of the correlation bandwidth Bc of the channel proceeds as follows: let $C(t, f)$ be the autocorrelation of the transfer function, i.e. the Fourier transform of the power of the impulse response. The equation for $C (t, f)$ is:

$$C(t,f) = \int_{\tau_0}^{\tau_3} P(\tau)e^{-2\pi jft} d\tau \tag{3.73}$$

The correlation bandwidth is defined as the frequency at which $|C(t, f)|$ is equal to X percent of $C(t, f = 0)$ (Parsons 1992). It describes the amplitude of the selective attenuation as a function of the separation in frequency. The correlation bandwidth is therefore the frequency at which the autocorrelation function of the transfer function falls below a given threshold.

For the analysis of experimental data, the ITU-R recommends using delay intervals in the case of thresholds 9, 12 and 15 dB below the peak value, delay windows in the case of a 50, 75 or 90 percent energy and a correlation bandwidth in the case of a 50 or 90 percent correlation.

The correlation bandwidth is connected to the delay spread by the following equation (Lee 1993):

$$Bc * Delay - spread = \frac{1}{2\pi} \tag{3.74}$$

Depending on the environment, the propagation channel will be more or less selective. The same applies to the form of the impulse response. In picocell environments, i.e. inside buildings, the decrease of power with delay follows an exponential law whereas in small cell and microcell environments important delays are very clearly observed. It should be pointed out that in Fig. 3.13 the time scales are not the same for picocell environment and for small cell and microcell environments.

Table 2 presents typical values of the delay spread in four different environments. These values, which are given here merely as an indication, have been obtained from the statistical analysis of broadband measurements at the 2200 MHz frequency.

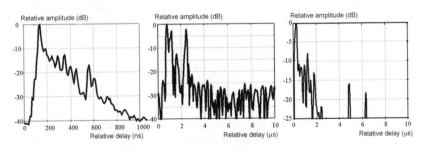

Fig. 3.14. Examples of impulse responses measured in various environments. From left to right: inside buildings, small cell and microcell environments

Table 3.3. Orders of magnitudes for the delay spread in different environments

	Delay spread at 50 percent	Delay spread at 90 percent
Inside buildings	70 ns	100 ns
microcell	300 ns	600 ns
Small cell	0.6 μs	1.3 μs
macrocell	few μs	a few μs

This table does not provide any precise value for the delay spread in macrocell environments, due to the fact that the results for such environments are not homogeneous from one location to another. For example, in mountainous environments, the delay spread might exceed 50 μs whereas in suburban environments it is generally no higher than 3 μs.

The characterisation of the channel must also allow for the analysis of new techniques. For instance, the utilisation of adaptive antennas and the Space Division Multiple Access (SDMA) are being intensively studied at present. These techniques should indeed enable to significantly increase the capabilities of existing and third generation systems (Swales 1990). Let us briefly recall that the principle at work here is to focus the gain of the base station onto a mobile and thus to treat with several lobes different mobiles located at different places but capable of functioning over the same temporal interval and at the same frequency. The design of systems of this type requires some additional data concerning the directions of arrival of the signals. Indeed, whereas in the absence of multi paths the energy arrives from a single direction defined by the transmitter (beam-forming techniques), in an urban area, the different multiple paths are not directed along the same direction. More detail concerning the evaluation of angles of arrival can be found in Appendix M which is more specifically devoted to this subject.

References

Barbot JP, Levy AJ, Bic JC (1992) Estimation of fast fading distribution functions. Com. URSI Commission F Open Symposium

Bello PA (1963) Characterization of randomly time-variant linear channels. IEEE Trans. Commun. Systems, vol CS-11 : 360-393

Boithias L (1987) Radiowave Propagation, Mc Graw-Hill, New-York

Braun WR, Dersch U (1991) A physical mobile radio channel. IEEE Trans. on Vehicular Technology, vol 40 2: 472-482

Clarke RH (1968) A statistical theory of mobile-radio reception. B.S.T.J. 957-1000

CODIT (1994) Final Propagation Models. Public Deliverable of the RACE 2020. CODIT Project R2020/TDE/PS/DS/040

Dersch U, Zollinger E (1994) Physical characteristics of Urban Microcellular Propagation. IEEE Trans. on Antennas and Propagation, vol 42 11:1528-1539

Failly M (1989) Final Report of COST 207, Digital Land Mobile Radio Communications. CEE Luxembourg

Fechtel SA, Meyr H (1993) Matched filter bound for treillis-coded transmission over frequency-selective fading channels with diversity. European Transactions on Telecommunications, vol 5 3:109-120

Gollreiter R (1994) Public Deliverable of the Race 2084. ATDMA project R2084/ESG/CC3/DS/P/029, Ed. Channel Issue 2

Guisnet B (1998) La propagation pour les services de mobilité. L'Echo des Recherches 170:15-24

ITU-R (1994) Propagation in Non-Ionized Media. PN Series Volume

Jouget M (1978) Ondes électromagnétiques. Propagation libre. Dunod, Paris

Kattenbach R, Fruchting H (1995) Calculation of system and correlation functions for WSSUS channels from wideband measurements. Frequenz 49, 3-4: 42-47

Lavergnat J, Sylvain M (1997) Propagation des ondes radioélectriques. Collection Pédagogique de Télécommunication, Masson, Paris

Lee WCY (1986) Mobile communications engineering, McGraw-Hill, New York

McNamara DA, Pistorius CWI, Malherbe JAG (1990) The Uniform Geometrical Theory of Diffraction, Artech, House, London

Parsons JD (1992) The Mobile Radio Propagation Channel. Pentech Press Publishers

RACE ATDMA Project (1994). Channel Models Issue 2. R084/ESG/CC3/DS/029/b1, ed R. Gollreiter

Rappaport TS, Sandhu S (1994) Radio Wave Propagation for Emerging Wireless Personal Communication Systems. IEEE Antennas and Propagation Magazine, vol 36, 5:14-23

Rappaport TS (1996) Wireless Communications. Principles and Practices. Prentice Hall PTR, New Jersey

Swales SC, Beach MA, McGrenham JP (1990) The performance enhancement of multibeam adaptative base stations antennas for cellular land mobile radio systems. IEEE, Trans. Veh. Tech. vol. 39, 1:56-57

Visoz R, Bejjani E, Kumar V (1998) Matched filter bound for equalization and antenna diversity over mobile radio channels. ICUPC, Florence

4 Ionospheric Links

4.1 Introduction

The propagation of high frequency radio waves over long distances is based on the reflection and refraction properties of the ionosphere. This medium, however, far from being stable, is submitted to a number of variations that may either attenuate or potentially reinforce the transmissions between two given points. Disturbances may arise suddenly with a high level, leading to the interruption of communications inside a given frequency band. As a consequence, the need arises to establish forecasts predicting the conditions for the propagation of radio waves: for a given link, these forecasts allow to plan and select antennas, as well as adequate frequencies and exploitation schedules.

4.2 Ionospheric Refraction

The study of the propagation of electromagnetic waves in the ionosphere leads to the magneto-ionic theory as developed by Appleton and Hartree (Budden 1961; Ratcliffe 1959).

The equation for the refractive index is:

$$\bar{n}^2 = \bar{\varepsilon}_r = \left(\mu - j\chi\right)^2 = \frac{c^2}{\omega^2}\bar{k}^2 = \frac{c^2}{\omega^2}\left(\alpha - j\beta\right)^2 \tag{4.1}$$

where μ is the real part of the refractive index, χ is the extinction index, α is the propagation coefficient and β is the attenuation coefficient.

In the ionosphere the equation for the refractive index is:

$$\bar{n}^2 = \bar{\varepsilon}_r = 1 - \frac{X}{1 - jZ - \dfrac{Y_T^2}{2(1 - X - jZ)} \pm \left[\dfrac{Y_T^4}{4(1 - X - jZ)^2} + Y_L^2\right]^{\frac{1}{2}}} \tag{4.2}$$

where:

$$X = \frac{Ne^2}{\varepsilon_0 m \omega^2} = \frac{\omega_p^2}{\omega^2} \tag{4.3}$$

$$Y_L = \frac{eB_L}{m\omega} = \frac{eB \cos \theta}{m\omega} = \frac{\omega_b \cos \theta}{\omega} \tag{4.4}$$

$$Y_T = \frac{eB_T}{m\omega} = \frac{eB \sin \theta}{m\omega} = \frac{\omega_b \sin \theta}{\omega} \tag{4.5}$$

$$Z = \frac{v}{\omega} \tag{4.6}$$

and:

- N is the electronic concentration,
- e and m are the charge and mass of the electron respectively,
- ε_0 is the permittivity of the vacuum,
- v is the collision frequency of the electrons,
- θ is the propagation angle, i.e. the angle formed by the wave vector and the magnetic field $\vec{B}_0$,
- ω_p is the plasma pulsation. $f_p = \omega_p/2\pi$ is the electronic plasma frequency and its numerical value is given by the equation:

$$f_p(MHz) \approx 9\sqrt{N(\text{électrons}/m^3)} \tag{4.7}$$

- ω_b is the electronic angular frequency. $f_b = \omega_b/2\pi$ is the gyrofrequency: it varies in space and has an average value equal to 1.25 MHz,
- indexes T and L refer respectively to the transverse (T) and longitudinal (L) components of the terrestrial magnetic field $\vec{B}$ compared to the normal to the electromagnetic wave with pulsation ω.

As can be seen from these equations, the index depends on the frequency, on the direction of propagation and on a double sign $\pm$. The plasma, being an ionised and magnetised medium, is dispersive, anisotropic and birefringent. This medium is furthermore absorbing.

The presence of the double sign $\pm$ corresponds to the presence of the magnetic field which allows the propagation of plane waves with two different types of polarisation. The propagation modes corresponding to the $-$ and $+$ signs are referred to as the ordinary mode and the extraordinary mode respectively. These two modes correspond to privileged polarisations where the medium is monorefringent.

The phenomenon of anisotropy is generated by the static magnetic induction $\vec{B}_0$ and disappears when $\vec{B}_0 = \vec{0}$.

The phenomenon of absorption, or the electromagnetic energy dissipation by Joule effect, is caused by collisions. If the ratio $Z = \nu/\omega$ is negligible compared to the unit, ε_r is real. If ε_r is positive, the wave number k is real and the medium is no longer absorbing.

The indexes are the same along the θ and $\theta+\pi$ directions, as indicated by the presence in the equation of the squares of the $\cos\theta$ and $\sin\theta$ trigonometrical functions.

The polarisation ratio is defined as the ratio between the E_x and E_y components of the $\vec{E}$ field and can be written in the following form:

$$\rho = \frac{E_x}{E_y} = -\frac{H_x}{H_y} = \frac{j}{2Y_L}\left[\frac{Y_T^2}{1-X-jZ} \pm \left(\frac{Y_T^4}{\left(1-X-jZ\right)^2} + 4Y_L^2\right)^{\frac{1}{2}}\right] \qquad (4.8)$$

The equation for the refractive index can be simplified in two different cases: when the wave propagates almost along the same direction than the direction of the magnetic field $\vec{B}_0$, and when it propagates along a perpendicular direction. These two situations are referred to as the quasi-longitudinal and quasi-transversal approximations respectively.

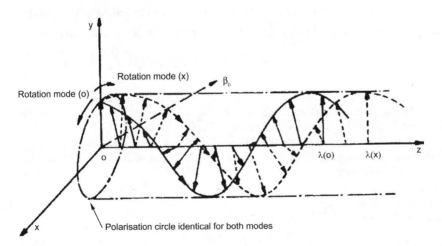

Fig. 4.1. Representation of the electric fields inside a magneto-ionic medium (longitudinal approximation) (Goutelard 1979)

In quasi-longitudinal approximation, Y_T is equal to zero. Neglecting collisions (i.e. assuming that $Z = 0$), we are presented with the equations:

$$\overline{n}^2 = \overline{\varepsilon}_r = 1 - \frac{X}{1 \pm Y_L} \tag{4.9}$$

$$\rho = \pm j. \tag{4.10}$$

The polarisation here is circular: the phase difference between E_x and E_y is equal to $\pm \pi / 2$, and $E_z = 0$.

In the ordinary mode, i.e. in left-hand polarisation where the wave travels away from the observer, the index ε_r is equal to zero if $X_0 = 1 + Y_L$, while in the extraordinary mode, i.e. in right-hand polarisation where the wave travels towards the observer, the index ε_r is equal to zero if $X_0 = 1 - Y_L$.

Assuming that the two modes (ordinary and extraordinary) are in phase when they penetrate inside a layer, the resultant of the two vectors will be linearly polarised, albeit out of phase, when the waves leave the layer: the extraordinary wave is delayed with respect to the ordinary wave. The phase difference is given by the equation:

$$\Omega = Cte \int_0^l N.B_0 \cos\theta dl \tag{4.11}$$

It may be noted that, in the case where $B_0 \cos\theta$ can be assumed to be constant, the penetration angle of the ionised layer is proportional in quasi-longitudinal approximation to the total electronic content $\int_0^l Ndl$ along the path L, i.e. to the number of electrons contained in a cylinder with unit section and with a length equal to the length of the path followed by the wave inside the ionised medium. This phenomenon is known as the Faraday effect.

In the quasi-transverse approximation, Y_L is equal to zero. Neglecting the collisions, i.e. assuming that $Z = 0$, we are presented with the two following systems of equations:

− in the ordinary mode:

$$\overline{n}^2 = \overline{\varepsilon}_r = 1 - X \tag{4.12}$$

The refractive index is therefore equal to zero for $X_0 = 1$.

$$\rho \approx 0 \qquad (4.13)$$

$E_x = 0$, the z-axis and the vectors $\vec{B}$ and $\vec{E}$ are coplanar.

− in the extraordinary mode :

$$\bar{n}^2 = \bar{\varepsilon}_r = 1 - \cfrac{X}{1 - \cfrac{Y_T^2}{1 - X}} \qquad (4.14)$$

$$\rho = -j\infty \qquad (4.15)$$

$E_y = 0$, the $\vec{E}$ and $\vec{B}$ vectors are perpendicular.

The refractive index is equal to infinite for $X_x = 1 - Y_T^2$ (pole) and to zero for $X_x = 1 \pm Y_T$ (zeros).

Polarisation is elliptic in all directions of propagation and depends on the mode (ordinary or extraordinary). The refractive index is equal to zero for three values of X irrespective of the value of the propagation angle:

− in the ordinary mode, if :

$$X_0 = 1 \qquad (4.16)$$

− and in the extraordinary mode, if :

$$X_x = 1 \pm Y \quad (Y^2 = Y_L^2 + Y_T^2) \qquad (4.17)$$

In Optics a spectral line is doubled in longitudinal propagation and tripled in transverse polarisation: this phenomenon is known as the Zeeman effect. Fig. 4.4 provides, for three different types of propagation inside a plasma and in the absence of collisions, a comparison between the normal Zeeman effect in Optics and the variations of the permittivity ε_r,

The following equation yields the absorption coefficient per length unit in the quasi-longitudinal approximation case:

$$\beta = \frac{\omega_p^2}{2c\mu\left[\left(\omega \pm \omega_b \cos\theta\right)^2 + v^2\right]} v \tag{4.18}$$

where $-\mu$ is the real part of the refractive index. The following conclusions can therefore be drawn:

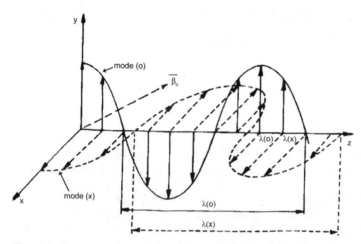

Fig. 4.2. Representation of the electric fields in a magneto-ionic medium (transverse approximation) (Goutelard 1979)

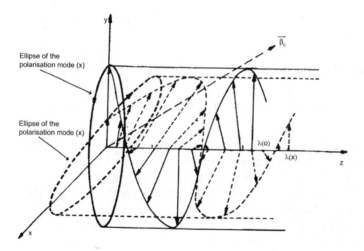

Fig. 4.3. Representation of the electric fields inside a magneto-ionic medium (general case) (Goutelard 1979)

- absorption is inversely proportional to ω^2, i.e. to the square of the frequency. This implies that high frequencies will propagate inside the ionosphere without loss.
- the presence of the $\omega_p^2 v$ term in the equation shows that absorption is a function of the collision frequency. A small increase in the electronic density will therefore increase the attenuation in proportion to the collision frequency. For example, the increase of the electronic density in the D layer during a solar flare induces in particular an important attenuation of the radio waves, which may result in the interruption of communications. In this case, absorption will be described as non-deviative.
- absorption is inversely proportional to the real part of the refractive index: in the presence of significant refraction effects, the attenuation increases. In this case, absorption is said to be deviative.

The following equation gives the phase velocity in the absence of both collisions and magnetic field:

$$v_\varphi = \frac{c}{\mu} = c\left(1 - \frac{\omega_p^2}{\omega^2}\right)^{-\frac{1}{2}} \tag{4.19}$$

The phase velocity is higher than the speed of light in vacuum and depends on the frequency. In this case, the medium is said to be dispersive: this means that two waves with slightly different frequencies will propagate at slightly different velocities, thereby causing interferences.

For a wave of the form $\cos(kz - \omega t)$, the group velocity is given by the following equation:

$$v_g = \frac{\delta\omega}{\delta k} \tag{4.20}$$

In a non-dispersive medium where ω/k is constant, $v_g = v_\varphi$, and the group refraction index μ' is defined by the equation:

$$\mu' = \frac{c}{v_g} = c\frac{\delta\omega}{\delta k} = c\frac{d}{d\omega}\left(\frac{2\pi}{\lambda}\right) = \frac{d}{d\omega}(\mu\omega) = \mu + \omega\frac{d\mu}{d\omega} = \mu + f\frac{d\mu}{df} \tag{4.21}$$

In the absence of a magnetic field, i.e. in a medium where $\mu^2 = 1 - \left(\dfrac{f_p}{f}\right)^2$, the

following equation holds:

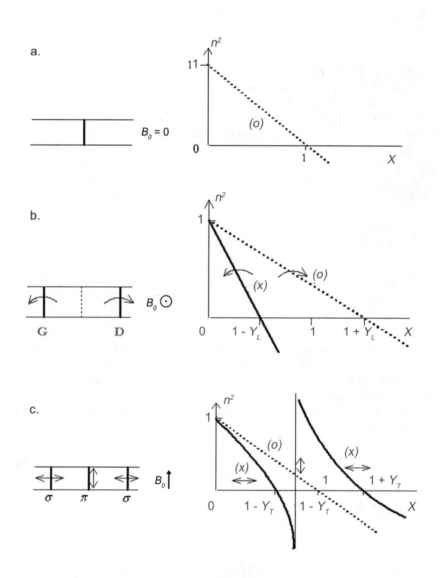

Fig. 4.4. Comparison between the Zeeman effect in Optics and the variations of the permittivity ε_r. *a.* in the absence of magnetic field, *b.* propagation perpendicular to the magnetic field, *c.* propagation parallel to the magnetic field

$$\mu' = \frac{d}{df}(\mu f) = \frac{1}{\mu} \qquad (4.22)$$

The group velocity is therefore given by the equation:

$$v_g = \mu c \qquad (4.23)$$

The angle α formed by the normal to the wave plane and the direction of the ray is defined by the equation:

$$\tan(\alpha) = -\frac{1}{v_\varphi}\frac{dv_\varphi}{d\theta} = +\frac{1}{\mu}\frac{d\mu}{d\theta} \qquad (4.24)$$

where θ is the angle beyond which the direction of the normal to the wave plane intersects with the reference axis.

To a maximum electronic density N_{max} corresponds a maximum frequency beyond which reflection is no longer possible. This frequency, referred to as the critical frequency, characterises the ordinary mode, and is determined by the following equation:

$$f_c(o) = f_p(N\max) \qquad (4.25)$$

The following approximate equation links the critical frequency, which characterises the extraordinary mode, to the frequency of an ordinary wave:

$$f_c(x) = f_c(o) + \frac{1}{2}f_b \qquad (4.26)$$

where f_b is the gyrofrequency. The real reflection height of a plane wave with frequency f is the altitude z_0 where the value of the real refractive index is equal to 0, i.e. $\mu_{(0)} = 0$ for an ordinary wave, or $\mu_{(x)} = 0$ for an extraordinary wave.

The following equation defines the reflection phase height:

$$h(f) = \int_0^{z_0} \mu dz \qquad (4.27)$$

The virtual reflection height is the height at which an impulse propagated at the speed of light would be reflected:

$$h'(f) = \frac{1}{2}c\tau(f) = \int_0^{z_0} \mu' dz \qquad (4.28)$$

where μ' is the group refractive index, while τ is the time required for an impulse to propagate forth and back between the origin and the reflection level in the ionosphere.

4.3 Trajectory Calculation

When a radio waves penetrates inside the ionosphere, the presence in this medium of charged particles such as ions and electrons, combined with the influence of the magnetic field, causes modifications in the characteristics of this wave. These effects shall be successively reviewed hereafter.

The wave is refracted, that is, it propagates along a curvilinear direction and at a group velocity lower than the speed of light in vacuum. The resulting delay in the propagation of the wave, or group propagation time, is a function of the frequency at which the wave has been emitted. At certain frequencies, a wave emitted from a point located at the surface of the ground might be reflected off the ionosphere and directed back to the surface. As the propagation media are birefringent, the ordinary and extraordinary components of the electric vector each undergo a different rotation, resulting in a rotation of the polarisation plane of the returning wave (assuming here that the wave is linearly polarised). This phenomenon is known as the Faraday effect.

The frequency of the received wave slightly differs from its frequency of emission: this difference is due to the movements of the layers within the ionosphere, and is described as the Doppler effect.

Further, in addition to these effects, which are due to the large scale gradients and to the small scale irregularities of the electronic concentration, the received signal is also affected by fluctuations in phase and in amplitude. These fluctuations are called scintillations.

A radio wave may also be absorbed along its trajectory, this being a consequence of the collisions occurring between the electrons, the ions and the neutral particles.

The very process of determining the path followed by a wave propagating in such a medium leads to considering some of these effects, in the form for instance of refraction, reflection, penetration and absorption phenomena. Fig. 4.5 schematically represents the propagation of an electromagnetic wave in the ionosphere. The different cases to be considered here are:

– if the frequency of the wave is too small (trajectory represented by curve a), the wave is absorbed inside the medium, more particularly inside the D layer,

- if the frequency of the wave is increased (trajectories represented by curves *b* and *c*), the wave is refracted and reflected at the E and F layers. This create the conditions for communication between a transmitter and a receiver,
- if the frequency of the wave is too high, the wave is refracted, though not to a sufficient degree for it to be reflected (trajectory represented by curve *d*). In this case, the wave is not directed back to the surface and instead disappears in the interplanetary medium. On the other hand, these frequencies are the very ones which allow radio communications between the Earth and satellites, either mobile or stationary.

It is therefore apparent that, in order to maintain an ionospheric link, the frequency cannot be too small, as the wave would then be absorbed, nor too high, for reflection would no longer be possible in this case. The two frequencies thus defined are referred to as the *lowest useful frequency* (*LUF*) and the *maximum useful frequency* (*MUF*) respectively. Hence, in order that a frequency be usable, the following formula must be satisfied:

$$L_{owest}U_{seful}F_{requency} < f < M_{aximum}U_{seful}F_{requency} \tag{4.29}$$

As the frequency decreases, the wave is submitted to a more or less strong absorption inside the D layer, where its energy is converted into heat. Therefore a lowest frequency referred to as the lowest usable frequency (*LUF*) exists. Below this frequency, the energy of the waves is too low for them to be used.

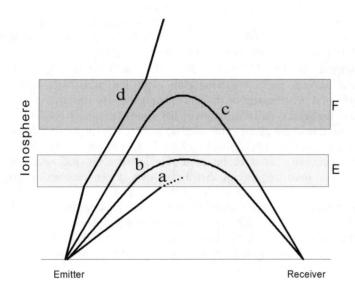

Fig. 4.5. Schematic representation of the propagation of an electromagnetic wave in the ionosphere

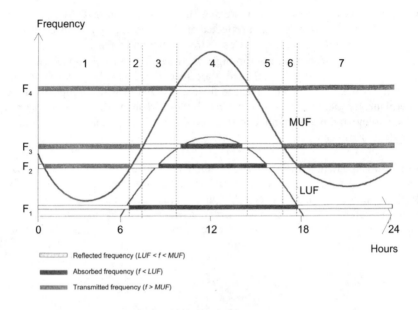

Fig. 4.6. Typical schematic representation of the daily variation of the usable frequency band for a given month, a given link and a given solar activity value. F1: timetables 1 and 7, F2: timetables 2 and 6, F3: timetables 3 and 5, F4: timetable 4

The maximum usable frequency (*MUF*) primarily depends on the electronic density of the medium, and therefore on the hour, the season, the geographical coordinates where the reflection takes place (and thereby on the geographical coordinates of the transmitter and the receiver) and the solar activity.

The lowest usable frequency (*LUF*) depends on the electronic density of the medium and therefore on the same parameters than the maximum usable frequency. In addition, it is dependent on the power emitted by the transmitter.

Fig. 4.6 presents the typical daily variation of the usable frequency band for a given month, a given link and a given value of solar activity. As can be seen from this figure, in order to continuously maintain a high frequency link (3-30 MHz), several different frequencies need to be used: for instance, in the situation presented in Fig. 4.6, four frequencies distributed over seven timetables are necessary.

Due to the great time and space variability of the ionospheric characteristics associated with the solar activity, the maintenance of an ionospheric link under satisfactory conditions requires that the usable frequency band be known. The need thus arises of setting up forecasts predicting the conditions of propagation along ionospheric paths of radio waves.

4.4 Ionospheric Forecasts

The conditions of propagation are obviously variable in time: therefore, different methods of forecast have to be developed depending on the duration chosen for the forecast. Forecasts are classified as short-term, medium-term or long-term forecasts depending on whether they are established over twenty-four hours, over a week or over more than a month.

4.4.1 Long-Term Forecasts

Forecasts of the Solar Activity

Long-term forecasts are valid over a one month period and are established from the ionospheric characteristics, from the forecast of the solar activity index and from statistics of the values of the ionospheric indices measured during previous similar situations. Forecasts are usually provided over periods equal to 30 or 90 per cent of the time, i.e. over periods of 10 or 27 days per month respectively.

The forecast of the solar activity is empirically conducted, using methods generally based on harmonic analysis, in order to account for the cyclic character of solar activity as well as for the relation between the maxima of solar activity and the duration of the ascending phase of the solar cycle. From these forecasts can be directly deduced forecasts predicting the ionospheric characteristics at each point of the Earth's surface and for each hour of the day.

Determination of the Propagation Modes

The different paths existing between a transmitter and a receiver can be determined using the complex equations provided by the magneto-ionic theory. The ray tracing technique provides an iterative procedure for the definition of the paths. However, since there exist but a very limited number of measurement sites, the characteristics of the medium itself are far from being known to an extent that would allow the application of this method. Further, the computation times would be far too long for a routine implementation of this method.

For these reasons, the forecast of the propagation of radio waves along a given link shall be obtained from a simplified calculation of the propagation, based itself on the introduction of certain simplifying assumptions, allowing to describe the paths followed by the waves in terms of successive multiple hops, i.e. alternate reflections between the Earth and one of the main ionospheric layers. These layers are the two regular layers E and F, and the sporadic layer Es which is of limited thickness but nonetheless presents high-density gradients at places where it exists. The propagation modes of practical interest are:

− modes propagating exclusively in the E layer, consisting of 1 to 3 hops,

- modes propagating exclusively in the F layer, consisting of up to 7 hops,
- mixed paths with two or three ionospheric reflections.

Fig. 4.7 illustrates some of the propagation modes usually taken into account as well as some of their possible configurations.

The geometrical characteristics of the link (the elevation angle, the skip distances, the coordinates of the reflections points at each layer, etc) are defined from the following models:

- the D layer is the layer with radius equal to $R + 80$ km, where R is the Earth's radius,
- the Es layer is the layer with radius equal to $R + 105$ km,
- the E layer is a parabolic thick layer characterised by $h_0 = 100$ km and $y_m = 25$ km. The maximum skip distance in this layer is equal to 2480 kilometres,
- the F layer is a parabolic thick layer characterised by $h_0 = 240$ km and $y_m = 96$ km. The maximum skip distance in this layer is equal to 3500 kilometres.

Determination of the Usable Frequency Band

The usable frequency band is limited on the one hand by the conditions of reflection at ionospheric layers and on the other hand by the necessity of obtaining a field with high enough amplitude at the reception point. If the frequency is too low, the wave is subjected to such an attenuation that it is no longer capable of reaching the receiver with the necessary signal level. On the other hand, if the frequency is too high, the wave is not refracted to a sufficient degree by the ionosphere, and therefore, instead of being reflected and transmitted back to the receiver, it passes through the ionosphere and disappears in the interplanetary medium.

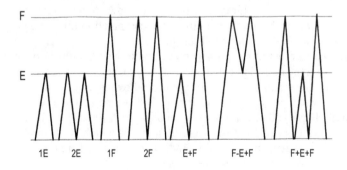

Fig. 4.7. Typical ionospheric propagation modes

The maximal useful frequency (*MUF*) is defined as the highest frequency allowing at a given time the propagation by ionospheric reflection of radio waves between two given terminal stations. This frequency is directly deduced from the values of the ionospheric characteristics.

The lowest useful frequency (*LUF*) is defined as the lowest frequency allowing at a given time the propagation by ionospheric propagation of radio waves between two given terminal stations and producing a field at the reception point of sufficient amplitude for maintaining a link of acceptable quality under specified operating conditions.

For each hour in universal time UT under consideration, the propagation of radio waves in the ionosphere will be analysed for different paths among those existing between the two extremities of the link and for different probabilities (30, 50 and 90 percent generally). The procedure can be described as follows:

- for each selected path *ch,* one calculates the maximum usable frequency MUF_{ch} and the lowest useful frequency LUF_{ch} both of which are dependent on the actual operational means. If the operational constraints are too important (for instance if the received signal is of too low an amplitude and does not allow to obtain the required signal-to-noise ratio) the LUF_{ch} may not be actually determinable,
- the paths where the MUF_{ch} is lower than the LUF_{ch} or where the LUF_{ch} is not calculable are eliminated,
- the lowest LUF_{ch} and the highest MUF_{ch} are selected among the MUF_{ch} and LUF_{ch} values of the remaining paths. These two values define the usable frequency band allowing, for a specified probability, the communications between a transmitter and a receiver.

Determination of the Maximum Useful Frequency (*MUF*). For a given link, the determination of the maximum useful frequency for each hour and each probability proceeds as follows:

- let P_i ($i = 1, n$) represent the reflection points along a path *ch.* At each of these points, the maximum useful reflection MUF_i for the hop *i* is calculated. The maximum useful frequency of the path is defined as the lowest such MUF_i.
- for a given hop *i*, the calculation of the associated MUF_i comes down to the evaluation of the upper limit $LimSup_i$ of the frequency of the ordinary mode, whence the value of the MUF_i can be directly derived using the following equation:

$$MUF_i = LimSup_i + \frac{1}{2} fh_i \qquad (4.30)$$

where fh_i is the value of the gyrofrequency at point P_i.
- the upper limit $LimSup_i$ of the frequency of the ordinary mode is evaluated in the following way:

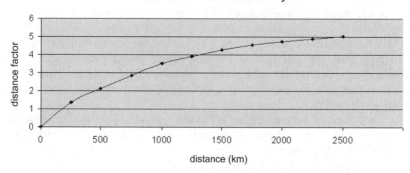

Fig. 4.8. Variation of the distance factor associated with the E layer

$$LimSup_i = f_{oi}X(d_i)$$ (4.31)

where f_{oi} is the upper limit of the frequencies reflected in vertical incidence at the point P_i. Its value depends on the layer under consideration (E, Es or F), on the hour, on the season, on the solar activity and on the selected probability. In statistical models, the 30, 50 and 90 percent time probabilities are more particularly considered.

Determination of the Upper Limit of the Frequencies reflected by the E layer. For each hop under the E layer, the following formula gives the upper limit of the reflected frequencies:

$$LimSup_E = f_o E * F_E(d)$$ (4.32)

where $f_o E$ is the monthly median value of the ordinary critical frequency of the E layer. Using the model described in the ITU-R, its value is determined from the geographical coordinates of the reflection point, the hour, the month, the solar activity and from a distance factor $F_E(d)$, which is itself a function of the skip distance in the E layer (see Fig. 4.8).

The values of $f_o E$ for the 90 and 30 per cent probabilities were obtained by multiplying the average dispersion coefficients $r = 0.95$ and $s = 1.02$.

Determination of the Upper Limit of the Frequencies reflected by the E_s layer. For each hop under the E_s layer, the upper limit of the reflected frequencies is given by the following formula:

$$LimSup_{Es} = f_o E_{Es} * F_{Es}(d)$$ (4.33)

where $f_o E_s$ is the monthly median value of the ordinary critical frequency of the E_s layer. The monthly median values foE_s (50 percent), the upper decile foE_s (90 percent) and lower decile foE_s (10 percent) at a given point and for a given hour are obtained from numerical charts. A description of each such chart can be provided in the form of a given set of coefficients U_{sk} for two different values of the solar activity: a low solar activity ($R = 10$, in 1954) and a high solar activity ($R = 180$, in 1958). Actually an interpolation has to be performed in order to obtain coefficients for the considered solar activity. This set of coefficients defines a double development, the first describing the time dependence in terms of Fourier series and the second describing the space dependence in terms of spherical harmonics.

The following equation, deduced from the median value and from the lower decile, gives the 30 percent value:

$$f_o E_s(30\%) = f_o E_s(50\%) + [f_o E_s(10\%) - f_o E_s(50\%)] * 0,3174$$ (4.34)

where $F_{Es}(d)$ is a distance factor and is a function of the skip distance in the E_s layer. The following equation yields it:

$$F_{Es}(d) = \frac{1}{Cos\alpha_{Es}}$$ (4.35)

where α_{Es} is the incidence angle at the Es layer.

The values of foE_s for the 90 and 30 percent probabilities are obtained from the multiplication of the average dispersion coefficients $r = 0.95$ and $s = 1.02$.

$$f_o E_s(90\%) = r * f_o E_s(50\%)$$ (4.36)

$$f_o E_s(30\%) = s * f_o E_s(50\%)$$ (4.37)

Determination of the Upper Limit of the Frequencies reflected by the F layer. For each hop under the F layer, the following equation gives the upper limit of the reflected frequencies:

$$LimSup_F = f_o F_2 * M(d)$$ (4.38)

where:

- $M(d)$ is a distance factor and a function of both the distance and the median value of the $M(3000)F_2$ factor,
- foF_2 is the monthly median value of the ordinary critical frequency in the F_2 layer. The monthly median values foF_2 and $M(3000)F_2$ for a given place and hour are given by numerical charts. These charts can be described by a set of coefficients U_{sk} specific to each month and considered solar activity. The values of foF_2 for the 90 and 30 percent probabilities are obtained from multiplication by the dispersion coefficients r and s, both of which depend on the season, on the hour and on the geomagnetic latitude.

$$f_o F_2(90\%) = r * f_o F_2(50\%),\qquad\qquad (4.39)$$

$$f_o F_2(30\%) = s * f_o F_2(50\%)\qquad\qquad (4.40)$$

Determination of the Lowest Useful Frequency (LUF). Once the different possible propagation modes have been identified, it is necessary, for each of these modes, to determine the lowest limit of the usable frequency band. This is imposed by the necessity for the field at the reception point to be of sufficient amplitude. More than usual, the radio noise, of natural or artificial origin, determines the practical limit beyond which the maintenance of a radio communication system with an acceptable quality is no longer possible.

For each path, each hour and each probability, the lowest useful frequency is determined as follows:

$$LUF_{ch} = Max\left(LUF_{chD}, LUF_{chA}\right)\qquad\qquad (4.41)$$

where LUF_{chD} is the lowest useful absorption frequency for the path under consideration and is determined by operational constraints. LUF_{chA} is the lowest useful screening frequency and is non-null for hop paths in the F layer. The E layer may screen off radio waves transmitted along these paths at frequencies not high enough. Let T_i (where $i = 1, n$) be the points where the path cut the E layers: at each of these points a limit LUF_{chAi} of the screening frequencies is determined. The lowest useful screening frequency LUF_{chA} is then defined as the maximum such LUF_{chAi}.

Determination of the Lowest Useful Absorption Frequency. The lowest useful absorption frequency is determined by assuming that, for a given probability, the loss transmission d_T, which is a decreasing function of the frequency, is equal to the maximum acceptable path loss dp.

In order to calculate the path loss occurring along a given link, the following parameters describing the transmission losses and gains are taken into account:

- the gains of the emitting and receiving antennas,
- the spatial attenuation, including the focusing of waves emitted under small elevation angles and waves propagating over very long distances,
- the absorption, with and without deviation,
- the auroral absorption,
- the attenuation due to reflections at the ground in the case of multiple-hop propagations.

In order to determine the lowest useful absorption frequency, the total attenuation of the signal during propagation is calculated. The total attenuation is the sum of:

- the spatial attenuation caused by the dispersion of the emitted flux at a surface, increasing with the distance from the emission antenna,
- the reflections losses in multiple-hop propagations, caused by the reflections at the ground,
- the ionospheric absorption losses generated by the collisions between the electrons and the neutral molecules.

The calculation of the lowest useful frequency comes down to the determination of the frequency at which the wave path loss is equal to a given maximum acceptable attenuation, determined by the operational performance requirements and by the level of local radio noise.

Determination of the Lowest Useful Screening Frequency. The following equation yields the lowest useful screening frequency (LUF_{occ}) at a point T where the path cuts the E layer:

$$LUF_{occ} = F_o E * S(d) \qquad (4.42)$$

where $S(d)$ is a distance factor depending on the skip distance in the F layer, on the propagation mode and on the transmission mode, while $F_o E$ is the median value of the critical frequency of the E layer.

In order to meet different types of requirements, different presentations of monthly statistical forecasts are available.

Normalised Geographical Area Forecasts

These forecasts are presented in the form of graphs indicating the approximate statistical limits of the usable operational frequency in the 2-35 MHz frequency band depending on different parameters: time, the localisation within a given delimited geographical area and the available operational means.

These forecasts are designed for radio channels whose extremities are located inside the same geographical area. They are still reasonably valid for links with length no larger than 3000 kilometres provided that the reflection point of the link is located inside this area.

Point-to-Point Forecasts

Point-to-point forecasts are presented in a form similar to that used for geographical area forecasts; these forecasts are suited to the practical situation of links operating between two given locations.

Fixed-mobile forecasts are developed for long distance links operating between a fixed station and a mobile whose geographical coordinates are known at every moment. A schematic chart indicates the positions of the mobile, represented by squares: the limits of the usable frequency band are calculated, for a given probability, at the centre of the square. The maximal useful frequency and the lowest useful frequency are then determined from the limits of the usable frequency band.

Different other presentations of these forecasts are possible, allowing for instance to search, among a list of frequencies for the frequency offering the highest probability of connection between a mobile and a fixed station, or between a mobile and several fixed stations. Forecasts can also be designed for links operating between a fixed station and a plane moving along a route at a given altitude.

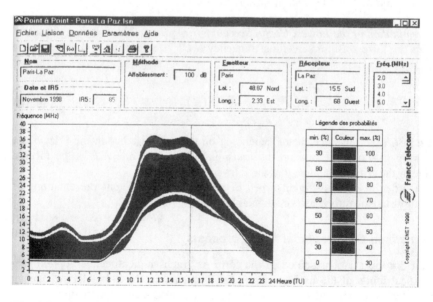

Fig. 4.9. Example of a forecast of the conditions of ionospheric propagation along a point-to-point link (ENST-Bretagne 2001)

Table 4.1. Example of a forecast in table form (ENST-Bretagne 2001)

Hour	S/B [dB]	Attenuation [dB]	Elevation [degrees]	Path	Group Time [ms]	Reliability [%]
0	59	77	9.8	2 F	14.9	95
1	59	77	10.2	2 F	15.9	95
2	59	77	10.6	2 F	15.0	91
3	60	77	10.8	2 F	15.0	85
4	60	77	10.4	2 F	15.0	82
5	61	77	8.7	2 F	14.9	85
6	58	80	7.0	2 F	14.8	89
7	64	75	6.1	F – E + F	14.7	95

Forecasts of the ionospheric propagation of radio waves are set out by different institutions worldwide: the ENST-Bretagne in France or the IPS in Australia can be mentioned here among others. As an example, Fig. 4.9 represents graphically a forecast established by the ENST-Bretagne. The corresponding values are presented in Table 4.1.

In the last recent years, a number of studies have been undertaken in order to improve both the propagation models and the associated ionospheric databases (ionospheric model) (COST 238 1999; COST 251 2001).

4.4.2 Medium-Term Forecasts

Medium-term forecasts are aimed at predicting the general conditions of propagation and more specifically the level of the maximum usable frequencies during the next week period. These forecasts are intended at three purposes:

- the correction and adaptation of long-term ionospheric forecasts with respect to season and to the evolution of solar and geomagnetic activities,
- the determination of the probability or imminence of events likely to the link, in order that all necessary arrangements for ensuring the continuity of the link be made in time,
- the dissemination of information concerning notable events which have occurred during the previous week period (sudden ionospheric disturbances, polar cap absorptions, significant auroral absorptions, etc.). This subject is addressed in more detail in Appendix F.

The forecasting method relies on a regression linear analysis over values observed in the past of the f_oF2 weekly medians for a given week and a given hour, and the corresponding value of the IR5 solar activity index. It might be recalled here that the IR5 index describes the solar activity over a five month period; more detail on the subject of solar activity indices can be found in Appendix A. In the analysis thus performed, the weeks presenting the highest level of magnetic disturbance are eliminated in order that the regression curves correspond to a calm ionosphere.

Solar activity forecasts allow determining the evolution during the next week, and for each hour of the day, of the values of the maximal useful frequency and to compare them with values defined over the long term.

The effects of ionisation may be included in the forecasts, since a high degree of correlation exists between ionisation and the magnetic activity.

The essential features of these forecasts are therefore their more accurate approach to seasonal variations and their better account of solar and magnetic activities.

4.4.3 Short-Term Forecasts

Short-term forecasts are provided over the next twenty-four hours, and are intended at determining over six hour periods the level of the usable frequency band in comparison with the usable frequency band defined over the long term. The method generally used for these forecasts relies on the research of precursor events of ionospheric disturbances among the solar and geophysics events constantly supervised in different observatories worldwide. The possible effects of such ionospheric disturbances on radio communications can then be predicted on the basis of similar situations in the past. Short-term forecasts are therefore intended at providing corrections on a daily basis of long-term forecasts over permanent areas.

4.5 Conclusion

Although the last decades have seen a remarkable development of satellite transmissions, ionospheric links still play an important part in radio communications: they require only a simple and inexpensive infrastructure and they allow low-rate transmissions of numerical data. These links find privileged fields of application in maritime communications and broadcasting services.

References

Budden KG (1961) Radiowave in the Ionosphere, Cambridge University Press, Cambridge

Ratcliffe JA (1959) The Magneto-ionic Theory and its Application to the Ionosphere. Cambridge University Press, Cambridge

Goutelard C (1979) Caractérisation du canal ionosphérique dans les transmissions ionosphériques hautes fréquences. Revue de CETHEDEC NC79-2

COST 238 (1999) PRIME (Prediction and Retrospective Ionospheric Modelling over Europe). Final report, Commission of the European Communities

COST 251 (1999) Improved Quality of Service in Ionospheric Telecommunication Systems Planning and Operation. Final report, Commission of the European Communities

ENST-Bretagne www-iono.enst-bretagne.fr

5 Terrestrial Fixed Links

5.1. Introduction

Terrestrial fixed links, also referred to as radio relay links, are based on the propagation of radio waves in the troposphere, that is, in the transition area extending between the terrestrial surfaces and the upper atmosphere. It might be recalled here that several significant meteorological phenomena occur in this region of the atmosphere : we may for instance mention the gradients of the refractive index and the presence in this region of hydrometeors, in particular rain and wet snow.

The study of terrestrial fixed links imposes to take into account a number of effects of different nature, such as attenuation, reflection, refraction, transmission, scattering or depolarisation phenomena. The aim of this chapter is to provide a survey of the main phenomena that the wave encounters during its propagation between the transmitter and the receiver.

The refractive index of the troposphere plays a most important part in this context. Within a vertical plane, the gradients of the refractive index induce the formation of ducting layers enabling the propagation of electromagnetic waves beyond the horizon. If their horizontal extent is large enough, these layers may induce significant variations of the level of the direct signal, represented by the curve 1 in Fig. 5.1, as well as variations in the angles of arrival or the appearance of multiple paths, represented by the curves 2 and 3 in this figure. The study of the refractive index and its effects in the case of a forward-scatter UHF experimental link shall be presented within this chapter.

The ground also induces different types of effects on the propagation of radio waves in the troposphere : waves may for instance be diffracted by high buildings or peaks located along the link, or reflected by the ground. The paths associated with these two phenomena are represented in Fig. 5.1 by the curves 4 and 5 respectively.

At frequencies higher than 10 GHz, electromagnetic waves interact with the neutral atmosphere and with various meteorological phenomena, for example with hydrometeors like rain, snow, or hail. These interactions generate an absorption and a diffusion of energy resulting in an attenuation of the transmitted signals. Some experimental results on this topic will be presented in the course of this chapter. Water vapour has a significant effect only at frequencies higher than 22

GHz, while the presence of hail and ice crystals creates but a limited amount of attenuation.

5.2 Radioelectric Parameters of the Atmosphere

5.2.1 Refractive Index

The study of the propagation of radio waves inside non-ionised media is essentially concerned with the lowest region of the atmosphere, the troposphere. This approximately ten kilometre thick layer is the region of the atmosphere where such meteorological phenomena as the wind or the rain appear. These phenomena depend mainly on the pressure, on the temperature and on the relative humidity of the air.

In radio electricity, the troposphere is considered as a dielectric medium with a refractive index close to unity. The variations of the refractive index of the troposphere are small but nonetheless play an important part. The following equation expresses the refractive index n:

$$n = \sqrt{\varepsilon\mu} \qquad (5.1)$$

where ε is the relative permittivity and μ is the relative permeability.

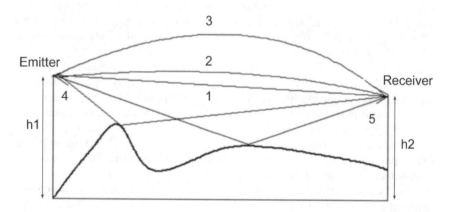

Fig. 5.1. Environment of a line-of-sight radio relay link

The average refractive index value n at ground level varies around a mean value of 1.0003. Since variations only appear at the fifth and sixth decimals, we shall hereafter consider the refractivity N which provides a practical value of the refractive index n. These two indices are connected to each other through the following formula (ITU-R 1996):

$$N = (n-1)10^6 \qquad (5.2)$$

Refractivity can be determined either by measuring it directly with a refractometer, or by calculating it from meteorological data using the following equation:

$$N = 77,6\frac{P}{T} + 3,73\times10^5\,\frac{e}{T^2} \qquad (5.3)$$

where T is the absolute temperature in K, P is the atmospheric pressure in hPa (or in mb), and e is the water vapour partial pressure in hPa.

Water vapour partial pressure is connected to humidity *(H)* and to saturation vapour pressure e_s (see Chap. 2). The first term, referred to as the dry component, is a function of both the pressure and the temperature, while the second term, called the wet component, is a function of both the humidity and the temperature. The dry component is the most important, and its contribution to the value of the refractivity lies between 60 and 80 per cent. The values assumed by the dry and wet components with respect to the different meteorological parameters are provided by abacuses (Boithias 1987).

In summer and at median latitudes, the wet component tends to increase while the dry component tends to decrease. Conversely, in winter, the wet component is weak whereas the dry component tends to grow because temperatures are low.

During summer months, anticyclonic high pressures generally result in high temperatures and in small humidity percentages, causing thereby an increase of the value of the refractivity. High pressure situations may also arise in winter, leading in this case to very low temperatures, and in arid areas, to small humidity percentages. Due to the low pressures which generally exist during the winter months, the humidity and the temperature tend to decrease during that time, and conversely, they tend to increase in summer (Bean 1966; Segal 1977).

5.2.2. Modified Refractive Index

For some applications, more particularly ray tracing, a modified refractive index M defined by the following equation may be used :

$$M = N + \frac{h}{a} \qquad (5.4)$$

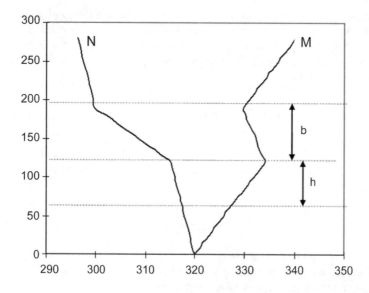

Fig. 5.2. Comparison between refractivity and the modified refractive index (b is the thickness of the super-refracting layer and h' is the height of the layer above the link)

where h is the altitude in metres of the point under consideration and a is the Earth's radius in thousands of kilometres.

This transformation reduces the problem under consideration to the problem of the propagation of radio waves above a plane Earth covered by an atmosphere with a refractivity equal to the refractive modulus M. The introduction of M leads therefore to the simplification of the equations used for the determination of trajectories. Although this method remains very useful for the resolution of ducting problems, the advances in data processing and numerical computation have significantly reduced its utility.

This leads to the following equation:

$$\frac{dM}{dh} = \frac{dN}{dh} + 157 \quad (M/km) \tag{5.5}$$

The standard value of the refractivity gradient N (- 40 N units/km) corresponds therefore to a gradient of the modified refractive index M equal to 117 M units/km. As will be seen later, in the presence of a tropospheric radio duct, then $\Delta N < -157$ N units/km or equivalently $\Delta M < 0$. A decrease with altitude of the modified refractive index M reveals the presence of a radio duct: therefore, unlike in the case of an N profile, where the slopes of the curves have to be determined, the presence of such a duct can be readily identified in a M profile by a negative slope coefficient, as can be seen in Fig. 5.2.

5.2.3 Reference Atmosphere for Refraction

In order to draw comparisons between different regions, it is essential to adopt a reference level, which means that the effects induced by altitude have to be eliminated. The basic profile proposed hereafter has been adopted by the ITU-R.

As altitude rises, the refractive index n generally decreases and tends towards the unit, while the refractivity N tends towards 0. However, this decay law is considerably disturbed in the troposphere by two factors:

– the dependence of the average decrease of n on the decrease of the atmospheric pressure,
– the variations around the average caused by irregularities in the temperature and the relative humidity.

Since the decrease of the atmospheric pressure follows an exponential law, the decay law of the refractive index is also exponential. This led to the introduction of a reference atmosphere for refraction, defined by the following equation (ITU 2000):

$$N(h) = N_0 \exp\left(-\frac{h}{h_0}\right) \tag{5.6}$$

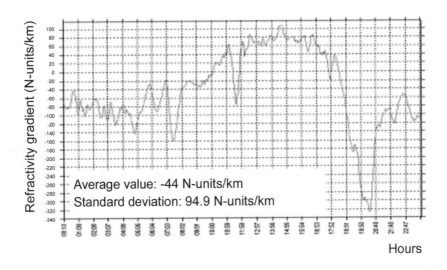

Fig. 5.3. Daily variation of the refractivity gradient observed on March 3, 1998 in Sélestat

where N_0 is the average value of refractivity extrapolated to the level of the sea (315 N-units), h is the height above the sea level and h_0 is the reference height (h_0 = 7.35 km in general).

5.2.4 Vertical Gradient of the Refractive Index

Like pressure, temperature and moisture, the refractive index of the air depends on the altitude, and this vertical variation of the refractive index is of great importance for the propagation of radio waves. Fig. 5.3 presents the daily variation of the refractivity gradient which was observed in Sélestat on March 3, 1998 (Blanchard 1999).

5.2.5 Variability of the Refractive Index

Space and Time Variations

In the first hundred metres of the atmosphere, the gradient of the refractive index in a given location may significantly vary around its standard value, depending either on time or on the altitude. This phenomenon is directly caused by the variations of the gradients of the temperature dT/dh and of the water vapour partial pressure de/dh, both of which depend on climatic phenomena like subsidence or night radiations and on the nature of the ground, for instance on the presence of water over its surface.

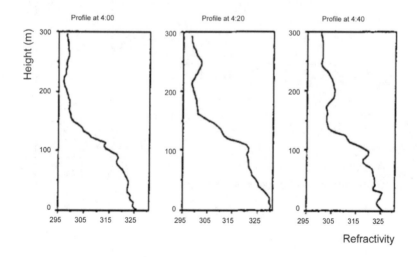

Fig. 5.4. Examples of vertical profiles

The profiles experimentally determined of the refractive index represented in Fig. 5.4 were recorded using a tethered balloon in the South West of France. These profiles show the great variability of the refractive index as well as its dependence on both the altitude and time. A refractivity pocket, resulting from an increase in water vapour, can be seen to appear at time 4:20 at the altitude of 200 metres and then move in the direction of the ground (Ishimaru 1982). The dependence on distance of the refractive index is particularly strong for mixed paths (ground-sea), due to the presence of islands along such paths (Chanh 1974).

Influence of Atmospheric Conditions

It is not unusual to take only into account the vertical component of the gradient of the refractive index. This component is dependent on the variations affecting the pressure, the humidity and the temperature. Under standard weather conditions, the following approximation holds true (Bean 1966; Boithias 1987):

$$\frac{dN}{dh} = 0,35\frac{dP}{dh} - 1,3\frac{dT}{dh} - 7\frac{dH}{dh} \tag{5.7}$$

where H is the humidity expressed in g/Kg in terms of mixing ratio, as defined by the equation:

$$H = 6,22\frac{e}{P} \tag{5.8}$$

This relation is valid in the lowest layer of the atmosphere.

As long as turbulences do not reach a degree high enough to cause a modification of the structure of the atmosphere, both the temperature and the humidity decrease with increasing altitude. In periods of calm weather, however, a temperature inversion and a strong moisture gradient may occur. As temperature inversions may occur over a large area for a relatively long time, they have a strong influence on the propagation of radio waves. Furthermore, by reducing turbulences, temperature inversions have a stabilising role, and they may therefore be an important factor in the appearance of strong gradients of humidity.

It is apparent from Eq. 5.7 that the variation of humidity is the most influencing factor on the variations of refractivity.

Different types of meteorological phenomena may induce the formation of layers of temperature inversion or the formation of layers with strong humidity gradients. As examples of such phenomena, we may mention here evaporation, advection, ground cooling by radiation and subsidence. Further detail on these different phenomena can be found in Chap. 2.

5.3 Refraction

5.3.1. Trajectories of Radio Waves

The trajectory followed by a radio wave is called a ray. The propagation of waves inside the troposphere is essentially a function of the value of refractive index and its gradient. In the case of a continuous medium, i.e. of a medium where the refractive index varies continuously with altitude, this law, known as the Snell-Descartes law, is written as follows:

$$n(h) \cos\phi(h) = \text{cste} \tag{5.9}$$

where $\phi(h)$ is the angle between the ray and the horizontal at the altitude h, as defined by the equation:

$$\varphi(h) = \frac{\pi}{2} - \theta(h) \tag{5.10}$$

In the presence of a spherical stratification, this equation must be adequately modified. The application of Snell-Descartes law to a medium with a spherical geometry is referred to as Bouguer's law. The geometry associated with this law is schematically represented in Fig. 5.6.
For a continuous medium, Bouguer's law states that:

$$n(h)r(h)\cos\phi(h) = \text{cste} \tag{5.11}$$

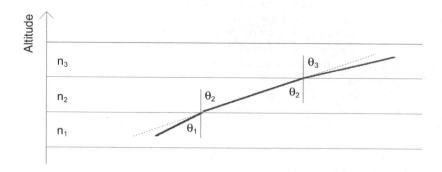

Fig. 5.5. Geometry associated with Snell-Descartes law

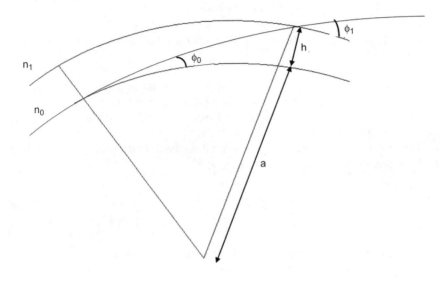

Fig. 5.6. Geometry associated with Bouguer's law

where $\phi(h)$ is the angle formed by the ray and the local horizontal plane at the altitude h, while $r(h)$ is equal to $(a+h)$. The parameter a in this expression is an arbitrary reference defined with respect to the spherical centre of symmetry. For instance, if the Earth's centre is considered as the centre of symmetry, the parameter a shall assume the same value than the radius of the Earth, i.e. approximately 6370 kilometres.

5.3.2 Radius of Curvature of the Paths of Radio Waves

As a radio wave propagates in the troposphere, the vertical gradient of the refractive index induces a bending of its path, which remains at every point of space contained within the vertical plane. The radius of curvature of the path is given by the following equation (ITU-R 2000):

$$\frac{1}{\rho} = -\frac{\cos \varphi}{n} \frac{dn}{dh} \tag{5.12}$$

where:

- ρ is the radius of curvature of the path,
- n is the refractive index of the atmosphere,
- dn/dh is the vertical gradient of the refractive index,
- φ is the angle formed by the ray path with the horizontal at the considered point.

The curvature of the path is defined as positive if its concavity is directed towards the surface of the Earth. This phenomenon does not significantly depend on the frequency as long as the gradient of the refractive index does not vary significantly over a distance equal to the wavelength.

If the antennas are located at a very short distance from the ground, then $\cos\varphi = 1$. Consequently, since $n \equiv 1$, we may write that:

$$\frac{1}{\rho} = -\frac{dn}{dh} \qquad (5.13)$$

It is thus apparent that the curvature of the trajectory is proportional to the refractivity gradient. Further, if the refractivity gradient is constant, the trajectories followed by radio waves are arcs of circle. For a reference atmosphere ($\Delta N = -40$N-units/km), the radius of curvature is equal to 25640 kilometres, approximately four times the Earth's radius of curvature. For a gradient equal to -157 N/km, the radius of curvature of the trajectories is equal to the Earth's radius: the trajectories themselves are therefore parallel to the surface of the Earth.

In order to go back to a plane Earth representation, it is often convenient to modify either the curvature of the ray path or the curvature of the Earth itself. This can be achieved by using either the concept of a modified refractive index or a representation in terms of rectilinear paths. By definition any of these two transformations preserves angles as well as the difference in curvature between the terrestrial surface and ray paths.

5.3.3 Effective Earth Radius

In order to achieve a representation in terms of rectilinear propagation, the notion of the effective Earth radius R_e may be introduced and shall be defined by the following equation:

$$\frac{1}{R_e} = \frac{1}{ka} = \frac{1}{a} + \frac{dn}{dh} \qquad (5.14)$$

where a is the real Earth's radius and k is the multiplying coefficient of the terrestrial radius (effective Earth radius factor), as defined by the equation:

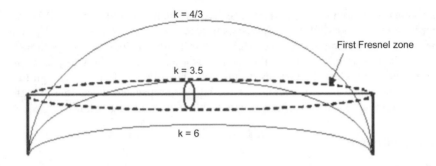

Fig. 5.7. Representation of the effective Earth radius (clearance of the first Fresnel zone)

$$k = \left(1 + a\frac{dn}{dh}\right)^{-1} \tag{5.15}$$

Thus, by measuring altitudes from the effective Earth radius, the trajectories followed by waves become rectilinear. For the reference atmosphere, k assumes a value of 4/3, while ka = 8500 kilometres = Re. Fig. 5.7 is the representation of a link according to the effective Earth radius model. The notion of the effective Earth radius has been found to be well suited for temperate climates and to the study of links operating at a short distance from the ground (Boithias 1987).

5.3.4. Effects of Refractivity Variations

In order to study the effects induced by variations in refractivity, we shall consider here an hypothetical Earth with a standard effective Earth radius. It might be noted that we could have considered as well a real or a hypothetical flat Earth: the only difference would have been the reference value ΔN, which is here equal to –39 N units/km, whereas it would have been equal to 0 N units/km in the case of the real Earth, and to -157 N units/km in the case of an hypothetical flat Earth.

Fig. 5.8 represents the trajectories followed by waves for these different values of the refractivity gradient, and according to these three possible representations (real Earth, hypothetical Earth with standard effective radius, hypothetical flat Earth) (Born 1964).

Subrefraction

If ΔN is larger than - 39 N units/km, the trajectories followed by radio waves undergo such a bending that the waves travel away from the terrestrial surface: this phenomenon is known as subrefraction. By adequately modifying the effective Earth radius, the possibility always exists to reason in terms of rectilinear trajectories. In the subrefraction case, the effective Earth radius is less than 8500 kilometres, while the effective Earth radius factor k is lower than 4/3: the Earth seems to be more curved while the distance to the radio horizon is reduced. Although this phenomenon may induce a fading of the signals propagated along line-of-sight paths, it does not have any influence in the case of forward-scatter links.

Superrefraction

The phenomenon of superrefraction occurs when ΔN is smaller than -39 N units/km. In this case, the values of the effective Earth radius factor k and the effective Earth radius are larger than 4/3 and 8500 kilometres respectively, with the consequence that the Earth seems to be flatter than it really is. Still considering an hypothetical earth with standard effective Earth radius, the curvature of the trajectories can be described as being more pronounced than the curvature of the effective terrestrial surface. The wave is therefore bent back towards the Earth and reaches points located beyond the radio horizon of the transmitter. In particular if $\Delta N < -157$ N units/km and if the wave is reflected by the ground, a propagation of the wave by multiple successive reflections will occur. This propagation mode is referred to as duct propagation.

Three different types of tropospheric radio ducts are to be distinguished: ground-base ducts or surface ducts, surface-based ducts and elevated ducts (ITU-R P.453-7). These three types are represented in Fig. 5.9. Since surface-ducts occur much less frequently than surface ducts, statistics common to these two types of ducts have been drawn up. Ducts are characterised by their intensity S_s in M units or E_s in M units and by their width $S_t(m)$ or $E_t(m)$. Two additional parameters are introduced for the characterisation of elevated ducts: the altitude E_b in metres of the lowest part of the duct, and the altitude E_m in metres where the duct presents a maximum value M. Charts for these different parameters can be obtained from the ITU-R bureau.

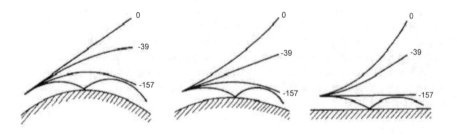

Fig. 5.8. Representation of horizontal trajectories with respect to the refractivity gradient for different representations: real Earth, effective Earth's radius equal to $4/3a$, flat Earth

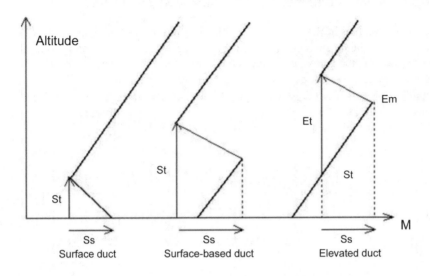

Fig. 5.9. Variation of the refractivity gradient for different types of ducts: surface duct, surface-based duct and elevated duct

5.4 Experimental Results on the Refractive Index

5.4.1 Refractivity

Fig. 5.10 represents the average daily variation of the ground refractivity observed in Sélestat from January 1998 to February 1999 (Blanchard 1999). Other examples are available throughout the literature (Bean 1966; Segal 1997; Blanchard 1997, 1999). As can be seen, some differences exist between the reported measured values and the values predicted by theoretical models, which shows the need of investigating more precisely meteorological data.

5.4.2 Refractivity Gradient

Fig. 5.11 represents the average daily variation of the refractivity gradient observed in Sélestat from January 1998 to February 1999 (Blanchard 1999). Here again, the reader will find further examples in the literature (Bean 1966; Segal 1997; Blanchard 1997, 1999).

5.4.3 Cumulative Distribution of the Refractivity Gradient

The representation of the refractivity gradient in the form of cumulative distribution curves is the most commonly used. These curves display interesting data, like for instance the median value of the refractivity gradient (50 percent of time) and the radio meteorological parameter β_0, defined as the percentage of time that the refractivity gradient is lower than - 100 N units/km. Some results concerning the distribution of the refractivity gradient over a few months at different observation sites have been collected by Blanchard (Blanchard 1999).

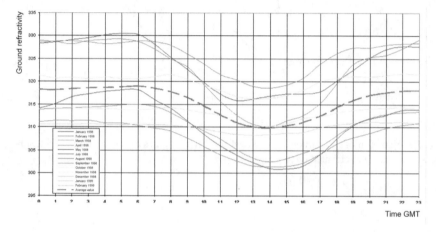

Fig. 5.10. Daily variation of the ground refractivity observed in Sélestat from January 1998 to February 1999

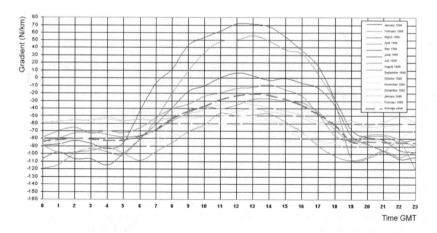

Fig. 5.11. Average daily variation of the refractivity gradient observed in Sélestat from January 1998 to February 1999

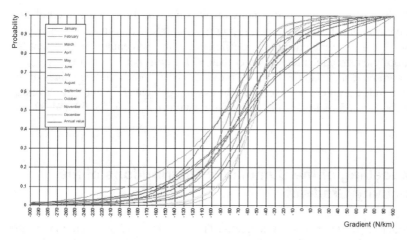

Fig. 5.12. Cumulative distribution in not exceeded values of the gradient of the refractive index during different periods

The observation of the reported distributions reveals the existence of certain differences related to season: in particular, the lowest and very negative values of the refractivity gradient, corresponding to a superrefraction phenomenon, can be observed in summer for much higher percentages of time. The same tendency was observed in almost all observation sites, and it may be noted as well as in the charts published by Bean and Dutton, or more recently by Craig and Hayton (Craig 1993). It may be further noted that the refractivity gradient assumes values in a broader range in summer than in winter, and that the slope of the curves is significantly steeper during the winter months.

Fig. 5.12 shows curves of the cumulative distribution in values not exceeded of the gradient of the refractive index, constructed from measurements realised from March 1998 to February 1999, i.e. over a whole year. In this figure, the y coordinate values represent the probability that the gradient be lower than the x coordinate values. The yearly distribution curve is plotted in thick dotted lines.

The observation of these curves shows that although the median value of the gradient of the refractive index is lower than the standard value, it does not deviate much from it.

5.4.4 Cartography

Different experiments have shown the lack of precision of the refractivity charts published first by Bean and Dutton. Further, with the advances realised in the field of data processing, it is interesting to have numerical charts of the different radio meteorological parameters, in order to integrate them directly into propagation models. For this purpose, different studies have been undertaken, for example at France Telecom R&D (Argos) or at the *Centre d'Etudes Terrestres et Planétaires*

(CETP). Fig. 5.13 presents cartography results obtained through the interpolation every five minutes of data at different points. This figure presents the values over the whole area of France of the β_0 coefficient i.e. the percentage of time that the gradient of the refractive index is lower than 100 N units per kilometre.

Recommendation ITU-R P. 453-8 provides charts indicating the values of the refractive index gradient not exceeded for a given percentage p during an average year in the first 65 metres of the atmosphere (with p = 1, 10, 50, 90 and 99 percent).

5.5 Modelling of the Cumulative Distributions of the Refractivity Gradient

The model presented in Recommendation ITU-R P.453-6 allows determining the cumulative distribution of the refractive index from a point of this distribution, which may be either the median or the β_0 coefficient (ITU-R P.453 1997). This model is composed from two submodels connected to each other by the median of the gradient of the refractive index.

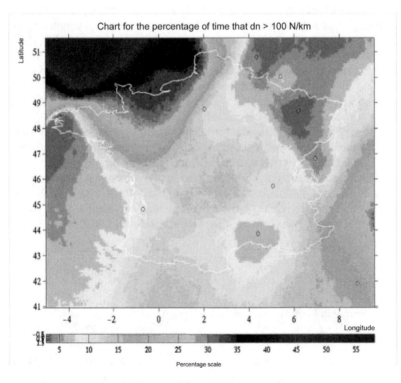

Fig. 5.13. Cartography of the β_0 coefficient from data collected by the CETP

5.5.1 Evaluation of the Median

If the time percentage P_0 that the gradient is equal to D_n is known, then the value of the median can be determined using the following equation:

$$Med = \frac{D_n + k_1}{\left(1/P_0 - 1\right)^{1/E_0}} - k_1 \qquad (5.16)$$

where $E_0 = \log_{10}\left(|D_n|\right)$ and $k_1 = 30$.

In the absence of available experimental data, the data presented in the Recommendation ITU-R P.453-6 for the β_0 coefficient (i.e. for $D_n < -100$ N units/km) can be employed.

5.5.2 Percentage of Time that the Gradient is lower than the Median

Let us assume here that the median value Med has been either experimentally measured or calculated as previously described. For each value $D_n < Med$ of the gradient of the refractive index, the following equation yields the percentage of time that the gradient is lower than D_n (Blanchard 1999):

$$P_1 = \frac{1}{1 + \left[\left(\frac{|D_n - Med|}{B} + k_2\right) * k_3\right]^{E_1}} \qquad (5.17)$$

where:

$$B = \left|\frac{0{,}3 * Med - N_s + 210}{2}\right| \qquad (5.18)$$

$$E_1 = \log_{10}(F + 1) \qquad (5.19)$$

$$F = \frac{2 * |D_n - Med|}{\left(\dfrac{B}{67}\right)^{6{,}5} + 1} \qquad (5.20)$$

$$k_2 = \frac{1.6 * B}{120} \tag{5.21}$$

$$k_3 = \frac{120}{B} \tag{5.22}$$

These expressions are valid for $Med > -120$ N-units/km and -300 N-units/km $< D_n < 50$ N-units/km

5.5.3 Percentage of Time that the Gradient is higher than the Median

For $D_n > Med$, the distribution of the gradient is given by the following equation, which can be seen as complementary to the previous one but for the coefficient k_3, which is here replaced by a coefficient k_4 (Blanchard 1999):

$$P_2 = 1 - \frac{1}{1 + \left[\left(\dfrac{|D_n - Med|}{B} + k_2\right) * k_4\right]^{E_1}} \tag{5.23}$$

where:

$$B = \left|\frac{0,3 * Med - N_s + 210}{2}\right| \tag{5.24}$$

$$E_1 = \log_{10}(F + 1) \tag{5.25}$$

$$F = \frac{2 * |D_n - Med|}{\left(\dfrac{B}{67}\right)^{6,5} + 1} \tag{5.26}$$

$$k_4 = \left[\frac{100}{B}\right]^{2.4} \tag{5.27}$$

These expressions are valid for $Med > -120$ N-units/km and -300 N-units/km $< D_n < 50$ N-units/km.

5.6 Main Propagation Mechanisms

The first researches on interferences confirmed that seven main mechanisms of propagation are to be considered in the prediction procedures of these interferences (ITU-R P.452). These propagation mechanisms can be classified into two categories:

- long-term mechanisms or normal propagation mechanisms: line-of-sight propagation, propagation by diffraction and tropospheric scatter.
- short-term mechanisms or abnormal propagation mechanisms: tropospheric radio duct, reflection and/or refraction at elevated layers of the atmosphere, line-of-sight propagation with possible enhancement of the signal, scattering by hydrometeors.

While interferences may occur through different propagation mechanisms, the prevalence of each of these mechanisms depends on several different parameters, including the climate, the frequency band used for the link, the time percentages, the length of the link and the topography. Here are the main propagation mechanisms:

- line-of-sight propagation: the most direct mode of propagation is line-of-sight propagation under homogeneous atmospheric conditions. An additional factor complexity may however appear in the presence of multiple paths creating constructive interferences, thus resulting in an increase in the signal level.

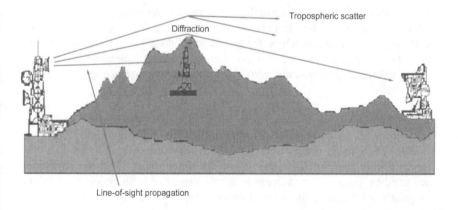

Fig. 5.14. Long-term propagation mechanism (ITU-R P.452)

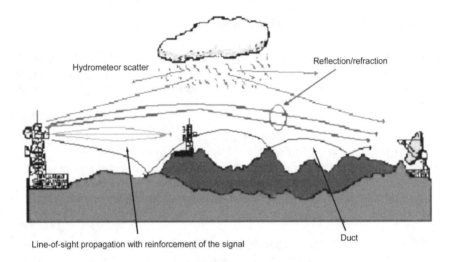

Fig. 5.15. Short-term propagation mechanisms (ITU-R P.452)

- diffraction: the effects induced by diffraction become prevalent beyond direct line-of-sight and in normal clear air situation. For services where problems of short-term abnormal propagation are of limited importance, the degree of accuracy to which diffraction can be modelled often determines the density of the radio systems which can be set up over a given area. Further, the effects induced by diffraction must be predicted in such diverse situations as a regular ground, the presence of discrete obstacles or an irregular ground.

- tropospheric scatter: scattering in the troposphere is the propagation mechanism which defines the background level of interferences for long paths (between 100 and 150 kilometres) where the diffraction field is low. However, with the exception of special cases of sensitive stations or strong transmitters (radar systems), interferences associated with tropospheric scatter are negligible due their low levels.

- tropospheric radio duct: this is the most important short-term interference mechanism above water surfaces or coastal regions with slight relief. This phenomenon may result in high amplitudes of the signals propagated over long distances (larger than 500 kilometres above the sea surface). If certain conditions are fulfilled, relating for example to the frequency or to the thickness and homogeneity of the duct, the level of these signals may be higher than the corresponding free-space level.

- reflection and/or refraction at elevated layers of the atmosphere: the analysis of reflection and refraction phenomena occurring at layers located at an altitude of a few hundreds of metres has a great importance since this propagation mode may, if a favourable path geometry exists, compensate for the attenuation induced by ground diffraction in the case of relatively long distances, ranging from 250 to 300 kilometres.

Two other propagation mechanisms which, they are unrelated to problems of clear air propagation, may nevertheless induce interferences, can be mentioned here:

- the diffraction induced by the ground and by buildings. Although this propagation mechanism was of very limited importance until recently, this may change in view of the development of high density networks (ITU-R P. 452-8 1997),

- the reflection of waves at airplanes: this mechanism is not negligible in areas with high air traffic density (CCIR 1990).

While all the propagation mechanisms discussed above are associated with variations of the refractive index, it is necessary here to introduce a distinction between propagation mechanisms created by large scale variations of the refractive index and propagation mechanisms induced by small-scale variations. For instance, phenomena like tropospheric radio ducts and spherical diffraction are the result of large scale variations of the refractive index, whereas phenomena like tropospheric scatter and reflection at elevated layers are caused by local variations of the refractive index.

5.6.1 Line-of-Sight Propagation

The simplest mode of propagation of radio waves is the propagation along line-of-sight paths. The concept of line of sight has its origins in geometrical Optics: in this context, the concept of a wave is replaced with the concept of a trajectory or a ray, while Maxwell's equations give way to simpler relations involving geometrical angles, like for instance the reflection law or Snell-Descartes law. In the terrestrial environment however, the presence along a given path of heterogeneous media, for example maritime or rural areas, may significantly modify the behaviour of the waves : indeed, in situations where geometrical optics would predict a null field, the propagation of the signals may still occur through diffraction phenomena involving obstacles present in these different media.

Propagation will generally be said to be in line-of-sight when diffraction phenomena are negligible, i.e. if there are no obstacles in the first Fresnel ellipsoid. As is described in more detail in Chap. 3, this ellipsoid delimits the region of space where almost all the energy is contained.

5.6.2 Tropospheric Scatter

Basic Principles of Tropospheric Scatter

During the 1940-1950 decade, the performances of radar and radio relay systems went through rapid and significant improvements due to such factors as the increase of the emission power in UHF, the increase of the antenna gains and the enhancement of the noise factors of the receivers. These technological advances led to the discovery of the existence, at a distance of several hundreds of kilometres, of fluctuating fields with an average amplitude significantly higher than the amplitude predicted for spherical diffraction.

Different theories were developed at this time in order to explain this phenomenon in terms for instance of atmospheric layers or whirlwinds. These theories were intended at constructing mathematical models for the prediction of the amplitude of the received fields, and at correlating the values thus predicted with actual measurements. However, this phenomenon could not be fully explained by any of these theories.

Unlike other phenomena, for example the phenomenon of ducting propagation described earlier in this chapter, tropospheric scatter is a quasi-permanent phenomenon. This property was therefore used for developing forward-scatter transhorizon radio links. Tropospheric scatter radio relay systems were thus set up in the early 1950s between stations distant from 200 to 350 kilometres, and even from 600 to 800 kilometres under favourable weather conditions. These systems enabled to provide rapidly and at a relatively low cost the radio coverage of such remote and almost inaccessible regions like the Far North or African deserts (Nemirovsky 1987).

In order to explain the propagation of radio waves over very long distances at very short wavelengths, it turned out necessary to consider the structure of the atmosphere at a significantly lower scale than is usual in meteorological studies. The very possibility of such a mode of propagation was demonstrated at relatively late a time since the corresponding attenuation, even though it is much lower than the attenuation that would result from the diffraction by the curvature of the Earth, still remains extremely high.

The different theories developed in order to account for this mode of propagation are based on the low amplitude of the air refractive index. According to some of these theories, the atmosphere, when considered at an adequate scale, behaves like a turbid medium and would therefore scatter in all directions a part of the energy which propagates inside it (scattering theories).

A different approach is based on the idea that the heterogeneities present in the atmosphere cannot be assumed to be isotropic: their dimensions are smaller along the vertical direction than along the horizontal direction. These heterogeneities would therefore act more or less like mirrors, i.e. reflect a part of the incident energy (partial reflection theory). This model actually seems quite plausible: the movements of the air in the atmosphere evidently occur at very different scales in the vertical and in the horizontal directions, and the same is probably true of the heterogeneities present in the atmosphere.

In practice, neither of these theories leads to results that would be directly usable for the prediction of propagation characteristics. As a matter of fact, these theories are based on the consideration of atmospheric magnitudes, like for instance the turbulence scale or the fluctuation spectrum, which are not routinely measurable at meteorological stations.

Without necessarily choosing one of these theories, the energy transmission system can be described as follows: the heterogeneities of the refractive index within the common volume of the antennas receive energy from the emission antenna and return a small part of this energy towards all directions, including the direction of the reception antenna. Since these heterogeneities fluctuate in time, the received level undergoes the same fluctuations. The study proceeds therefore in two steps, first the study of the average level of the received signal, then the study of the fluctuations about this mean level.

Fluctuations of the Scattering Field

Seasonal fluctuations. In areas with temperate climates, seasonal fluctuations of the scattering field can be observed, which very often result in higher degrees of attenuation in winter. The difference between the worst month average value and the annual average value may reach the order of 12 dB: this difference decreases with increasing distance, and it might be reduced to 3 dB for links with lengths from 500 to 1000 kilometres.

In areas with desert climates, the opposite phenomenon can be observed: in summer the average monthly values of attenuation may be 20 dB higher than the averages annual values. Due to these important fluctuations, the unavailability rates may be very different from one season to another, depending on the margin selected for maintaining the link.

A consequence of these variations is the recording time necessary for validating a radio link or a radio system. This difficulty indicates the importance of being able to rapidly perform a reliable evaluation of the annual average attenuation.

Daily Fluctuations. These fluctuations of the scattering field are associated with traditional meteorological variations. In areas with temperate climates, attenuations are generally minimal during the morning and maximal during the afternoon. While these variations reach very high levels in desert climates, they are small in equatorial zones.

Hourly variations follow approximately a lognormal law, with a σ standard deviation dependent on the length of the link. Indeed, the longer the link, the highest the probability that superrefraction phenomena may compensate the effects induced by subrefraction. The values which can be admitted from a statistical point of view are $\sigma = 9$ dB for links with a length of approximately 100 kilometres and $\sigma = 3$ dB for links with a length of several hundred kilometres (Boithias 1987).

The 80 and 99.9 percent attenuations can be deduced from the value found for σ using the two following relations:

$$a(80\ \%) = a(50\ \%) + 0.84\sigma \qquad (5.28)$$

$$a(99.9\ \%) = a(50\ \%) + 3.10\sigma \qquad (5.29)$$

Fast Fluctuations. The fast variations of the received field which can be observed are the consequence of the chronic instability of the atmospheric layers and whirlwinds. The resulting wave is the vector combination of several waves with no significant difference in amplitude, yet randomly out of phase. The fast fluctuations of the field therefore follow a Rayleigh law (10 dB per time decade), and several successive fades of the signal per second can be observed in UHF.

As far as interferences are considered here, it might be stressed here that scatter propagation takes place over long distances and along all directions, including directions towards areas where the signal is not desired, although the amplitude of the signal in this case is low compared with other abnormal propagation phenomena. When studying interferences problems, the interest lies more specifically in small percentages of time, contrary to the case where an operating time close to 100 percent is desired.

Scattering Geometry

In scattering problems, an important parameter is the scattering angle formed by the pointing directions of the antennas. This angle depends on the distance along the great circle and on the angles of sight α_1 and α_2. The geometry of a tropospheric link and the parameters to be considered in this context are represented in Fig. 5.16. The scattering angle in this figure is equal to:

$$\theta = \alpha + \alpha_1 + \alpha_2 \qquad (5.30)$$

where the altitude and the distance to the horizon are expressed according to the modified Earth radius factor ka. After some simplifications, the following equation for the angles of sight α_1 and α_2 is obtained:

$$\alpha_i = \frac{h_i' - h_i}{d_i} - \frac{d_i}{2ka} \qquad (i = 1,2) \qquad (5.31)$$

The equivalent distance used in attenuation calculations is defined by the following equation:

$$DE = ka\theta = ka\ (\alpha + \alpha_1 + \alpha_2) = D + ka(\alpha_1 + \alpha_2) \qquad (5.32)$$

Path Loss

Since the field may significantly vary in time, the need arises of considering its annual median value. Different models have therefore been developed in order to determine the annual median value of the field. In these models, the following parameters are used:

− the average refractive index along the vertical of the scatter area,
− the geographical distance,
− the scattering angle or the equivalent distance,
− the frequency, generally with a 30 log(f) law accounting for free-space attenuation.

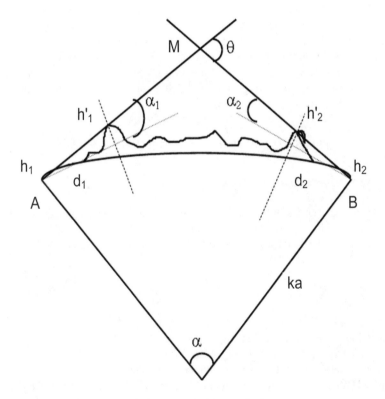

Fig. 5.16. Geometrical parameters of a tropospheric link

The correlation between climatic data and the average field is not always easy to achieve since only in temperate areas are data like the ground refractivity known with a relatively high degree of precision. More generally, although many models require that the ground refractivity be known, data for this parameter are unfortunately difficult to obtain.

Early models. The approximations that were used during the 1950s were extremely simple. As an example, here is the formula recommended at this time by RCA:

$$A_P = A_0 - 104 + 10 \log_{10}(f) + 60 \log_{10}(DE) \qquad (5.33)$$

where A_0 is the attenuation in free space, f is the frequency in MHz, and DE is the equivalent distance in kilometres.

A more elaborate model, developed on the basis of a large number of experimental data, can be found in the Technical Note 101 of the National Office of Standards (Rice 1967). Although this model can be theoretically applied to a large number of different climates, it nonetheless has the disadvantage of using a function of the product θD which requires the introduction of curve networks parameterised with a dissymmetry factor and dependent upon the ground refractivity.

For $N_s = 301$ and $\theta d < 10$, this method nonetheless provides a simple formula which can still be used. Including free-space attenuation, the median annual transmission loss is given by the equation:

$$A_P = 135,8 + 30 \log_{10}(f) + 30 \log_{10}(\theta) + 10 \log_{10}(D) + 0,34\,\theta D \qquad (5.34)$$

where θ is expressed in radians, D is in kilometres and f is in MHz,

BTRL-YEH Model. This solution recommended by the COST 210 (COST 210 1991) has been developed from the Yeh model by British Telecom Laboratories Radiocommunications (BTRL). In this model, the median transmission loss is given by the following equation:

$$A_P = 190,1 + K(f) + 20 \log_{10} D + 0,573\,\theta - 0,15\,N_0 + A_g - C_s \qquad (5.35)$$

where D is in kilometres and θ is in milliradians, while:

- A_g (in dB) is a term accounting for atmospheric absorption,
- C_s (in dB) is a corrective term introduced in order to take into account the dissymmetry of the path. The following equation yields the factor S which characterises the asymmetry of the path :

$$S = \frac{\alpha_1 + \dfrac{\alpha}{2} + \left(\dfrac{h_1 - h_2}{d}\right)}{\alpha_2 + \dfrac{\alpha}{2} - \left(\dfrac{h_1 - h_2}{d}\right)} \tag{5.36}$$

where the angles are expressed in milliradians, the heights are in metres and the distance d is in kilometres. C_s is then defined as follows:

$$C_s = \left(1,25 |Ln(S)|\right)^{1,58} \tag{5.37}$$

The COST 210 recommends considering the value for the refractivity at the sea level in order to use the data published by Bean $et~al$ (Bean 1966). In addition, the selected value should be the value at the middle of the path.

For the estimation of interferences, the attenuation not exceeded during a percentage of time p ranging between 0.001 and 50 percent can be determined from the median attenuation by using the following relation:

$$A_P(p) = A_P(50) - 10,125 \left[-\log_{10}\left(\frac{p}{50}\right) \right]^{0,7} \tag{5.38}$$

This equation shows that the level of the signal at the reception may increase from approximately 20 dB during 0.1 percent of the time, thereby inducing interferences.

Other Models. Different other models allowing the determination of the path loss could have been discussed here, among which we should in particular mention the model developed by the CCIR 1990 or the CNET model empirically developed by Boithias from the study of propagation data should be mentioned. A survey of these different models is provided by Deygout (Deygout 1994). The indicative curves given by Boithias for the values path loss observed in different climates, either for different percentages of time, or for the worst month, are also of interest (Boithias 1987).

5.6.3 Duct Propagation

If the gradient of the refractive index is lower than -157 N units/km, then the curvature of the trajectories followed by waves will be higher than the curvature of the terrestrial surface. A region of the atmospheric where these superrefraction conditions exist is referred to as a ducting layer. Ducting layers have significant effects on line-of-sight paths as well as on transhorizon links. In the presence of a duct, the very notion of radio horizon has no longer any precise meaning, since points located beyond the horizon can be reached. Further, the level of the received signal at the time of such a phenomenon may reach and even exceed its free-space level (Rana 1993; Shen 1995; Vilar 1988). Therefore, ducting layers are among the main causes of interferences occurring between communication services using the same frequency band. If the ducting layer extends at a relatively low altitude and if the ground is sufficiently reflective, then the duct is referred to as a surface-based duct. If the layer extends on the contrary at a high altitude, waves are bent successively towards the Earth and towards space, and are therefore caught between two altitudes: if these waves do not touch the ground, the duct thus formed is referred to as an elevated duct.

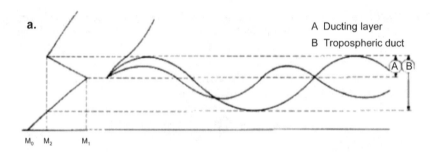

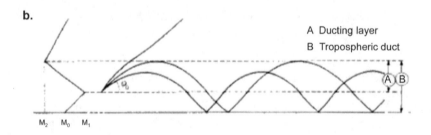

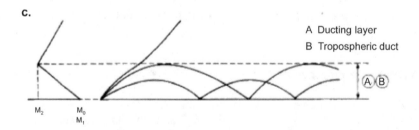

Fig. 5.17. Different types of ducting layers *a.* elevated duct *b.* surface-based duct *c.* surface duct (Boithias 1987)

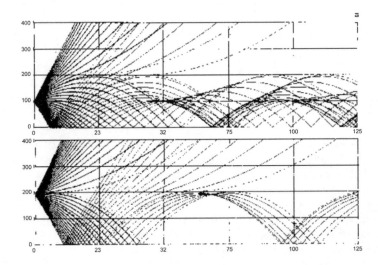

Fig. 5.18. Effects induced by the presence of a ducting layer (ray tracing) (Boithias 1987)

Ducting layers cause waves emitted by a given transmitting antenna to cross each other at certain points of space. This leads to the appearance of interference zones where waves associated to multiple paths cross each other and of zones where almost no waves propagates and where the signal level is low, the latter being referred to as radio holes. The boundary between these two zones constitutes a caustic along which the level of the signal is very high. As the refraction conditions are variable in time, a given point of space may thus be alternatively in one or the other of these zones, which results in abrupt fluctuations of the level of the received signal.

The effect of the height of the emission antennas can be observed in Fig. 5.18: waves are trapped inside the ducting layer and, assuming the antenna to be located within this layer, only a very limited amount of energy leaks away. If on the contrary the antenna is located above the ducting layer, a leakage of energy into the region extending above the ducting layer occurs.

While the thickness of a duct seldom exceeds a few hundred metres, ducting layers may extend over several hundred square kilometres, especially above coastal regions or very humid regions. The evaporation ducts which may appear due to the large negative gradients of water vapour near the sea surface have a thickness of the order of only ten metres, but they remain present during a high percentage of time.

When an emission antenna is located inside a radio duct with horizontal layers, only waves emitted at very small angles of elevation can be trapped inside the ducting layer. In the simplified case of a normal refractivity profile above a surface duct, and assuming the gradient of the refractive index to be constant, the critical angle at which waves are trapped is given by the following equation (ITU-R P.453 1997):

$$\alpha = \sqrt{2.10^{-6} \left| \frac{dM}{dh} \right| \Delta h}$$ (5.39)

where dM / dh is the vertical gradient of the modified refractive index and Δh is the thickness of the duct, corresponding to the height above the emission antenna of the upper boundary layer of the duct.

The maximal trapping angle increases rapidly as the refractive index gradients falls below - 157 N units/km or when the thickness of the duct increases. Recommendation ITU-R P.834 presents curves giving the maximum trapping angle depending on the refractive index and on the thickness of the duct (ITU-R 1995).

It can therefore be seen that the presence of a propagation duct influences the level of the received signal: although the multiple paths cause a fading of the signal, however, as far as interferences are concerned, the most important property of this mode of propagation is the possible enhancement of the signal.

5.6.4. Reflection at Elevated Layers

The previously described case of duct propagation could be reduced to the consideration of two main types: surface ducts and elevated ducts. However, the complexity of atmosphere results in the possibility for multiple layers to exist. Although the refractive index varies on average with the atmospheric pressure in altitude, local variations of the refractive index, caused by irregularities in the temperature and in the relative humidity of the air, may also appear. These irregularities are themselves due to the temperature inversion occurring at sunrise and to anticyclonic subsidence. The thin reflective layers thus created are also called sheets. The reflective capacity of these layers depends on the gradient of their refractive index. These layers may extend over several tens of square kilometres, while their thickness may reach an order of 100 metres. As an example, the altitude at which these layers form in the United Kingdom is approximately 1.4 km (Lane 1965).

The probability for such layers to exist is higher than the probability of a tropospheric radio duct. This results from the fact that while a strong variation of the refractive index may be at the origin of these layers, an additional condition for the formation of a tropospheric duct is that the thickness of the layers must be sufficient for the waves to be guided.

From a macroscopic point of view, the atmosphere exhibits an horizontal stratification: accordingly, tropospheric layers can be regarded as being roughly plane surfaces. These surfaces, however, can be deformed through a number of different mechanisms, for instance the vertical gradients of the wind, turbulences or the movement of vertical layers. The irregularities thus superimposed to the plane surface are of two types:

- primary irregularities, with a size of the order of the metre, which are generated by microscopic flows in the vicinity of the surface,
- secondary irregularities, with a size of the order of ten metres, i.e. an order of magnitude larger than the primary irregularities.

Over very broad layers, sinusoidal irregularities with wavelengths of a few kilometres can be sometimes observed (Lane 1965).

The reflection of waves at layers of this type may occur in two different ways:

- either by specular reflection if the separation surface between the layers does not present any irregularity,
- either by scattering reflection in the more frequent case where the interface is irregular.

Small variations in the characteristics of these layers, either in the quality of their surfaces or in their position, generate fluctuations of the level of the received signal. The signature of this propagation mechanism reveals the existence of both slow and fast variations, to which may be added scintillations due to tropospheric scatter.

The combination of reflection and tropospheric refraction may result in relatively high signal levels beyond the radio horizon. Refraction induces effects on the curvature of the ray paths, which result in the reduction of the incidence angle of the rays with the reflective layer and enhance the conditions of reflection. As regards the problem of interferences, it can therefore be seen that a transmitter may reach points located beyond the geometrical horizon.

Several different theories have been developed in order to account for the phenomena of reflection at these layers. The simplest such theories are very similar to Fresnel theory with the difference however that they take into account the irregularities that these layers present. For this purpose a reduction factor p is introduced through the following definition (Boithias 1987):

$$p = \exp(\frac{-8\pi^2 (\Delta h \sin \varphi)^2}{\lambda^2})$$
(5.40)

where Δh is the standard deviation of the height distribution of the irregularities inside the layer, which is assumed to be a Gaussian distribution, while φ is the incidence angle of the ray with the average surface.

An effect of the combination of reflection and refraction is that when the angle of incidence decreases, the p reduction factor tends towards the unit and accordingly the higher the reflection factor becomes. Even though the theoretically predicted value of the reflection factor is no higher than 1, this mechanism of propagation still represents a real threat in the case of relatively important percentages of time.

5.6.5 Diffraction

Basic Principles of Diffraction

The presence of an obstacle, and more specifically of the Earth, between a transmitter and a receiver may result in a received signal with a relatively high power. This is due to the fact that the direct signal and the signal emitted back by the obstacle may both be received at the reception point. This phenomenon, which cannot be explained by geometrical optics, is known as diffraction.

Although the study of diffraction phenomena is primarily concerned with boundary condition problems, the surface complexity of actual obstacles makes this study extremely difficult.

In the case of waves propagating at a close distance from the Earth, the main obstacle is the Earth itself. The higher the frequency is, the higher is the influence exerted by the irregularities of the ground. In diffraction studies, two well-studied models for the diffraction by obstacles exist: single knife-edge diffraction and rounded obstacle diffraction. These two models can then be combined for describing more complex cases. There exist other models of diffraction intended at describing more complex forms of diffraction, in particular the models developed by Deygout, Millington, Epstein and Peterson (Boithias 1987). More detailed information on this subject can be found in Appendix K devoted to the methods used for determining diffraction.

The term of spherical diffraction propagation is applied to the propagation of waves along paths without any significant diffraction edge. In this case, the amplitudes of the received signals are in general much lower than the amplitudes of the received signals in the case of propagation by knife-edge diffraction. However, due to the stability of signals propagated by spherical diffraction along favourable paths, these signals may still have a significant amplitude in the absence of ducting, i.e. for instance for percentages of time higher than 20 percent.

A link where the transmitter and the receiver are in direct visibility is not necessarily a line-of-sight link: this essentially depends on the value of the refractive index. More specifically, and as represented in Fig. 5.19, when the conditions for subrefraction are fulfilled, waves may graze the ground and be diffracted by possible obstacles present along the path.

In order to study diffraction phenomena, Fresnel introduced a division of space into different regions which have since been referred to under the name of Fresnel ellipsoids. This subject is addressed at more depth in Chap. 3. Let us here simply recall that if the first Fresnel ellipsoid, or at least 60 percent of its surface, is clear, then it is not necessary to take into account the effects induced by diffraction.

As in the case of tropospheric reflection, the combination of diffraction and refraction is a possible cause of interferences which is not to be neglected. Fig. 5.20 presents an example of a realistic path which clearly shows that the attenuation due to diffraction is inversely proportional to the gradient of the refractive index.

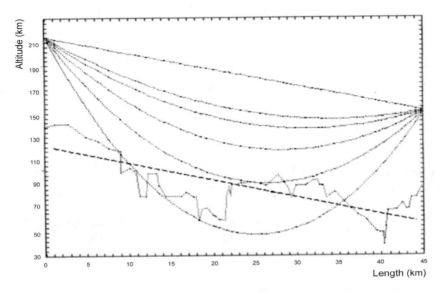

Fig. 5.19. Refraction and path profile: occultation of the radio wave in subrefraction

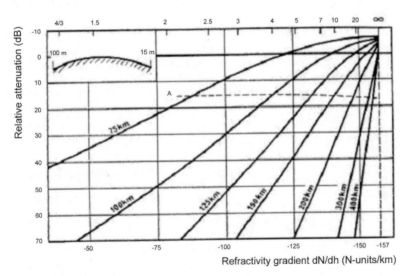

Fig. 5.20. The attenuation due to spherical diffraction at the 4 GHz frequency as a function of the gradient of the refractive index (CCIR 1990)

Under the assumption of a spherically stratified atmosphere with a vertical gradient of the refractive index, the effects induced by refraction can be represented by using the effective Earth radius, since the latter is indeed dependent on the gradient of the refractive index. The trajectories followed by waves can thus be represented by straight lines.

The effective Earth radius is given by the equation:

$$\frac{1}{R_e} = \frac{1}{ka} \qquad (5.41)$$

where k is the effective Earth radius factor, while a is the effective Earth radius.

For a standard atmosphere, $\Delta N / \Delta h = -39$ N units/km, while $k = 4/3$ and $Re = 8500$ kilometres.

Fig. 5.21 shows the representation of the radio horizon obtained by using the effective Earth radius. If h_t is the height of the emitting antenna, then the distance to the radio horizon is given by the following equation:

$$D_h = \sqrt{2R_e h_t} = \sqrt{2kah_t} \qquad (5.42)$$

Hence, the larger the k factor is, the larger is the distance to the radio horizon. Signals obtained in the conditions of spherical diffraction increase and decrease without undergoing any significant fluctuations. As the level of the signals is not as high as in the case of ducting propagation, a phenomenon of tropospheric scatter may still often appears, which means that a superposition of the effects induced by these two propagation mechanisms occurs.

The probability that such a phenomenon occurs is higher than the probability of a duct, especially in the case of short terrestrial links.

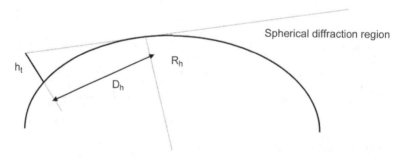

Fig. 5.21. Spherical diffraction

6.5.2 Diffraction Models

A number of different models developed for the resolution of diffraction problems can be found in the literature. These models will not however be described here: a detailed survey of these models can be found for instance in the works by Boithias and by Deygout (Boithias 1987; Deygout 1994). Among the most commonly used diffraction models, we may mention here the Bullington model, the Longley-Rice model or the Deygout model. These models derive from the researches conducted by Van Der Pol, Bremmer and Norton. Furthermore, these models are continuously improved in order to approach as closely as possible the reality, leading for instance to the development of multiple-edge models.

Nomograms are provided by the ITU-R for the estimation of attenuation caused by spherical diffraction depending on the polarisation, on the frequency and on the distance.

Elementary diffraction models also exist, for instance for the diffraction by an edge, a cylinder or a multi-cylinder. The difficulty in applying these models to the study of diffraction lies in the fact that the surface of the ground must be reduced to these elementary geometrical forms: as an example, hills are often represented by simple edges.

5.7 Experimental Results

In order to characterise these long-distance propagations over the Earth's surface, a series of measurements was conducted over a one hundred kilometre length in the plain of Alsace in north-eastern France at the 400, 900 and 2200 MHz frequencies. These measurements were aimed at determining the prevalent propagation mechanisms in the case of a terrestrial forward-scatter link and at drawing statistics for the enhancements of the signal, i.e. for the occurrence of abnormal propagation.

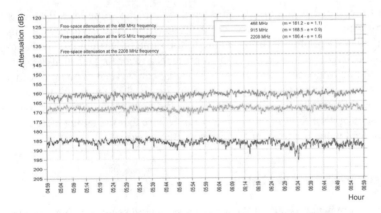

Fig. 5.22. Example of a phenomenon of tropospheric scatter at the 468 MHz, 915 MHz and 2208 MHz frequencies

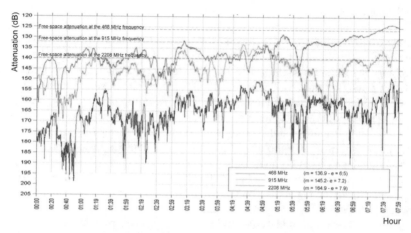

Fig. 5.23. Example of a recording showing a signal enhancement by tropospheric duct at the 468 MHz, 915 MHz and 2208 MHz frequencies

5.7.1 Tropospheric Scatter

The phenomenon of tropospheric scatter is associated with the fluctuations of the refractive index. The received signal is the sum of the signals scattered by the turbulences present in the common volume of the antennas. In Fig. 5.22 is presented a recording in the presence of a tropospheric scatter phenomenon (Blanchard 1999).

5.7.2 Duct Propagation

The level of the signal in the case of duct propagation can sometimes be higher than the corresponding level in free-space propagation. The signature of this mechanism reveals a relatively strong fading of the same order than the levels observed in line-of-sight propagation. Fig. 5.23 presents a recording taken at the time of appearance of a tropospheric radio duct: the signal can be seen to be high with a few major fading effects which range from 10 to 20 dB. The fading frequency is far less important than the scintillations owing to the fact that duct propagation is generally a stable process. It might be simply noted here that the fading effects are more pronounced when the frequency increases (Blanchard 1999).

5.7.3 Reflection at Elevated Layers

The reflection at elevated layers of the atmosphere can be seen as acting as a signal repeater. Points located beyond the radio horizon may be reached by reflection at such layers, which are not necessarily present along the entire link.

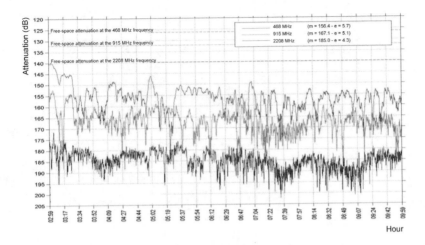

Fig. 5.24. Example of a recording showing a phenomenon of reflection at elevated layers at the 458 MHz, 915 MHz and 2208 MHz frequencies

The received signal depends on the incidence angle of the waves with these layers as well as on the quality of their surfaces. A small variation of these layers, either in the quality of their surfaces or in their position (altitude, slope, etc) may induce very strong variations of the signal level (Blanchard 1999).

5.7.4 Spherical Diffraction and Superrefraction

Under normal conditions, spherical diffraction has but a limited influence on the level of the path loss in transhorizon propagation. In the conditions of superrefraction however, i.e. when the gradient of the refractive index becomes much lower than - 40 N units/km, the distance to the radio horizon increases, thereby reducing the effects induced by diffraction.

The existence of such a phenomenon is revealed by a regular and slow decrease of the path loss. Scintillations due to tropospheric scatter are very often superimposed to the diffraction envelope since the signal level is not high enough to prevail over scintillations. A superposition of these two propagation mechanisms therefore occur (Blanchard, 1999).

5.7.5 Daily Variation of the Path Loss

Fig. 5.26 represents the daily variation of the path loss that was observed for each day of July 1998 at the 915 MHz frequency. The diurnal tendency of the path loss as well as its variability from day to day can be seen in this figure (Blanchard 1999).

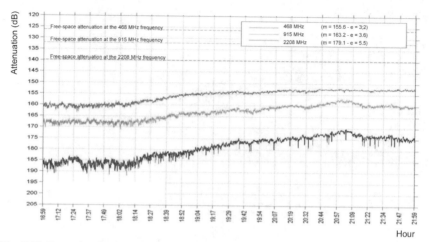

Fig. 5.25. Example of spherical diffraction combined with superrefraction at the 458 MHz, 915 MHz and 2208 MHz frequencies

5.7.6 Seasonal Variation of the Path Loss

Fig. 5.27 shows the daily average variations of the path loss observed at three different frequencies in 1998. The sliding averages for each considered frequency over a 30 day period have also been reported in this figure. As can be observed, the signal is approximately 8 dB higher during the summer months than during the winter months. The same tendency had already been observed by other authors (Boithias 1987). A strong correlation between the various curves can be noted in this figure (Blanchard 1999).

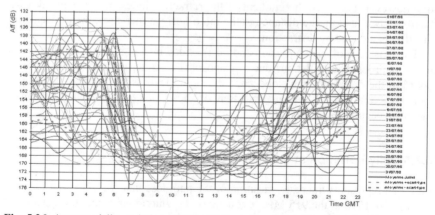

Fig. 5.26. Average daily variation of the path loss observed at the 915 MHz frequency in July 1998

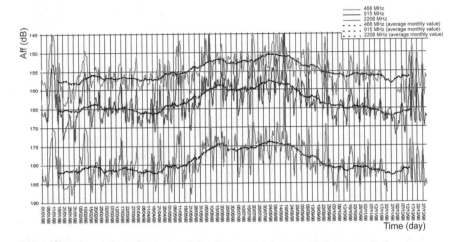

Fig. 5.27. Seasonal evolution of the path loss at the 458 MHz, 915 MHz and 2208 MHz frequencies

5.7.7 Statistical Distribution of the Path Loss

Annual Distribution

Cumulative Distribution. Fig. 5.28 presents the cumulative distributions of the path loss observed for three different frequencies in 1998. The corresponding free-space levels ($EL_{frequency}$) have also been reported in this figure in order to both provide a reference and show the abnormal character of the propagation occurring along such a link: the path loss attenuation is lower than the free-space attenuation. The representation used here has been selected is such a way as to show the small percentages of time which are to be considered when studying interference problems.

The intersection of the distribution curves with the lines representing free-space attenuation indicates the percentage of time that the path loss is lower than the free-space reference level. As can be seen, free-space attenuation was exceeded only at the lowest frequencies. At the 2.2 GHz frequency, the path loss tended strongly to the value of the free-space attenuation without however exceeding it.

At the 468 MHz frequency, free-space attenuation was exceeded during approximately 0.05 percent of time, which amounts to a total duration of 4 hours and 22 minutes over one year. At the 915 MHz frequency, this percentage of time is approximately 0.1 percent, i.e. 8 hours and 45 minutes. It should be borne in mind that this value represents the time cumulated over one year, and that the maximum duration ever observed over one day was of the order of approximately half an hour. These values show the abnormal character and the variability of the propagation along long distance links.

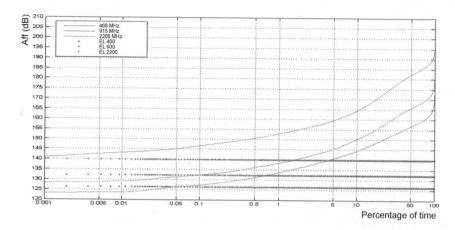

Fig. 5.28. Cumulative distribution of the path loss observed in 1998 at the 458 MHz, 915 MHz and 2208 MHz frequencies

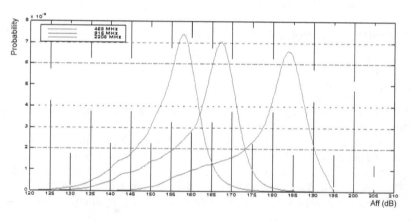

Fig. 5.29. Probability distribution curve of the path loss. From left to right, 458 MHz, 915 MHz and 2208 MHz frequencies

Probability Density. Fig. 5.29 represents the probability distribution of the path loss observed during the year 1998. It provides information on the median value of the path loss, which can be seen to be close to the most probable value, and equal to it in the Gaussian case. The observation of these curves allows determining what type of law does the path loss attenuation follows: the probability density curves appear to be close to a Gaussian curve but for the dissymmetry which can be noticed in the lowest part of the curves (Blanchard, 1999).

Daily Distribution

In Fig. 5.30 is displayed a three-dimensional representation of the evolution of the cumulative distribution of the path loss (percentage of time that the path loss attenuation is lower than a selected value of attenuation) at the 468 MHz frequency during an average day in 1998. For the sake of visualisation, the scale of the percentages of time is linear and not logarithmic. As can be seen in this figure, the highest values of attenuations are observed in the middle of the day: the three-dimensional representation shows the progressive recess with time of the distributions curves to increasing values of attenuation and the return in the evening to the values observed in the morning (Blanchard 1999).

5.7.8 Experimental Results on Propagation Mechanisms

Cross-Channel Experiment

In order to study a trans-horizon propagation link, a 110 kilometre link was established for several years between France and England (Shen 1995). Two transmitters at a frequency close to 16.147 GHz with an instantaneous frequency deviation of 1 MHz were set up in France near Le Havre. The two sites were separated by a distance of one kilometre and were located at different altitudes. The signals emitted by each transmitter could be received by the two receivers, located near Portsmouth at different altitudes and distant from 7 kilometres.

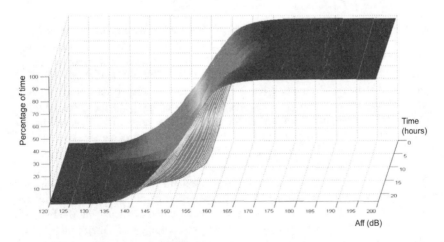

Fig. 5.30. Three-dimensional representation of the hourly cumulative distribution curves at the 468MHz frequency

Table 5.1. Percentages of time for different propagation mechanisms

	Cross-Channel experiment	Terrestrial link
Tropospheric scatter	from 73 to 87 %	from 56 to 78 %
Ducting	from 6 to 8 %	from 4 to 20 %
Reflection at elevated layers	from 2 to 3.5 %	from 10 to 20 %
Diffraction	from 5 to 14 %	from 2 to 5 %

Terrestrial Link Experiment

A terrestrial long-distance forward-scatter link with a length of 105 kilometres was established by France Telecom R&D in the plain of Alsace between Strasbourg and Mulhouse, at the altitude of 42 metres and 38 metres respectively. Three frequencies, 400, 900 and 2200 MHz, were considered in this experiment, which was conducted over a fourteen month period.

Experimental Results

The two previously described experiments allowed to obtain statistical results concerning the percentage of time that the phenomena of tropospheric scatter, ducting, reflection at elevated layers and diffraction are present (Blanchard 1999; Shen 1995). These results are presented in Table 5.1.

5.8 Influence of Rain on Propagation

The emergence of new telecommunication needs, associated with the increasing congestion of the radio spectrum, have led radio system designers to become interested in increasingly high frequencies and more specifically to frequencies in the millimetre wavelength range, ranging from 30 to 300 GHz. Given the bandwidth available in this frequency band, considerably increased data transmission rates exceeding some hundreds of Mbits/s can be obtained. At these frequencies however, atmospheric precipitations in the form of rain, snow or hail cause important disturbances on the propagation of radio waves. System designers need therefore   prediction methods predicting the effects induced by these disturbances in order to offer reliable transmission supports under any meteorological conditions.

Two different experiments were thus conducted at France Telecom R&D in order to study the influence of rain on the propagation of radio waves (Gloasguen 1993; Veyrunes 2000). The first experiment was intended at measuring the bistatic scattering of electromagnetic waves by rain particles along a 312 metre path at the 94 GHz frequency in vertical polarisation (Gloasguen 1993). This experiment was conducted using two transmitting and receiving rotating antennas, which could both be rotated independently in the horizontal plane, thereby allowing the measurement of bistatic scattering angles ranging from - 120° to +120°. The size distributions of raindrops were measured by a disdrometer at a close distance from the propagation path (Gloasguen 1995). Since narrow beam antennas were employed for this experiment and the path length was relatively short, the scattering volume was small and the rain statistics could be assumed to be homogeneous over the scattering volume. The authors demonstrate that at the 94 GHz frequency the first order multiple scattering approximation suffices for the description of the lateral scattering by rain. The computations performed using the first order multiple scattering approximation and Mie theory for spherical particles were found to be in agreement with experimentally measured values (Gloasguen 1996a, 1996b).

The second experiment was intended at simultaneously measuring the meteorological conditions and the variations of the radio field at the 30, 50, 60 and 94 GHz frequencies along an 800 metre path in line-of-sight. The statistical study of the resulting experimental data, concerning for instance the interactions occurring between electromagnetic waves and different phenomena of atmospheric and meteorological nature, or the development of propagation prediction models, will be described hereafter.

5.8.1 Experimental Device

The experimental device, set up in rural medium near the town of Belfort in 1998, consisted of a ensemble of four transmitters and four receivers operating at the 30, 50, 60 and 94 GHz frequencies in vertical polarisation and of a ensemble of meteorological sensors distributed along the radio propagation channel in order to allow an accurate characterisation of the propagation channel.

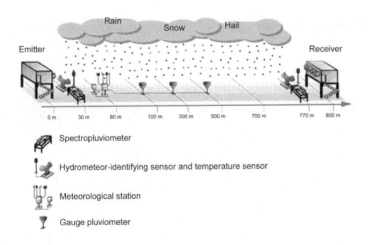

Fig. 5.31. Schematic representation of the experimental setup

Theoretical Results

By their very nature, wave propagation problems are essentially electromagnetism problems, and in theory their determination supposes the resolution of Maxwell's equations. Unfortunately, due to the complexity of natural propagation media, the resolution of these equations can only be performed once the problem has been simplified, sometimes to the point of being far removed from reality. An ideal model describing the attenuation by rain of electromagnetic waves has thus been developed (Veyrunes 2000).

Fig. 5.32 presents a comparison between the values of attenuation predicted by this model and the experimentally measured values at the 30 and 50 GHz frequencies during a very intense rain event.

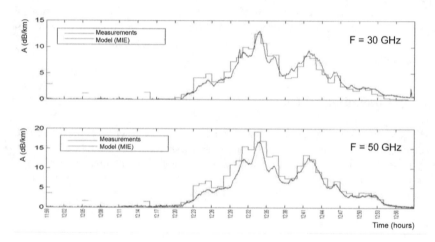

Fig. 5.32. Comparison between the predicted and the measured values of specific attenuation at the 30 and 50 GHz frequencies

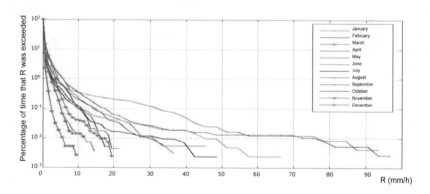

Fig. 533. Monthly cumulative distributions of rain precipitation intensities

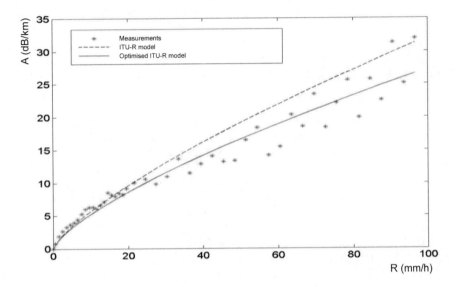

Fig. 5.34. Comparison between experimentally measured, ITU-R predicted and optimised values of the specific attenuation in dB/km at the 94 GHz frequency as a function of the rainfall rate in mm/h

Statistical Results

System designers are interested in the statistical description of disturbances induced by hydrometeors for a determined link in space and time. The statistical study of the experimental results was therefore concerned with different problems which are discussed hereafter.

Statistical Distribution of Rain Intensities. An essential characteristic of the rain regime in a given place as regards telecommunication services is the statistical distribution of the rain intensities R (mm/h). The determination of the probability of rain occurrence, expressed as the probability for a given threshold R to be reached or exceeded, is essential in this context. As an example, Fig. 5.33 presents the monthly cumulative distributions of rain precipitation intensities (Veyrunes 2000).

Attenuation due to Rain Intensity. A most fundamental parameter to consider for the understanding of the characteristics of the attenuation due to hydrometeors is the relation between the specific attenuation A (dB/km) and the rainfall rate R (mm/h). The comparison of experimentally measured values of attenuation with the values predicted by the ITU-R model ($A = kR^{\alpha}$) led to the conclusion that the parameter k was underestimated and the parameter α overestimated at the frequencies considered by this model. The optimisation of these parameters leads to enhanced performances, more specifically in the case of rain intensities higher than 20 mm/h (Veyrunes 2000).

Several different models can be found in the literature. A synthesis of these models was carried out within the COST255 framework (COST255 1999). The most powerful such models are described in Appendix I devoted to rain attenuation.

Frequency Scaling. The determination of frequency scaling law of rain attenuation along a propagation path may turn out to be necessary for extending long term measured attenuations statistics to other frequencies at which the necessary data cannot be collected, either out of time constraints or for economic reasons. The aim of the long term frequency scaling law is to derive a long term cumulative distribution of attenuation at the considered frequency from the distribution available at another frequency. A frequency scaling model of the attenuation due to rain and snow was developed in the frequency band ranging from 30 to 94 GHz.

Fig. 5.34 presents a comparison between the frequency scaling of rain attenuation at the 94 GHz frequency and rain attenuation at the 30 GHz frequency (Veyrunes 2000).

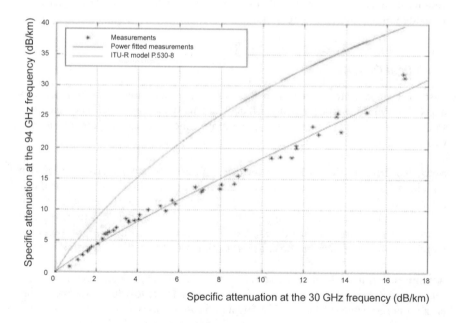

Fig. 5.35. Modelling of the correlation between the specific attenuation at the 94 GHz frequency and the specific attenuation at the 30 GHz frequency. Comparison with the ITU-R P.530-8 model

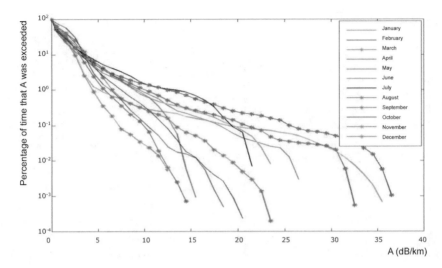

Fig. 5.36. Monthly cumulative distributions of the specific rain attenuations at the 94 GHz frequency

Statistical Distribution of Attenuation. A most important parameter in the development of communication systems is the statistical distribution of attenuation. This parameter indicates the percentage of time that the selected value of attenuation is reached or exceeded. In order to investigate this parameter, hourly, monthly, annual and worst month distributions were studied and modelled.

Fig. 5.36 represents the evolution of the experimentally determined monthly cumulative distributions of the specific rain attenuations at the 94 GHz frequency (Veyrunes 2000).

Statistics for Fade Duration. The dynamic characteristics of the attenuation due to rain were also considered in the course of this experiment. Although these characteristics have been little investigated in the past, radio system designers are now becoming increasingly interested in their study, which concern the following parameters:

- the fade durations. This parameter provides information about the distribution of the cut-off times for the link. Fig. 5.37 presents an example of seasonal cumulative distributions of rain fade durations exceeding the attenuation thresholds 2, 4, 6, 8 and 10 dB/km at the 60 GHz frequency (Veyrunes 2000).
- the intervals between individual fades. This parameter allows determining the availability of the signal, i.e. its correct operating time with respect to its specifications. Fig. 5.38 presents an example of seasonal cumulative distributions of the times between individual rain fades at the 94 GHz frequency (Veyrunes 2000).

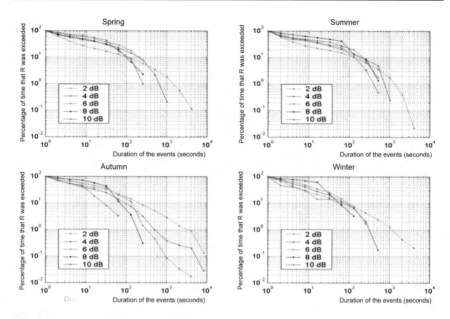

Fig. 5.37. Seasonal cumulative distributions of the rain fade durations exceeding the attenuation thresholds 2, 4, 6, 8 and 10 dB/km at the 60 GHz frequency

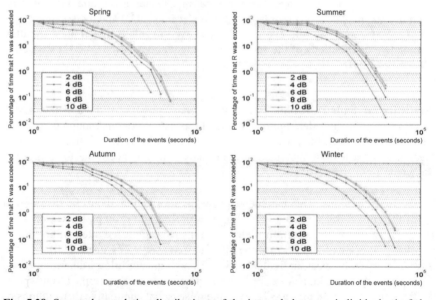

Fig. 5.38. Seasonal cumulative distributions of the intervals between individual rain fades exceeding the attenuation thresholds 2, 4, 6, 8 and 10 dB/km at the 94 GHz frequency

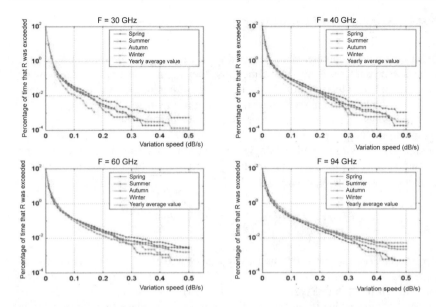

Fig. 5.39. Seasonal cumulative distributions of the fade rates associated with rain attenuation at the 30, 50, 60 and 94 GHz frequencies

- the fading rates. The study of the variations of this parameter allows evaluating the desired speed of the error control systems used for the link. Examples of seasonal cumulative distributions of the fading rates associated with rain attenuation at the 30, 50, 60 and 94 GHz frequencies are presented here in Fig. 5.39 (Veyrunes 2000).

The modelling of these different cumulative distributions has led to a significant improvement of the quality of prediction in the 30-94 GHz frequency band.

These studies lead to a better understanding of the complexity of propagation phenomena in natural media and to the possibility of drawing statistical distributions of the occurrence of the parameters involved in these phenomena. Theses statistics address the new needs of communication system designers, and should therefore enable the design of reliable transmission supports for high rate binary proximity links.

The analysis of the measurements was conducted over a 20 month period. Conducting similar experiments in regions with different rain characteristics should allow to compare the experimental results thus obtain and to extend the models.

The propagation models described here were developed for relatively short paths along which precipitations are almost homogeneous. This condition however is rarely fulfilled in practice, especially in the case of links with a length of a few kilometres. It would therefore be of interest to conduct experiments on longer links in order to develop a model for an average effective length and thus extend the results obtained for 800 metre links to longer links.

5.9 Propagation Modelling

The transmission channel shall be considered here as a linear filter and as a time invariant characterised by three functions: the impulse response in the temporal domain, the transfer function in the frequency domain and the complex transfer function, the latter being the Laplace transform of the impulse response.

The impulse response $h(t)$, which is a real-valued function, is the channel response to a Dirac impulse. The transfer function $H(\omega)$ is the Fourier transform of the impulse response and is a complex-valued function. This function may be represented either by his real and imaginary parts, or by his module and phase, as shown by the following equation:

$$H(\omega) = R(\omega) + jX(\omega) = |H(\omega)| e^{-j\Phi(\omega)} \tag{5.42}$$

$R(\omega)$ and $|H(\omega)|$ are even functions, while $X(\omega)$ and $\Phi(\omega)$ are odd functions. Furthermore, for a causal system, $h(t) = 0$ for $t < 0$: the functions $R(\omega)$ and $X(\omega)$ therefore turn out in this case to be related by a Hilbert transform (Roubine 1970).

The complex-valued transfer function $H(p)$ is the Laplace transform of the impulse response with $p = \sigma + j\omega$, while the transfer function $H(\omega)$ is the restriction of $H(p)$ to the imaginary axis. The transmission channel being passive, $H(p)$ is a continuous function and is derivable in the half-plane $\sigma > 0$, within which it has no pole : the filter presents a minimum phase difference and the functions $H(\omega)$ and $\Phi(\omega)$ are related to one another through Kramers-Krönig dispersion relations, also referred to as Bayard-Bode relations.

In the situation of multiple paths, it shall be assumed that several propagation paths extend between the transmitter and the receiver: the received signal is defined as the sum of the signals which have followed different paths. Each such signal is characterised by an amplitude a_k, a delay τ_k and a phase difference φ_k which itself is a function of both the directions of the paths and the reflections off the ground and the atmospheric layers.

Neglecting the phase difference, the transfer function is defined by the following equation:

$$H(\omega) = \sum_{k=1}^{k=N} a_k e^{-j(\omega\tau_k)} \tag{5.43}$$

while the corresponding impulse response is defined as:

$$h(t) = \sum_{k=1}^{k=N} a_k \delta(t - \tau_k) \tag{5.44}$$

The impulse response expresses the reception of N echoes of the emitted signal. The differences between the propagation times, i.e. the delays τ_k, are due to the different lengths of the paths. The attenuations a_k are induced by such propagation phenomena as the focusing or the defocusing of waves, or by effects associated with the antennas, for instance the dependence on the antenna gain on the departure and arrival angles of the wave.

Taking into account the phase differences φ_k, the transfer function becomes:

$$H(\omega) = \sum_{k=1}^{k=N} a_k e^{-j(\omega\tau_k - \varphi_k)} \qquad (5.45)$$

while the corresponding impulse response becomes:

$$h(t) = \sum_{k=1}^{k=N} a_k \left[\cos\varphi_k \delta(t-\tau_k) - \sin\frac{1}{\pi} P\left(\frac{1}{\tau_k}\right) \right] \qquad (5.46)$$

As can be seen, this equation is much more complex than the previous relation, and its solution requires the introduction of distributions.

The number of paths is too large to allow these different paths to be easily measurable. Similarly, the previously described transfer functions do not lend themselves to a practical representation of the propagation channel as regards to the determination of its quality, which is a fundamental property of a link.

There exists however three mathematical models which correctly represent the transfer functions of a transmission channel up to the 55 MHz bandwidth: the first order complex polynomial model (Greenstein 1980) the two-ray model with fixed delay (Rummler 1979) and the normalised two-ray model. All these models are models with three parameters.

5.9.1 First Order Complex Polynomial Model or Greenstein Model

The transfer function is represented in this model by the following mathematical function:

$$H(\Omega) = R_0 - (R_1 - jX_1)\Omega \qquad (5.47)$$

where Ω is the pulsation measured at the centre of the analysis band ($\Omega = \omega - \omega_c$).

5.9.2 Two-Ray Model with Fixed Delay

In this model, the transfer function is represented by the following mathematical function:

$$H(\Omega) = a\left(1 - be^{-j(\Omega-\Omega_0)\tau}\right) \qquad (5.48)$$

where Ω is the pulsation measured at the centre of the analysis band ($\Omega = \omega - \omega_c$), while the constants a, b and Ω_0 are parameters of the model and τ is equal to $1/6B$, where B is the frequency bandwidth.

5.9.3. Normalised Two-Ray Model

In this model, the transfer function is represented by the following mathematical function:

$$H(\Omega) = 1 - be^{-j(\Omega t-\varphi)} \qquad (5.49)$$

where Ω is the pulsation measured at the centre of the analysis band ($\Omega = \omega - \omega_c$) and b, t and φ are parameters of the model.

For each of these models, joint probability laws of the different parameters have been set by Lavergnat and Sylvain (Lavergnat 1985).

5.10 Performance Prediction

5.10.1 Quality Prediction

Microwave links are subjected to momentary deteriorations of theirs conditions of propagation in the form of aperiodic and selective fading effects, distortions of the group delay, field enhancements or polarisation decoupling phenomena (Leclert 1984). This results in the appearance of errors in the transmitted numerical data and may even leads to the total cut-off of the transmissions if a given error rate threshold has been exceeded. These clear air phenomena affect all frequencies: the effects induced by these phenomena on the transmissions are directly proportional on the length of the link and on the frequency. Interruptions shorter than 10 seconds concern the quality of the link, while interruptions longer than 10 seconds concern its availability.

Therefore, in order to predict the performances of a given link, both the total duration of short cut-offs, i.e. shorter than 10 seconds and long cut-offs, i.e.

longer than 10 seconds have to evaluated. These two parameters enable the prediction of the quality and the availability of the link.

The constitutive paths of the link have very diverse characteristics with respect to their length, to the roughness of the profile, to the orientation of the antenna beam, to the height of the antenna or to the climate. The conditions of propagation are therefore variable from a link to another. The prediction for a given link of these two cut-off times supposes an understanding of the conditions of propagation along these paths and of the behaviour of the systems under different conditions of propagation. The characterisation of the systems should include all the different parameters which are necessary for the evaluation of the error probability, like the frequency, the emission power, the modulation, the emission and reception filters or the transmission rates.

In order to maintain a link with a satisfactory transmission quality, even in the case where an additional attenuation due to rain is present or in clear air propagation, it is necessary that the emission power be significantly higher than the power that would be necessary if the conditions of propagation were always normal. The need therefore arises of considering a margin: the raw margin is defined as the difference between the nominal received power and the power corresponding to the error rate of cut-off.

The unavailability time due to propagation can be easily determined from the statistics of the fading depths and the raw margin The fading effect induced by rain events simply causes a decrease of the signal to noise ratio at the reception point : the behaviour of the system can therefore be easily characterised.

The prediction of the quality of a link is a more complex procedure, due to the selective character of the fading effects associated with multipath propagation, which result in distortions of the transfer function and in a lower signal to noise ratio. The prediction procedure is conducted according to two criteria: error seconds and severe error seconds. These errors appear when the lower limit of the signal to noise ratio of the transmission channel is reached. The error rate 10^{-3} has been selected as the criterion used for characterising severe error second situations in order to evaluate the quality of transmission (ITU-T G.821). A practical approach consists in assimilating the selective fading effect to a thermal noise: the fading statistics are then considered at a fixed frequency. The margin to be examined in this case is not the raw margin but a reduced margin referred to as the dispersive fade margin. The deterioration due to multipath interference can be represented through the difference between the raw margin and the dispersive fade margin. A margin reduction is said to occur due to fading.

It might be noted here that selective fading is not the only factor that might cause a margin reduction: interferences resulting from the presence of adjacent channels, or of adjacent paths in the case of a nodal station, may also induce margin reductions. As the sum of the deteriorations is not linear, a new margin, referred to as the clear margin, is introduced, which can then be used for determining the cut-off time for the link, provided however that fading statistics at fixed frequency are available.

France Telecom R&D has developed a method, known as the Prequahn method, for the prediction of the transmission quality of numerical radio relay systems (Martin 1993). This method can be used for determining the cut-off time of any radio relay link for the worst month. These calculations can be performed on the basis of the predicted values of attenuation at a fixed frequency, and for a selected quality standard based on a certain value of the binary error rate. This method is valid either with or without use of numerical equalisers, and neither does it depend on the use of space diversity, angular diversity and/or frequency diversity methods. Several applications are possible, such as for instance:

- the study of the influence on the quality of transmission of the height above the ground level of the main antennas, either in a diversity configuration or not,
- performance predictions, allowing for instance to determine the better configurations of the systems, either with or without the use of diversity methods.

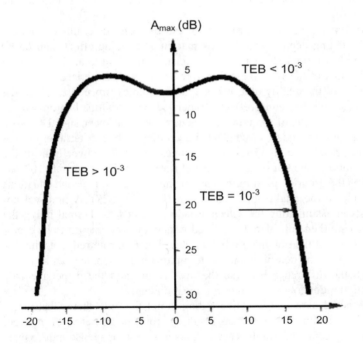

Fig. 5.40. Signature of equipment for an error rate equal to 10^{-3} per bit (Leclert 1984)

5.10.2 Signature of a Radio System

The measure of the robustness against selective fading of a radio relay system is a necessary and essential step in the prediction process of the performances of a link. This robustness should be characterised without referring to specific propagation statistics. The measurement criterion will therefore only take into account the quality of the radio relay system, leading to the possibility of comparing different systems with respect to the modulation techniques, to the synchronisation methods and to the signal analysis systems used in order to compensate for the selective fading effects.

The most commonly employed robustness measure is the signature tracing of the system, represented here in Fig. 5.40. This operation can be performed only conditionally with respect to a given propagation model, i.e. an analytical representation of the distortions due to propagation, without needing necessarily the statistics for the parameters of the model.

In this model, the two following parameters are considered:

- the frequency decentring Δf of the fading compared to the central frequency of the channel,
- the fading depth.

The signature is defined as the set of points whose coordinate values lead to a bit-error rate (BER) equal to 10^{-3} in the absence of noise. The signature thus separates the plane into two subregions (Leclert 1984).

A broader definition for the notion of signature was introduced by Emschwiler (Emschwiler 1978). Let us here consider an N parameter propagation model. For each N multiple values of the parameters of the model and the specified equipments, the transfer function of the propagation channel can be determined, as well as a performance of the channel, defined in terms of error probability for a given modulation and in terms of signal to noise ratio in the case of an analog modulation. Using a criterion (BER = 10^{-3} in the case a digital link) for separating correct and incorrect operations, the signature can then be defined as the space field D of the values of the parameters of the models at which the operation is incorrect (Levy 1989). The signature depends on both the propagation model and the choice of a correct operation, and a change in any of these two elements results in a modification of the signature itself. Propagation models adapted to multi-path channels are generally three-parameter models. Accordingly, the signature is a volume limited by a surface, and can be represented in the form of a network of curves, constituted by the intersections of the limiting surface with planes corresponding to fixed values of some parameter of the model.

Signatures can be either measured in laboratory using a simulator designed in accordance with the propagation model considered, or determined using analytical methods. Independently of the propagation characteristics of the link, a system with a signature entirely located beneath the signature of another system will be considered as being more robust than the latter. For more detailed information concerning the measurement and calculation of signatures, as well as their applications to the different propagation models (Rummler, Greenstein, etc.), the reader is referred to the very detailed communication presented by Levy *et al.* (Levy 1989).

The signature of a system can be used for the determination of the cut-off times of a link and for comparing two different systems with respect to a given link: assuming that a given propagation model has been selected, that the probability law of the parameters of the model have been defined and that the signature has been calculated, then the cut-off times can be determined per integration of the probability law on the field D of the signature (Levy 1989). The comparison between the different systems under consideration then directly proceeds from the determination of the cut-off times respectively associated with them.

References

Bean BR, Dutton EJ (1966) Radio Meteorology, US. Department of Commerce National Bureau of Standards Monograph

Bertel L (1994) Introduction aux Radiocommunications. Première partie. Cours de DEA Université de Rennes 1

Blanchard L (1999) Contribution à l'étude des propagations longues distances en UHF : Mécanismes, mesures de champ et brouillage. Ph.D. thesis, Université de Rennes

Blanchard L, Sizun H (1997a) Refractivity over France – First results of the ARGOS experiment. COST 255, Prague, Czech Republic

Blanchard L, Sizun H (1997b) Condition de refraction en France. Cartographie et comparaison avec un modèle de répartition cumulée du gradient du coïncide proposé à l'UIT-R. Troisièmes journées d'études sur la propagation électromagnétique dans l'atmosphère terrestre du décamétrique à l'angström, Rennes

Blanchard L, Sizun H (1997c) Comparison with ITU Cumulative distribution model of the refractivity gradient. COST 255, Brussels, Belgium

Blanchard L, Sizun H (1998a) First results on a terrestrial transhorizon link. COST 255, Vigo, Portugal

Blanchard L, Sizun H(1998b) Correlation studies of long distance propagation and refractivity conditions. URSI, Aveiro, Portugal

Blanchard L, Sizun H (1998c) Correlation studies of long distance propagation and refractivity conditions. COST255, Noordwijk, Netherlands

Blanchard L, Sizun H (1998d) Results of measurements of refractivity on towers in France. Improvement of the ITU-R P.453-6 Rec. cumulative distribution model of refractivity gradient. CLIMPARA'98, Ottawa, Canada

Blanchard L, Sizun H (2000) Signal levels statistic for a terrestrial over-the-horizon link at UHF band. AP2000, Davos, Switzerland

Blanchard L, Morucci S, Sizun H (1999) Transhorizon measurements compared with ITU P. 452 Rec. URSI, Toronto, Canada

Boithias L (1987) Radiowave Propagation, Mc Graw-Hill, New-York

Bye GD (1988) Radio-Meteorological Aspects of Clear-Air Anomalous Propagation in NW Europe. Br Telecom J vol 6, 3

CCIR, "Evaluation des Facteurs de Propagation liés aux Problèmes de Brouillage entre Stations Situées à la Surface de la Terre aux Fréquences Supérieures à 0,5 GHz", Rec. 569-4, 1990.

Claverie J, Hurtaud Y (1992) Validation de modèles de propagation des ondes centimétriques et millimétriques au-dessus de la mer. L'Onde Electrique vol 72: 42-46

Claverie J, Klapisz C (1985) Meteorological Features Leading to Multipath Propagation, Observed in the PACEM 1 Experiment. Annales des Télécommunications vol 40 11-12 : 660-671.

COST 210 (1991) Influence of the Atmosphere on Interference between Radio Communications Systems at Frequencies above 1 GHz. Commission of the European Communities

Craig KH, Hayton TG (1993) Investigation of β_0 Values Derived From Ten Years Radiosonde Data at 26 Stations CP 182 COST 235

Craig KH, Tjelta T (1996) Clear Air Climatic Parameters in Multipath and Interferences Models. Proceedings of URSI Commission F Workshop on Climatic Parameters in Radiowawe Propagation Prediction, Oslo, pp 8-9

Deygout J (1994) Données Fondamentales de la Propagation Radioélectrique. Les cours de l'Ecole Supérieure d'Electricité, Eyrolles, Paris

Du Castel F, Misme P, Voge J (1959) Réflexions Partielles dans l'Atmosphère et Propagation à Grande Distance. Annales des Télécommunications, Tome 14 1-2 : 33-40

Emsshwiller M (1991) Characterization of the performance of PSK digital radio transmission in the presence of multipath fading. IEEE ICC Paper 47-3

Fernandez E, Mathieu M (1991) Les Faisceaux Hertziens Analogiques et Numériques. Collection Technique et Scientifique des Télécommunications, Dunod, Paris

Gloasguen C (1993) An experiment for propagation studies at 94 GHz. Proc. ICAP93, Edinburgh, United Kingdom, pp 406-409

Gloasguen C, Lavergnat J (1995) Raindrop size distribution near Paris. Electron. Lett. 31 (5): 405-406

Gloasguen C, Lavergnat J (1996) Attenuation due to hydrometeors at 94 GHz: experimental results and comparison with theory. IEE Proc. Microwave Antennas Propagation vol 143, 1

Gloasguen C, Lavergnat J (1996b) 94 GHz Bistatic Scattering in Rain. IEEE Transactions on Antennas and Propagation vol 44, 9

Greenstein LJ, Czekaj BA (1980) A polynomial model for multipath fading channel responses. Bell. Syst. Tech. J. 59, 7: 1197-1225

Gossard EE (1981) Clear Weather Meteorological Effects on Propagation at Frequencies Above 1 GHz. Radio Science vol 16: 589-608

Hautefeuille M (1980) Anomalies de Propagation des Ondes Centimétriques au Sénégal. Revue Technique Thompson CSF vol 12 : 927-957

ITU-R P.369 (1995) Reference atmosphere for refraction. Rec. ITU-R P. 369-6

ITU-R P.581 (1995) The concept of 'worst month' Rec. ITU-R PN 581-2

ITU-R P.834 (1995) Effects of tropospheric refraction on radiowave propagation. Rec. ITU-R P. 834-1

ITU-R P.841 (1995) Conversion of annual statistics to worst-month statistics. Rec. ITU-R PN 841

ITU-R (1996) Manuel de Radiométéorologie

ITU-R P.452 (1997) Prediction procedure for the evaluation of microwave interference between stations on the surface of the Earth at frequencies above about 0.7 GHz. Rec. ITU-R P.452-8

ITU-R P.453 (1997) The radio refractive index: its formula and refractivity data. Rec. ITU-R P.453-6

Lane JA (1965) Some Investigation of the Structure of Elevated Layers in the Troposphere. J.Atmos.Terr.Phys. vol 27: 969-978

Lavergnat J, Sylvain M (1985) Statistique de la fonction de transfert par multi trajets pour des largeurs de bande moyennes. Application à la prévision de la qualité. Annales des Télécommunications 11-12: 604-616

Lavergnat J, Sylvain M (1997) Propagation des ondes radioélectriques. Masson, Paris

Leclert A, Martin L, Druais B, Vandamme P (1984) Propagation dans l'atmosphère. Première partie: influence sur les faisceaux hertziens numériques à grande capacité. L'Echo des Recherches 117

Levy AJ, Sylvain M (1989) Signature et modèles de propagation pour faisceaux hertziens numériques. Ann. Télécommun., 44, 7-8

Levy AJ (1992) Propagation Troposphérique en Air Clair. L'Onde Electrique vol 72: 28-34

Lystad S (1995) Investigation of Surface Refractivity and Refractive Gradients in the Lower Atmosphere of Norway. CP 121, COST 235

Martin L, Giraud B, Bouidene A (1993) La Prévision de la Qualité de Transmission sur les Faisceaux Hertziens Numériques. L'Echo des Recherches 153 : 51-60

Melin L, Ronnlund R, Angbratt R (1994) Radio Wave Propagation : A Comparison Between 900 and 1800 MHz, IEEE

Nemirovsky AS (1987) Les Faisceaux Hertziens à Diffusion Troposphérique. Journal des Télécommunications vol 54 : 449-454

Ravard O (1994) Etude et Modélisation de Canaux de Transmissions VHF Intermittents. PhD thesis, Université de Rennes 1

Rana D (1993) Etude Théorique sur la Caractérisation du Canal de Propagation Hertzien en Présence de Trajets Multiples. PhD thesis, Université de Paris 7

Rana D, Webster AR, Sylvain M (1993) Origin of Multipath Fading on a Terrestrial Microwawe Link. Annales des Télécommunications vol 48 : 557-566

Rice PL, Longley AG, Norton KA, Barsis AP (1967) "Transmission Loss Predictions for Tropospheric Communication Circuit. Technical Note 101 of National Bureau of Standards

Rotheram S (1989) Clear Air Aspects of the Troposphere and their Effects on Propagation Mecanisms from VHF to MILLIMETRE Waves. IEE Electomagnetic Waves Series 30, 9: 150-172

Roubine E (1970) Introduction à la théorie de la communication, Tome 1, Masson, Paris

Rummler WK (1979) A new selective fading model : application to propagation data. Bell. Syst. Tech. J. 58, 5: 1037-1071

Segal B, Barrington RE (1977) La Radioclimatologie au Canada: Atlas du Coïndice de Réfraction Troposphérique pour le Canada. Centre de Recherche sur les Communications, Rapport du CRC 1315-F, Ottawa

Shen XD (1995) Study of the Propagation Mechanisms Present in Transhorizon Links. PhD thesis, Portsmouth University

Tawfik A (1991) Experimental and Statistical Studies of X-Band Transhorizon Radio Link over the Sea. PhD thesis, Portsmouth University

Touati M, El Zein G, Citerne J (1994) Etude Expérimentale d'une Liaison Hertzienne à 15 GHz. Mesures de Propagation sur une Bande de 500 MHz. Société des Electriciens et des Electroniciens, 2èmes journées d'études, session 4, Perros-Guirec

Veyrunes O (2000) Influence des hydrométéores sur la propagation des ondes électromagnétiques dans la bande 30-100 GHz: Etudes théoriques et statistiques, PhD thesis, Université de Toulon et du Var

Vilar E, Spillard C, Rooryck M, Juy M (1988) Observations of Troposcatter and Anomalous Propagation Signal Levels at 11.6 GHz on 155 km Path Over the Sea. Electronique Letters vol 24: 1205-1207

6 Satellite Links

6.1 Introduction

Satellite links, operating either between the Earth and a satellite or the other way round, are characterised by an oblique direction of propagation. For this reason, compared with terrestrial links operating between two points located at the surface of the Earth, the study of the propagation of radio waves in this case is to some extent simplified. For instance, the influence of the ground can be neglected, thus allowing for the elimination of ground reflection and diffraction phenomena. Further, since waves generally propagate inside atmospheric layers at angles higher than a few degrees, atmospheric paths can be for the most part eliminated.

Therefore, the study of radio wave propagation between the Earth and a satellite comes down, besides free-space attenuation, to the study of phenomena related to the refractive indexes inside the troposphere and the ionosphere, to the absorption due to atmospheric gases, oxygen and water vapour in particular, and to the attenuation caused by hydrometeors like clouds, rain, fog, snow or ice.

6.2 Free-Space Attenuation

Attenuation in free-space, also referred to as transmission loss, is due to the dispersion of energy which takes place as the wave travels away from the transmitter. It is defined by the equation:

$$A_0 = -20\log_{10}\left(\frac{\lambda}{4\pi d}\right) = 32.4 + 20\log_{10}(f) + 20\log_{10}(d) \qquad (6.1)$$

where A_0 is the free-space attenuation (dB), λ is the wavelength in kilometres, d is the distance in kilometres travelled between the transmitter and the receiver and f is the frequency in MHz.

Distances being generally large, the attenuation values for space links are relatively high compared with values obtained in the case of line-of-sight radio relay systems. The difference is about 50 dB, as can be seen in Fig. 6.1.

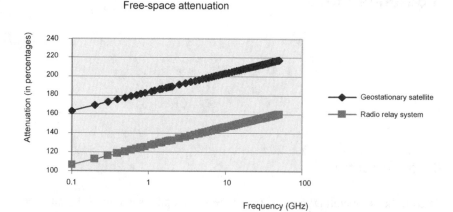

Fig. 6.1. Attenuation in free-space for a geostationary satellite (d = 36 000 km), and a radio-relay system (d=50 km)

6.3 Phenomena associated to the Refractive Indexes

The refractive indexes of the media (troposphere, ionosphere, plasmasphere, etc.) where an electromagnetic wave successively travels vary along the direction of propagation. Accordingly, an electromagnetic wave will propagate along a curvilinear direction: in this case the wave is said to be refracted. From this phenomenon originate a number of different other effects: a lengthening of the path, changes in the propagation velocity or the angle of arrival, frequency variations, scintillations, etc.

Before turning to the study of these various phenomena, an understanding of the refractive indexes of the crossed media (troposphere, ionosphere) is necessary.

6.3.1 Troposphere

The refractive index n in the troposphere is defined from the meteorological parameters (pressure, temperature, humidity). It assumes a value close to unity and is generally characterised by his refractivity N connected to n by the following equation (see Chap. 3, Sect. 2.1):

$$N = (n-1)10^6 = \frac{77,6}{T}\left(P + 4810\frac{e}{T}\right) \text{ (N-unit)} \qquad (6.2)$$

where n is the refractive index, T is the temperature in K, P is the pressure (in hectopascals or in mb) and e is the water vapour partial pressure in hectopascals.

For further information on the subjects of modified refractive indexes, the standard atmosphere, space and temporal variability and the influence of atmospheric conditions, the reader is referred to Chap. 5 devoted to terrestrial fixed links.

6.3.2 Ionosphere

The refractive index in the ionosphere may assume different forms depending on whether the influence of the geomagnetic field and the influence of the collisions (absorption) between the electrons and atoms composing the propagation medium are taken into account. Different cases are therefore to be successively considered:

Non-absorbing medium not subjected to the influence of the terrestrial magnetic field

The equation for the refractive index in this case is:

$$n = \left(1 - \frac{Ne^2}{\varepsilon_0 m\omega^2}\right)^{1/2} = \left(1 - \frac{\omega_0^2}{\omega^2}\right)^{1/2} = \left(1 - \frac{f_p^2}{f^2}\right)^{1/2} \tag{6.3}$$

where:

- E is the electron charge,
- ε_0 is the permittivity of the medium,
- m is the mass of the electron
- N is the electronic density, i.e. the number of electrons per m^3,
- ω is the pulsation of the wave,
- $\omega_0^2 = \dfrac{Ne^2}{m}$ is the plasma pulsation,
- f_p is the plasma frequency in Hertz,
- f is the frequency, expressed as follows :

$$f_p = 9\sqrt{N} \tag{6.4}$$

Absorbing medium not subjected to the influence of the terrestrial magnetic field

The equation for the refractive index in this case is:

$$n = \left(1 - \frac{\omega_0^2}{\omega(\omega - jv)}\right)^{1/2} \tag{6.5}$$

where v is the collision frequency.

Non-absorbing medium subjected to the influence of the terrestrial magnetic field

The equation for the refractive index becomes:

$$n = \sqrt{1 - \frac{\omega_0^2}{\omega^2 - \dfrac{\omega^2 \omega_H \sin^2 \theta}{2(\omega^2 - \omega_0^2)} \pm \sqrt{\omega^2 \omega_0^2 \cos^2 \theta + \dfrac{1}{2}\left(\dfrac{\omega^2 \omega_H^2 \sin^2 \theta}{\omega^2 - \omega_0^2}\right)^2}}} \tag{6.6}$$

where:

- $\omega_H = \dfrac{eB}{m}$ is the gyrofrequency,

- θ is the angle of propagation, i.e. the angle formed by the direction of propagation and the geomagnetic field.

Absorbing medium subjected to the influence of the terrestrial magnetic field

In such a medium, the equation for the refractive index is:

$$n = \sqrt{1 - \frac{\omega_0^2}{\omega(\omega - jv) - \dfrac{1}{2}\cdot\dfrac{\omega^2 \omega_H \sin^2 \theta}{\omega(\omega - jv) - \omega_0^2} \pm \sqrt{\omega^2 \omega_0^2 \cos^2 \theta + \dfrac{1}{2}\left(\dfrac{\omega^2 \omega_H^2 \sin^2 \theta}{(\omega - jv) - \omega_0^2}\right)}}} \tag{6.7}$$

Unlike in the troposphere, the refractive index here is lower than unity and is dependent upon the frequency. The refractive index thus defined is the phase index n_φ.

The following equation gives the group index, which is higher than unity:

$$n_g = \frac{1}{n_\varphi} = \frac{1}{\sqrt{1-\left(\dfrac{f_p}{f}\right)^2}} . \tag{6.8}$$

6.3.3 Refraction

The surfaces with equi-index $n(r)$ are concentric around the Earth, since the refractive index $n(r)$ is dependent on the distance from the Earth's centre. The paths followed by radio waves will therefore be defined by the following equation (see Fig. 6.2):

$$n(r).r.\cos\varphi = C^{te} \tag{6.9}$$

where r is the distance to the Earth's centre, φ is the refraction angle complementary to the incidence angle i, and $n(r)$ is the refractive index in the atmosphere.

The following equation defines the curvature at a point contained within the vertical plane:

$$\frac{1}{\rho} = \frac{-\cos\varphi}{n}\frac{dn}{dr} \tag{6.10}$$

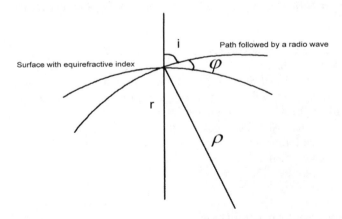

Fig. 6.2. Geometrical elements of the path followed by a radio wave

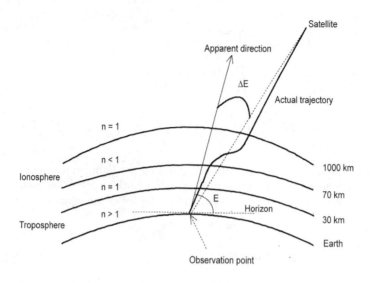

Fig. 6.3. Standard path between a satellite and the Earth

where:

- ρ is the radius of curvature of the ray path,
- n is the refractive index of the atmosphere,
- $\dfrac{dn}{dr}$ is the vertical gradient of the refractive index,
- φ is the angle to the horizontal direction of the ray path at the point under consideration.

The curvature of a ray is defined as being positive when its concavity is directed towards the Earth's surface.

In the troposphere, the vertical refractive index gradient is negative. Accordingly, the curvature of the paths will always be oriented along the same direction, and its concavity will be directed towards the Earth's centre.

In the ionosphere, the vertical refractive index gradient changes sign as the ray travels through the ionisation maximum: at this point, the curvature of the paths changes direction, as represented in Fig. 6.3.

6.3.4 Delay and Propagation Time Distortion

The propagation time from the satellite to the Earth is longer than the time calculated in free-space: indeed, the path actually followed by the wave is not rectilinear and the wave propagates with a group velocity lower than the speed of light in vacuum. The resulting delay, known as the group delay or group propagation time, induces an error in the estimation of the distance from the source, in this case from the satellite. The following equation expresses the apparent lengthening of the path as a function of the geometrical path:

$$\Delta L = \int_0^S (n-1)ds \qquad (6.11)$$

where s is the curvilinear coordinate on the path, n is the refractive index, and S is the distance separating the two extremities of the path (the terrestrial station and the satellite).

In the troposphere the path lengthening does not depend on frequency, but is affected by the meteorological conditions and decreases very rapidly with the angle of elevation. For elevation angles higher than ten degrees, the length difference can be expressed by the following equation (ITU-R P.834-2):

$$\Delta L = 0.00227 * P + \frac{1.79V}{T\sin\theta} \qquad (6.12)$$

where:

- ΔL is the length difference in metres
- P is the atmospheric pressure in hPa or in mb,
- T is the temperature at ground level in K,
- θ is the elevation angle,
- V is the total water vapour content expressed in kg/m^2 or in millimetres of precipitable water.

Values for ΔL vary from 2.2 to 2.7 metres at sea level along the zenith direction. The reader will find statistics for V in Recommendation ITU-R P.836, along with a method for extracting the value of the total atmospheric water vapour content from radiometric measurements.

In the ionosphere the path lengthening depends on both the frequency and the total electronic content, as described by the following equation:

$$\Delta L \approx \frac{40}{f^2} \int_0^S N ds \qquad (6.13)$$

where f is the frequency in MHz and N is the number of electrons (el/m^3).

In the case of a total electronic content (TEC) of $30*10^{16}$ el/m^2, the approximate lengthening at the 100 MHz frequency is of an order of 1200 metres.

Fig. 6.4 represents the variation of the path lengthening inside the ionosphere as a function of the frequency for different values of the total electronic content (10^{16}, 10^{17} and 10^{18} el/m^2). The dependency of the path lengthening on the frequency leads to a distortion of the propagation time and consequently to a deformation of the transmitted signals. The delay the wave undergoes is written in the form:

$$\Delta t = \frac{d\left(\dfrac{\Delta L}{c}\right)}{df} = \frac{80\Delta f}{cf^3}\int_0^S Nds \,. \tag{6.13}$$

The wave phase, being dependent on the total electronic content, varies in time, thereby inducing a Δf frequency shift:

$$\Delta f = \frac{40}{cf}\frac{d}{dt}\int_0^S Nds \tag{6.14}$$

The frequency of the wave received at ground level differs from the frequency emitted by the satellite. The apparent variations of frequency due to the ionosphere generally lie between 0.1 Hz (1.6 GHz) and a few Hertz in the case of low orbiting satellites emitting at 150 MHz.

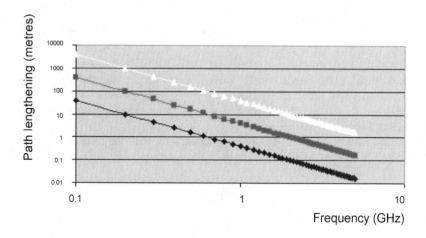

Fig. 6.4. Path lengthening in the ionosphere expressed as a function of the frequency for different values of the total electronic content: 10^{16}, 10^{17} and 10^{18} el/m^2

6.3.5 Direction of Arrival

The radio electric direction of the satellite differs from the geometrical direction by a ΔE value defined as follows:

$$\Delta E = \frac{(L + r \sin E) r \cos E}{h_i (2r + L) + (r \sin E)^2} \frac{\Delta L}{L} \tag{6.15}$$

where:

- L is the distance to the satellite,
- r is the Earth's radius,
- E is the apparent angle of elevation of the satellite, i.e. the angle as measured,
- h_i is the altitude of the electronic content average (300-450 kilometres).

6.3.6 Rotation of the Polarisation Plane

Due to the presence of the terrestrial magnetic field, the medium where the wave travels through is birefringent. Each component (ordinary and extraordinary) of the electric vector undergoes a different rotation. Consequently, a rotation of the polarisation plane of the linearly polarised wave can be observed as the wave reaches the boundaries of the ionosphere (Faraday effect):

$$\Delta\Omega = \frac{2,36 * 10^{-5}}{f^2} \int_0^S NB \cos\theta \sec\chi \, ds \tag{6.16}$$

where N is the total electronic content (TEC), B is the intensity of the terrestrial magnetic field, θ is the angle formed by the magnetic field with the direction of propagation, and χ is the solar zenith angle.

The values for $\Delta\Omega$ generally lie between 1 and 500 radians. Measurements conducted at several ground stations of the rotation of the polarisation plane of a wave emitted by a satellite have permitted the determination of the characteristics of itinerant ionospheric disturbances, such as their amplitude, their period, etc (Sizun 1979).

6.3.7 Scintillations

As it travels inside the ionospheric medium, an electromagnetic wave undergoes rapid variations in its amplitude, phase and directions of arrival. These variations, called scintillations owing to the presence of irregularities inside the medium, are characterised by their depth, period and speed variation.

Different indices (S2, IF, etc.) have been defined for the study of scintillations:

$$S_2 = \frac{\overline{X^2} - \left(\overline{X}\right)^2}{\left(\overline{X}\right)^2} \qquad (6.17)$$

6.4 Attenuation by Atmospheric Gases

The transmission attenuation caused by atmospheric gases results from the molecular resonance of oxygen and water vapour.

An oxygen molecule has a single permanent magnetic moment. At certain frequencies, its coupling with the magnetic field of an incident electromagnetic wave causes resonance absorption. In particular, at frequencies around 60 GHz a coupling occurs between the intrinsic moment of the electron, its spin, and the rotational energy of the molecule, generating a series of absorption lines quite close to each other in the spectrum. These absorption lines come to merge, thus forming a single and broad absorption band. Fig. 6.5 represents the specific attenuation due to oxygen at different altitudes in the 50-70 GHz frequency range: the lower the pressure, the higher the resolution of the bands. Fig. 6.6 represents the specific attenuation coefficient due to atmospheric gases.

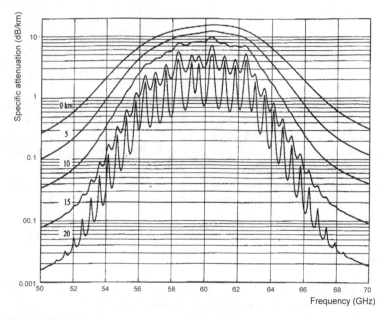

Fig. 6.5. Specific attenuation in the 50-70 GHz range at various altitudes (0, 5, 10, 15 and 20 km) (ITU-R P.676)

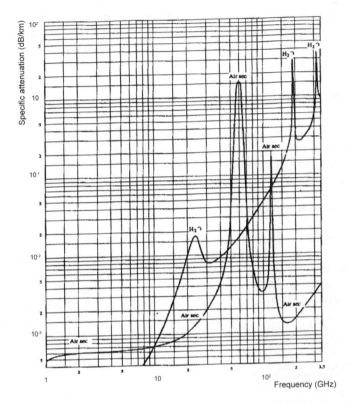

Fig. 6.6. Specific attenuation due to atmospheric gases (ITU-R P.676)

A water vapour molecule behaves like an electric dipole. The interaction of such a molecule with an incident wave disorientates the molecule by generating an additional internal potential energy. The attenuation maximum reached around the 22 GHz frequency is due to the resonance of the water molecule which starts to rotate while absorbing a high proportion of the incident electromagnetic energy.

The most accurate method for evaluating the attenuation due to atmospheric gases is by taking into account the contribution of all the absorption lines of oxygen and water vapour and the continuous spectrum of the absorption due to water and ice. Several models can be found in the literature (Liebe 1993; ITU-R 1999 Rec. ITU-R P.476-4; Salonen 1990; Gibbins 1986; Konefal et al. 1999).

The reference model, developed by Liebe et al., is known as the MPM93 model (Liebe 1993). This model allows determining the refractive index related to atmospheric oxygen and water vapour as well as the attenuation related to each of these components for frequencies up to 1000 GHz (Liebe 1981, 1985, 1989, 1993). The input parameters of this model are the pressure, the temperature, the relative humidity observed over a vertical profile of the Earth's atmosphere and the frequency (COST255 1999; ITU-R P.676-4).

As a numerical application of this model, the specific attenuation due to atmospheric gases in the case of an average atmosphere (7.5 g/m^3) was found to be equal to approximately 0.2 dB/km and 15 dB/km at 20 and 60 GHz respectively.

6.5 Hydrometeor Attenuation

The transmission attenuation due to hydrometeors like clouds, snow, fog or rain is caused by two factors: the energy absorption by Joule effect by hydrometeors and the wave diffusion induced by the particles.

6.5.1 Attenuation due to Clouds and Fog

Attenuation due to clouds and fog is determined on the basis of the total water content per volume unit. At frequencies around 100 GHz and at higher frequencies, fog attenuation may reach significant levels. The liquid water concentration is typically equal to approximately 0.05 g/m^3 inside a moderate fog (visibility of the order of 300 metres) and of 0.5 g/m^3 inside a thick fog (visibility of the order of 50 metres).

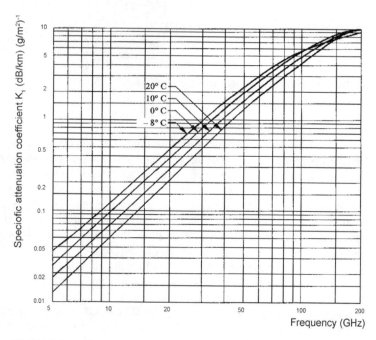

Fig. 6.7. Variation of the specific attenuation inside the cloud as a function of frequency (see Rec. ITU-R P.840)

In the case of clouds and fogs consisting entirely of very small droplets with diameter inferior to 0.01 centimetres on an average, the Rayleigh approximation is valid at frequencies lower than 200 GHz. Attenuation can therefore be expressed as a function of the total water content per volume unit (g/m^3).

The following equation yields the specific attenuation in clouds or fogs with such characteristics:

$$\gamma_c = K_l \, M \text{ (dB/km)} \tag{6.18}$$

where γ_c is the specific attenuation inside the cloud in dB/km, K_l is the specific attenuation in dB/km per g/m^3 and M is the concentration of liquid water in clouds or in fog in g/m^3.

Fig. 6.7 presents values for K_l at frequencies ranging from 5 to 200 GHz and temperatures varying between $-8\ °C$ and $20°\ C$. For attenuation due to clouds, it is appropriate to use the curve corresponding to a $0°\ C$ temperature.

In the case of a moderate fog ($0.05\ g/m^3$), the orders of magnitude for attenuation are 0.002 and 0.1 dB/km at the 20 and 60 GHz frequency ranges respectively, while for a thick fog ($0.5\ g/m^3$) they reach 0.02 and 1 dB/km respectively.

In order to determine the attenuation due to clouds for a given probability value, it is necessary to know the statistics of the total content of liquid water L in kg/m^2 contained in a column or, equivalently, the total content of precipitable water in millimetres for a given site. The equation for attenuation can thus be written in the form:

$$A = \frac{L\ K_l}{\sin \theta} \text{ (in dB)} \quad \text{with } 90° \geq \theta \geq 5° \tag{6.19}$$

where θ is the angle of elevation, and the values for K_l are as reported in Fig. 6.7.

Statistics for the total content of a liquid water column can be obtained either from radiometric measurements or by launching radio-probes. If there are no local measurements available, it is advised, in order to calculate the attenuation due to clouds, to use the values of the total content of liquid water in a cloud column (normalised at $0°\ C$) which can be collected from ITU-R data (ITU-R Rec. P.676). These data give the normalised content of liquid water in a cloud column (kg/m^2) exceeded during 20, 10, 5 and 1 percent of time over a year.

Let us describe a numerical application here: for a normalised total content of liquid water in a cloud column equal to $0.6\ kg/m^2$, a value exceeded during 1 percent of time over a year (average value in France), the zenith attenuation is equal to 0.25 and 1.2 dB at 20 and 60 GHz respectively. At an elevation angle of 45 degrees, the zenith attenuation is equal to 0.3 and 1.7 dB respectively, while at an elevation angle of 5 degrees it reaches 3 and 13 dB respectively.

6.5.2 Rain Attenuation

The determination of the attenuation due to rain relies more particularly on precipitation intensities (rainfall rates). Fig. 6.8 represents an example of the variation rate of the path loss in dB/km expressed as a function of the rainfall rate in mm/h obtained over an 800 metre horizontal link in Fontaine in France on May 31 1998, at the 30, 50, 60 and 94 GHz frequencies (Veyrunes 2000).

The specific attenuation γ_R (dB/km) is deduced from the rain rate R (mm/h) using the following power law equation:

$$\gamma_R = kR^\alpha \qquad (6.20)$$

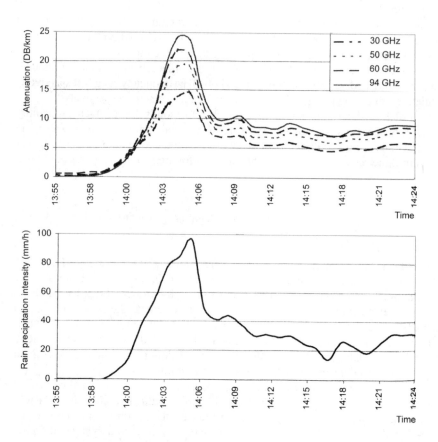

Fig. 6.8. Event of May 31, 1998: temporal rain attenuation (upper graph), temporal rainfall rate (lower graph)

The k and α coefficients depend on both the frequency and the polarisation (ITU-R P.837). Values for k and α can be obtained at different frequencies using a logarithmic scale for the frequency, a logarithmic scale for k and a linear scale for α.

For a 20 mm/h rain rate, a value exceeded during 0.1 percent of time in Belfort, rain attenuations is of the order of 2 and 8 dB/km at the 20 and 60 GHz frequencies respectively,. For a 40 mm/h intensity, a value exceeded during 0.01 percent of time, attenuation values are of the order of 3.2 and 13 dB/km respectively.

The Recommendation ITU-R, P.837 provides a model for the determination of the rainfall rate Rp, exceeded for any given percentage p of an average year, and for any given site. The same recommendation also gives examples of world charts for rain intensity (mm/h) exceeded during 0.01 percent of an average year. The data files were established on the basis of data collected for fifteen years by the *European Center of Medium-range Weather forecast* (ECMWF).

For long-term statistics concerning the calculation of rain attenuation over an oblique path for any given site, the reader is referred to Recommendation ITU-R P.618 §2.2.1.1. The input parameters are the point rainfall rate on the site over 0.01 percent of an average year (mm/h), the height above the average sea level of the ground station, the elevation angle, the latitude of the station, the frequency and the effective Earth radius. The method thus described is generally employed for the calculation of the scaling attenuation with parameters like the elevation angle, the polarisation and the frequency.

Concerning the problem of the long-term scaling in frequency and polarisation of the statistics for rain attenuation, the reader is referred to Rec. ITU-R P.618 §2.2.1.2.

A number of different models can be found in the literature. The reader will find in the COST255 final report, in addition to the ITU-R P.618 model, the most powerful models for rain rate forecasts, among which the Brazil model, the Bryant model, the Crane global model, the Crane two-component model, the DAH (Dissanayake-Alnutt-Haidara) model, the Excell model, the Flavin model, the Garcia model, the Karasawa model, the Leitao-Watson showery model, the Matricciani model, the SAM model, the Sviatogor model, the Assis-Einluft model or the Misme-Waldteufel model may be mentioned. Some of these models are detailed in Appendix I on rain attenuation.

Under the assumption that the rain to be homogeneous along the whole link, the attenuation corresponding to a given time percentage is obtained by multiplying the specific attenuation associated to the corresponding rain intensity for the same time percentage by the length of the link.

However, precipitation is generally intense only over relatively limited areas, at least in temperate climates. For this reason, rain cannot be assumed to be homogeneous. Rain attenuation could be calculated by integration from the distribution of the specific attenuation along the path using the equation:

$$A_{dB} = \int\limits_{Path} \gamma(x)dx \qquad (6.21)$$

Unfortunately neither the variation of rain intensity nor the specific attenuation along the path can actually be determined. Different practical methods have been proposed in order to account for the heterogeneity of rain. The most physical method consists in modelling intense rain areas, or rain cells. The implementation of this method requires a statistical representation of the dimensions of the cells as a function of rain intensity and even possibly to define a rain profile inside each cell, from its centre to its periphery. This physical model leads to relatively complicated calculations and would require a statistical knowledge of the structure of the rain which is far from being presently available.

The method universally employed at present is more empirical in character than the preceding, and is based on the equivalent path length notion. Assuming that the specific attenuation has already been determined, the attenuation along a path with length L is written in the form:

$$A_{dB} = \gamma kL = \gamma L_e \tag{6.22}$$

where L_e is defined as the equivalent length.

The equivalent length is always shorter than the geometrical length. This arises from the fact that attenuation forecasts are established over very small time percentages corresponding, due to the use of an equiprobability method, to high rain intensities, and the extension of such intensities is always inferior to the length of the link. The corrective factor k (or reduction coefficient required for passing from the effective distance to the hypothetical distance along which an uniform rain would fall) depends on the length of the link (with a value tending towards to unit for very short links where rain is almost homogeneous), and on the structure of the rain, i.e. on its intensity, as the latter is the only parameter allowing to characterise it. The following equation must therefore be satisfied:

$$k = k \, (L, R). \tag{6.23}$$

Other relations have been suggested where the rain rate R is replaced by the considered time percentage P, leading therefore to the expression:

$$k = k(L, P). \tag{6.24}$$

As an illustration, we indicate hereafter two different formulas proposed for the determination of the corrective factor (Lavergnat 1997):

– the equation advanced by Lin on the basis of American data is of the first type and is written in the form:

$$k(L,R) = \frac{1}{1 + L\frac{(R-6.2)}{2636}} \qquad (6.25)$$

- the equation advanced by Boithias (Boithias 1987) on the basis of European data is written in terms of time percentage rather than of rain intensity:

$$k(L,P) = \frac{1}{1 + 0.014\log(2/P)^{1.7} L^{0.9}} \qquad (6.26)$$

6.6 Depolarisation Attenuation

Using orthogonal polarisations, two independent information channels with the same frequency can be transmitted along the same link. While in theory these two orthogonally polarised channels are completely isolated from each other, in reality a certain level of interference inevitably arises between them, due to the fact that the polarisation characteristics of the antennas are not perfect and to the depolarisation effects along the propagation path. Absorption and diffusion by hydrometeors are the main causes for the cross-polarisation that takes place at centimetric and millimetric wavelength ranges. These hydrometeors can be raindrops, ice, snow or hail. In the case of oblique satellite paths, all these different types of hydrometeors may be found along the path, either simultaneously or at different times and places. All these hydrometeors share the property of existing in a non-spherical form (for instance, the flattened shape of the large raindrops), thus creating different propagation characteristics (phase coefficient and specific attenuation) along the two principal axes. The two axes can generally be considered as orthogonal. The differential phase and attenuation effects due to hydrometeors thus modify the polarisation state of the radio waves propagated through the medium.

For instance, a linearly polarised wave whose polarisation plane is not aligned on any of the principal planes of the anisotropic medium will turn into an elliptically polarised wave. This means that a component perpendicular to the original wave, albeit out of phase with respect to it, has been produced. Therefore, a part of the energy will be transferred from the original polarisation (copolar channel) unto the orthogonal polarisation (cross-polar channel). The energy transfer which has thus taken place is measured by the discrimination ratio or decoupling polarisation factor (XPD) or by the cross-polarisation isolation factor (XPI). The definition and modelling of these different parameters are addressed at more depth in Appendix D devoted to the cross-polarisation caused by the atmosphere.

Table 6.1. Building penetration loss for different types of materials at the 17 and 60 GHz frequencies

Materials	17 GHz (V)	17 GHz (H)	60 GHz (V)	60 GHz (H)
Meshed glass	1.6 dB	0.03 dB	4.3 dB	0.6 dB
Glass	1.4 dB	0.9 dB	1.8 dB	2.2 dB
Plywood (3/4" thick)	5.6 dB	5.8 dB	12.7 dB	10.9 dB
Plasterboard (1/4" thick)	0.9 dB	0.8 dB	0.8 dB	2.6 dB
Two sheets of plasterboard	1.9 dB	1.8 dB	7.3 dB	6.4 dB
Thermolite block	54.7 dB	46.0 dB	56.8 dB	51.4 dB
Aluminium sheet (1/8" thick)	48.8 dB	43.2 dB	51.9 dB	42.3 dB

6.7 Building Penetration Loss

The building penetration loss is the power attenuation that an electromagnetic wave undergoes as it propagates from outside a building towards one or several places inside this building. This parameter is generally calculated by comparing the external field and the field present in different parts of the building where the receiver is located.

The value of the building penetration loss is influenced by a number of different physical parameters, whose effects intermingle most of the time. Among these different parameters the following are traditionally distinguished:

- the near environment : a distinction is drawn between districts with high towers more or less separated from each other and more traditional districts with buildings of average height,
- the reception depth in buildings: the amplitude of the field decreases as the mobile moves from the front of the building towards a room located inside it, while the influence of the inhomogeneities decreases as the penetration depth inside a building increases. Waves penetrate more easily through window panes than through brick walls. Accordingly, the paths followed by radio waves will be more or less attenuated, and might even be occulted.
- the incidence angle, which determines the reflection and transmission coefficients at a surface,
- the frequency,
- the nature of the materials : Table 6.1 presents numerical values of the attenuation of electromagnetic waves obtained for different types of materials at the 17 and 60 GHz frequencies and in horizontal *(H)* and vertical *(V)* polarisations respectively (Fiacco 1998). Further detail on this subject can be found in the final report by the RACE *Mobile Broadband System* project (RACE-MBS R2067 1999).

6.8 Attenuation due to the Local Environment

Depending on the angle of elevation, the local environment, for example in the form of buildings in suburban and urban media or of trees along roads in rural media, may prevent the propagation of radio waves.

6.8.1 Effects of Buildings

On the subject of the shadowing effects of buildings along roads, the reader is referred to Recommendation ITU-R P.681 4.2 where a model for the statistical probability of building shadowing with respect to the elevation angle, the average heights of the buildings, the height of the mobile and the distance between the mobile and the street, is presented.

6.8.2 Effects of Vegetation

An empirical model for representing the shadowing effect induced by trees along roads has been made available for the 2-30 GHz frequency range by the ITU-R (ITU-R P.681 4.1). In this model, the influence of the trees is represented by the percentage of optical shadowing resulting from their presence for a 45 degree elevation angle along the direction of the source of the signal.

In the case where the environment is known with an adequate level of precision (for example, if geographical base contour data are available), ray models can be employed for the determination of the coverage area of a transmitter at any frequency.

The attenuation due to vegetation appears to increase with the frequency. However, while the ITU-R provides a few attenuation models in the UHF band, data available at frequencies higher than 10 GHz are few and scarce. This lack of results has been repeatedly stressed in the last recent years (Grindrod 1997; Seville 1997). Further, the existing results present quite different values for this parameter, depending on the type of vegetation under consideration.

An ancient ITU-R model gave the following equation for the excess attenuation due to vegetation:

$$A= \alpha f_{MHz}^{\beta} d^{\gamma}$$ (6.27)

where f is the frequency, d is the distance travelled inside vegetation and α, β, and γ are three parameters of the model (CCIR 1986).

A comparison between this equation and experimental measurements with the coefficients indicated by the CCIR yields a 22 dB standard deviation error (Seville 1997). Adjusting the coefficients permits to bring down the error to 11 dB. Taking into account the region simultaneously illuminated by the emitting and receiving antennas, the standard deviation error can be further reduced down to 8 dB (Seville 1997). In the aforementioned study, the author points out the great diversity of results, which depend on the type of vegetation: for instance, whereas for a spruce the transmitted wave attenuation is on an average equal to 10 dB, it is equal to 22 dB for a ficus at 40 GHz.

An experimental and theoretical study based on the energy radiative transfer theory (Schwering et al. 1988) was conducted at the 9.6, 28.8 and 57.6 GHz frequency ranges. This study reveals that the vegetation attenuation expressed in dB as a function of the distance d inside the vegetation increases linearly and rapidly, with rates ranging from 1.3 to 2 dB/m when d is relatively small, i.e. smaller than 30 metres. As d increases, the attenuation due to vegetation passes through a transition phase beyond which it still linearly increases, albeit at the slower rate of 0.05 dB/m. The explanation advanced is that this phenomenon results from the combination of the rapidly attenuated path with the more slowly attenuated diffused paths. If the diffused paths are predominant, the depolarisation of the waves can reach a high degree.

In some propagation models, the vegetation in urban areas is regarded as a shadowing region which prevents the propagation of electromagnetic waves (Correia 1996). At the 60 GHz frequency, the authors indicate values ranging from 6 to 8 dB for the attenuation due to vegetation (Correia et al. 1994).

A few qualitative results have been presented in a study realised at frequencies near 28 GHz by the US-WEST. According to this study, the attenuation due to vegetation may reach the order of several tens of dB. The excess attenuation due to vegetation and shadowing effects in suburban areas compared to free-space attenuation was found to have an average value, of interest from a statistical point of view, equal to 4 dB/km with a 10 dB standard deviation at the 28 GHz frequency (US-WEST 1995).

The presence of vegetation is also the cause of the fast fading of the received field, correlated with the speed of the wind. In narrow band, an adjustment of this fast fading can be performed using a Rice-type law (US-WEST 1995). The study conducted by US-WEST shows that in this case the Rice parameter is correlated to a reasonable extent with the additional attenuation due to vegetation.

Measurements of the attenuation of the signal have also been carried out at the 60 GHz frequency in forest areas along roadside trees whose branches extend over the road (Grindrod *et al.* 1997). The transmitter is therefore only partially in line-of-sight. Up to a distance equal to 20 metres, the attenuation due to the foliage hanging over the road is no higher than 5 dB. Beyond this distance, strong variations of signal can be observed, which are explained by the authors by the interferences arising between the direct path attenuated by the leaves and the diffused paths. When the transmitter is placed directly behind a group of conifers, the attenuation reaches values ranging from 30 to 50 dB, corresponding to distances inside vegetation ranging from 60 to 80 metres.

In addition, the crossing of vegetation causes a depolarisation of the waves in direct proportion to the additional attenuation that it induces. For an initial isolation value of 24 dB between the two horizontal and vertical polarisations in free-space propagation, experiments carried out by the US-WEST have shown that the isolation of the two polarisations decreases by 1 dB when the additional attenuation due to vegetation rises by 3 dB. It suggests that the depolarisation effect due to vegetation is much more important than the depolarisation due to rain.

For further detail on the evaluation of attenuation due to vegetation, the reader is referred to Appendix J on vegetation attenuation.

References

Boithias L (1987)) Radiowave Propagation, Mc Graw-Hill, New-York

COST 255 (1999) Radiowave propagation modelling for Satcom services at Ku-band and above. COST255 Final Wokshop, Bech, Luxembourg

Fiaco M, Parks H, Radi H, Saunders SR (1998) Indoor Propagation factors at 17 and 60 GHz. Final report by the Centre for Communication Systems Research on behalf of the Radiocommunications Agency

Foulonneau B, Glangetas E, Bic JC (2000) Characteristics and reflection scattering profile of building materials in the 60 GHz band for indoor propagation prediction tools. AP2000, Davos

Gibbins CJ (1986) Improved algorithms for the determination of specific attenuation at sea level by dry air and water vapour in the range 1-350 GHz. Radio Science 21 6: 945-954

Grindrop EA, Hammoudeh A (1997) Performance characterisation of millimetre wave mobile radio systems in forests. 10th ICAP pp 2.391-2.246

ITU-R P.618 Propagation data and prediction methods required for the design of Earth-space telecommunication systems

ITU-R P.676 Attenuation by atmospheric gases

ITU-R P.833 Attenuation in vegetation

ITU-R P.834 Effects of tropospheric refraction on radiowave propagation

ITU-R P.837 Characteristics of precipitation for propagation modelling

ITU-R P.839 Rain height model for prediction methods

Lavergnat J, Sylvain M (1997) Propagation des ondes radioélectriques. Introduction. Masson, Paris

Liebe HJ (1983) Modeling attenuation and phase of radio waves in air at frequencies below 1000 GHz. Radio Science 16:183-1199

Liebe HJ (1985) An updated model for millimetre wave propagation in moist air. Radio Science 20 5: 1069-1089

Liebe HJ (1989) MPM - An atmospheric millimetre-wave propagation model. Int. Journal of infrared and millimetre waves, 10: 631-650

Liebe HJ, Hufford GA, Cotton MG (1993) Propagation modelling of moist air and suspended water/ice particles at frequencies below 1000 GHz. AGARD 52nd Specialists meeting of the EM wave propagation panel, Palma de Maiorca

RACE-MBS (1999) Final report on propagation aspects. R067/IST/2.2.5/DS/P/070.b1

Salonen E *et al.* (1990) Study of propagation phenomena for low availabilities. ESA/ESTEC Contract 8025/88/NL/PR, Final report

Schwering FK, Violette EJ, Espeland RH (1988) Millimetre wave propagation in vegetation: experiments and theory. IEEE Trans. On Geoscience and remote sensing vol 26 3: 355-367

Seville (1997) Vegetation attenuation : modelling and measurements at millimetric frequencies, 10th ICAP: 2.5-2.8

Sizun H (1979) Les perturbations ionosphériques itinérantes de moyenne échelle, étude et recherche de leur source. PhD Thesis, Université de Rennes

US WEST Advanced Technology Inc, HP Company; (1995) Final Report of the LMDS Phoenix Field Trial. Report T-09_02-004963-01.00

7 Mobile Radio Links

7.1 Introduction

Compared with the three previously described links (ionospheric, terrestrial and Earth-satellite links), mobile radio links are based on the concept of a non line-of-sight propagation between the transmitter, i.e. the base station and a mobile receiver. This is imposed by the fact that, although the radio coverage must be ensured in different types of environments (urban areas, suburban areas, rural areas or indoors), the base station cannot always be set up on a high point like for instance a tower, a hill or a mountain.

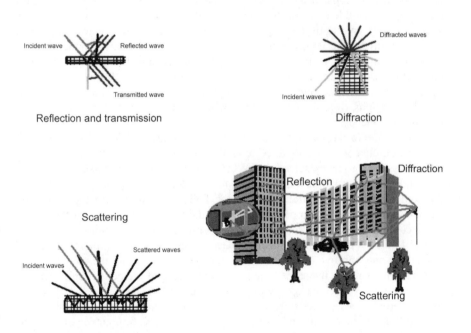

Fig. 7.1. Propagation mechanisms

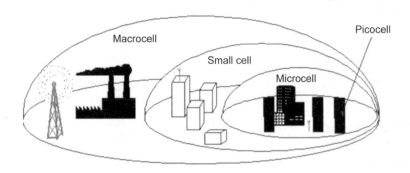

Fig. 7.2. Types of cells

In this context, radio waves will no longer be propagated only in line-of-sight but through a variety of propagation mechanisms: reflection, for instance at mountainsides or at walls, diffraction at edges, either horizontal (roofs) or vertical (corners of buildings), scattering by vegetation or guiding in street canyons. These different modes of propagation are here schematically summarised in Fig. 7.1. This results in a multitude of elementary paths. All these paths are characterised by an attenuation, an amplitude and a phase difference, and interfere when they arrive unto the receiver. These interferences can be either constructive when the different paths arrive in phase, leading to a reinforcement of the signal, or destructive, causing in this case a fading of the signal. Further the mobile itself moves inside this figure of interferences, and therefore propagates successively through luminous and dark regions (interference fringes), which results in a major fading or fast variations of the signal.

Further, due to the multiple paths and to the displacements of the mobile, the propagation channel presents three fundamental properties: attenuation, temporal variability and frequency selectivity. Whereas for analog communication systems, the concept of attenuation was sufficient for the study of the propagation channel, for numerical communication systems, the fading resulting from the variability and the selectivity of the propagation channel induces a deterioration of the quality of the communication which occurs independently of the attenuation.

Due to the ever increasing demand, telecommunication operators have been led to increase and improve their network, which means increasing the number of the base stations while at the same time reducing the size of the cells. Let us here recall that a cell is the area covered by a base station. As represented in Fig. 7.2, four different types of cells are commonly distinguished: macrocell, small cell, microcell and picocell, depending on the physical base station antenna and on its geographical coverage area. The characteristics of these cells are dependent on the location, on the power and on the height of the base station antenna height as well as on the geographical environment.

The largest cell is the macrocell, with an activity radius of the order of several ten kilometres. The environment of cells of this type is generally rural or mountainous, and the base station antenna is located at an elevated point: the typical height of the base station is 15 metres on a mast and 20 metres on top of a building. The geographical coverage area is predominantly rural and induces for a number of paths important delays (up to 30 µs). Further, due the limited number of diffusers and the distance between them, no significant fast fading occurs.

With the increase in users, in urban areas for the most part, the dimensions of the cells had to be decreased in order to reduce the reuse distance of the allocated frequencies. The most current urban cell is the small cell. Its coverage area has a radius lower than a few kilometres and the base station antenna is located above roof level, i.e. from 3 to 10 metres above ground level. The maximum duration of the impulse response is 10 µs.

In very dense urban areas, small cells are replaced by microcells with an activity radius of a few hundreds metres. The antennas are located below roof level, and the waves are guided by the streets. The maximum duration of the impulse response is 2 µs.

Finally, the smallest cell is the picocell, with radius of only of a few tens metres, corresponding to communications in the buildings where the base stations antennas are located. The maximum duration of the impulse response is 1 µs.

7.2 Types of Models

7.2.1 Theoretical Models

Theoretical models are based on the fundamental laws of physics combined with adequate approximations and with atmosphere and land models. These models lead to complex mathematical relations and require the resolution of Maxwell's equations through the use of different methods: finite element and finite difference methods, parabolic equation method, physical and geometrical optics methods, etc. A more detailed survey of these methods can be found in Appendix O. One of their main drawbacks however is a relatively high computation time which is often incompatible with operational constraints, especially for engineering purposes. These models can nevertheless be used as reference models in some specific cases, in order for instance to evaluate the approximations introduced in other models.

Since the variables used in such models are in general deterministic variables, these models are generally referred to as deterministic models. They may however take into account random variables characterised by their distribution.

7.2.2 Empirical or Statistical Models

Empirical models are based on the statistical analysis of a large number of experimental measurements conducted with respect of several different parameters like the frequency, the distance, the effective heights of the base station antenna and of the mobile, etc.

The best known such model is the Okumura-Hata model, which is based on the statistical analysis of a large number of experimental measurements conducted inside and near Tokyo with respect to different parameters like the frequency or the distance.

The path loss is a function of the frequency f, which may range from 50 MHz to 1500 MHz, of the effective height h_b (between 30 and 200 meters) of the base station antenna and the effective height h_m (between 1 and 10 meters) of the mobile, on the distance d to the transmitter, which may range from 1 to 20 kilometres, and finally on three corrective coefficients depending on the nature of the environment (low urbanised areas, moderately urbanised areas and highly urbanised areas).

In this model, the attenuation A_p is given by the following equation:

$$A_p = 69.55 + 26.16 \log_{10} f - 13.82 \log_{10} h_b + (44.9 - 6.55 \log_{10} h_b) \log_{10} d - a(h_m) \qquad (7.1)$$

where:

- for small and medium-sized towns, $a(h_m) = (1.1 \log_{10} f - 0.7) h_m - (1.56 \log_{10} f - 0.8)$
- for large cities with $f \leq 200$ MHz, $a(h_m) = 8.29 (\log_{10} (1.54 h_m))^2 - 1.1$
- for large cities and $f \geq 400$ MHz , $a(h_m) = 3.2 (\log_{10} (11.75 h_m))^2 - 4.97$

The attenuation formula given in Eq. 7.1 was deduced from measurements carried out near Tokyo. Deviations have been noted when this formula is applied to towns with characteristics significantly different to those of Tokyo (COST231 1999).

The Okumura-Hata model for small and medium-sized towns was extended to the 1500-2000 MHz frequency band by the COST231 research groups (COST231 1999), leading to the so-called COST231-Hata model:

$$A_p = 46.3 + 33.9 \log_{10} f - 13.82 \log_{10} h_b + (44.9 - 6.55 \log_{10} h_b) \log_{10} d - a(h_m) + C_m \qquad (7.2)$$

where:

- $a(hm)$ is as in the expression defined higher,
- $C_m = 0$ dB in small and medium-sized towns and in urban areas,
- $C_m = 3$ dB in large cities.

Empirical models can be implemented rapidly without requiring any extremely accurate or expensive geographical databases. Further, these models are robust since they are as a general rule developed from a large number of measurements. However, empirical models are unsuited to the study of propagation over short distances, and they lack precision for selective analyses of environments with a specific topography. These models are also unsuited to microcell and picocell environments.

7.2.3 Semi-Empirical Models

Semi-empirical models combine the analytical formulation of physical phenomena like reflection, transmission, diffraction or scattering (Deygout 1966, 1991; Epstein 1953; Vogler 1982) with a statistical fitting by variables adjustment using experimental measurements. This method is more robust than purely empirical methods since it avoids the improbability of independent variables. The best known such model is the COST model which relies on a multiscreen diffraction. The statistical optimisation is based on traditional linear regression techniques, but also to more elaborated techniques such as neuronal networks (Balandier 1995; Deldique 1997), nonparametric regression by binary decision tree (Gerome 1997a), classification (Gerome 1997b), etc.

7.3 Uses of Models

Propagation models are used in the conception and the design of radio interfaces in order to optimise their performances and subsequently during the actual field deployment of radio systems in order to determine the radio coverage.

In the first case propagation models are implemented in software in order to simulate the transmission chain: this process allows to identify and reproduce the relevant characteristics of the propagation channel and to evaluate systems, more specifically in terms of quality and of error rate. These models are based on the consideration of the impulse response and its evolution in space and time, and rely on generic or typical environments rather than on geographical databases (Biro 1995).

In the second case propagation models are implemented in engineering tools for the prediction different parameters useful for the field deployment of systems, for the study of the radio coverage (selection of the emission sites, frequency allocation, powers evaluation, antenna gains, polarisation) and for the definition of the interferences occurring between distant transmitters. These models are heavily dependent on geographical databases which contain data relating to the topography and the land use. For more detail on geographical databases, the reader is referred to Appendix N.

7.4 Macrocell Models

Macrocell models are semi-empirical models generally based on the analysis of the transmitter-receiver vertical profile and of the different obstacles, like for instance hills or forests, which are present along this profile. These models generally rely on geographical datasets, in the form of digital terrain model and land use maps. An adjustment of the variables is then performed on the basis of experimental measurements: this procedure is necessary due to the scarcity of available geographical data (estimation of the land use over a 100 metre mesh) and to the simplicity of the computation algorithms. Two different types of macrocell models exist: models developed for rural and moderately urbanised areas, and models for mountainous areas.

7.4.1 Rural Models

Rural models only take into account the ground vertical profile between the transmitter and the receiver for the determination of attenuation, and more specifically the attenuation due to diffraction by obstacles (ground or land use obstacles). These models can be used for the estimation of the radio coverage with a precision given by a standard deviation equal to 6 to 7 dB from the mean error between the measured and the predicted values. The performances of these models can be significantly improved in urban environments through the use of geographical data with a higher degree of precision. Performances deteriorate when hills and valleys are more pronounced, especially in mountainous areas, due to the fact that multipath effects tend to become predominant in this case.

CAPADOCE Model

This model can be applied to the study of the propagation of waves at frequencies ranging from 30 MHz to 30 GHz in flat, undulating and mountainous terrains, in rural and in moderately urbanised areas. It takes into account the physical parameters of the actual environment as represented by geographical databases: digital terrain model (DTM), the land use, the relative permittivity and conductivity of the ground, the refractive index, the roughness, which is a reduction factor of the reflection coefficient, and the polarisation. The following parameters are considered in this model:

- free-space propagation,
- reflection on smooth, spherical surfaces,
- diffraction from flat surfaces,
- multiple edge diffraction,
- tropospheric refraction,
- tropospheric scatter.

Applied to urban environments, this model presents an average error lower than 1.5 dB and a standard deviation close to 6 dB.

UHF Model

This model, developed by France Telecom R&D, is valid for distances ranging from 200 metres to 50 kilometres and at the 400, 900 and 1800 MHz frequencies. It is based on geographical mesh databases representing the relief and the land use, and takes into account the following parameters:

- free-space propagation,
- multiple edge diffraction, using the Deygout method,
- land use attenuation,
- corrective terms of additional attenuation in order to account for such parameters as the effective height of the antennas, the value of the Fresnel-Kirchhoff parameter, the number of points present within the first Fresnel ellipsoid or the height differences.

From a point of view concerned with performances, this model is centred, i.e. the average value between the measured and the predicted values is equal to zero, and its standard deviation is approximately of the order of 7 dB. It has been optimised for different environments (rural areas, maritime areas, urban areas, etc) and can be adapted to different databases (different types of DTM files, different mesh sizes, different topics, etc). This model is gauged automatically using a specific module which generates a neuronal structure for the local optimisation of the model.

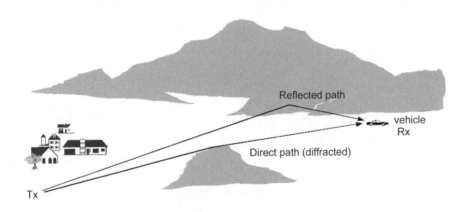

Fig. 7.3. Schematic representation of a mountainous environment

7.4.2 Mountainous Model

This model considers in addition to the direct line-of-sight path the paths followed by waves reflected by the sides of the neighbouring mountains (Hashemi 1993a; Kirlin 1994). In order to identify the reflection areas, more complex algorithms have to be employed for the determination of the attenuation, which leads to longer computation times than for instance with the UHF model.

The optimisation of computation times for the prediction of the attenuation has been carried out through the use of a variety of different methods, including:

- the search for reflected paths confined within an ellipse with foci collocated with the emitter and the receiver and with a maximum delay allowed Δt_{max} equal to 25 μs (see Fig. 7.4)
- the consideration of a single reflection on mountainsides in direct line-of-sight of both the emitter and the receiver,
- the consideration of the size of the mesh. It has been found that a 400 metre mesh is an acceptable compromise between the computation times and the accuracy of the predictions.
- the assembly of individual meshes into facets in order to obtain a single profile between the reception point and the barycentre of the mesh facet.
- the pre-processing of geographical data in order to determine the line-of-sight visibility between the meshes. This method avoids extracting a large number of profiles.

The standard error deviation between the measured and the predicted values is of the order of 6.4 dB. As can be seen when comparing Figs. 5 and 6, the improvement brought by this model as compared with the rural model is significant, not only in the case of high mountains like the Alps or the Pyrenees, but also for medium-sized mountains like the Vosges in France (Bourdeilles 1997).

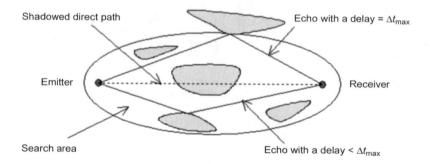

Fig. 7.4. Schematic representation of the different paths considered in mountainous media

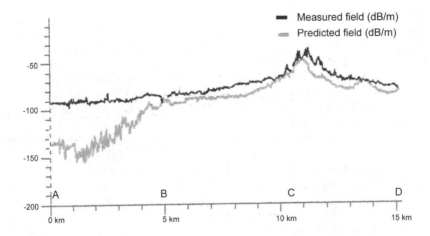

Fig. 7.5. Comparison between prediction and measurement of the field strength in a mountainous environment using a model allowing for the paths reflected on mountainsides

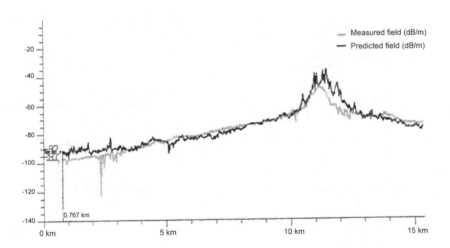

Fig. 7.6. Comparison between prediction and measurement of the field strength in a mountainous environment using a model allowing for the paths reflected on mountainsides

7.5 Small Cell Models

7.5.1 COST-CNET Model

The integration of a profile mode and different multiple diffraction algorithms over a succession of edges significantly improves statistical models for small suburban and urban cells. These methods, which have already been tested for the prediction of the field strength in rural environments, offer the advantage of taking into account the influence of the land use and the topography through the use of geographical databases with a more or less high degree of precision.

The best known such model, developed within the Action COST 231 framework, is a two-dimensional semi-empirical model well suited to small cells. The profile is located within the vertical plane containing the transmitter and the receiver. It is more particularly suited for GSM (900 MHz) and for DCS 1800 (1800 MHz) engineering with an emitting antenna situated at a height between 5 and 15 metres above roof level. The waves propagate essentially above roof level before finally diving down into the street where the mobile moves. In order to extend the applicability of this model to less common configurations, and more specifically in order to take into account the relief, an additional term for the diffraction by a dominant edge, referred to as the main edge, was introduced. Due to the distances and to the fact that the variations in altitude are rarely much significant within a given town, this term is determined using the Deygout method generally employed at France Telecom R&D.

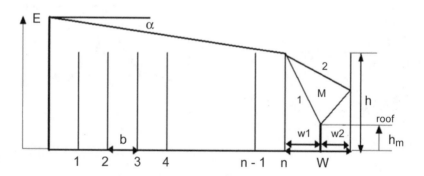

Fig. 7.7. Schematic representation of the profile between the emitter and the receiver in a small cell context

The method used for the determination of the attenuation derives from the researches carried out by Walfish and Bertoni on multiple diffraction by aligned and uniformly distributed edges (Walfish 1988; Xia 1992; Maciel 1993), and by Ikegami for the specific consideration of the diffraction by the last edge, i.e. the edge located immediately before the mobile (Ikegami 1984). As represented in Fig. 7.7, two paths are taken into account: the direct line-of-sight path between the last edge and the mobile and the path that would be followed by a wave reflected at the building located behind the mobile in the vertical emitter-receiver plane.

The resulting total losses *(L)* are decomposed into four main terms representing respectively the losses associated with the distance between the transmitter and the receiver, the losses due to multiple edge diffraction, the losses associated with the last diffraction and reflection at buildings and behind the mobile respectively and finally the losses due to diffraction by a main edge:

$$L = L_0 + L_{msd} + L_{rts} + L_{\deg} \tag{7.3}$$

Losses associated with the Distance between the Transmitter and the Receiver (*L₀*)

Let A_0 be the attenuation due to free-space propagation between two antennas separated by a distance *d*. The following equation yields it:

$$A_0 = \left(\frac{\lambda}{4\pi d}\right)^2 \tag{7.4}$$

Hence,

$$L_0 = -10 \log_{10}(A_0) \tag{7.5}$$

Losses due to Multiple Edge Diffraction

These losses can be estimated using the Walfish and Bertoni method as revised either by the COST-231 or by France Telecom R&D (*L_{msd}*). These two methods will both be described hereafter.

COST-231 Model

$$L_{msd} = L_{bsh} + k_a + k_d \log_{10} d + k_f \log_{10} f - 9 \log_{10} b \tag{7.6}$$

where:

$$L_{bsh} = -18\log_{10}\left(1+\left(h_b - h_{toit}\right)\right) \quad \text{if} \quad h_b > h_{toit} \tag{7.7}$$

$$L_{bsh} = 0 \quad \text{if} \quad h_b \leq h_{toit} \tag{7.8}$$

$$k_a = 54 - 0.8\left(h_b - h_{toit}\right)\frac{d}{0.5} \quad \text{if} \quad h_b \leq h_{toit} \text{ and } d < 0.5km \tag{7.9}$$

$$k_a = 54 - 0.8\left(h_b - h_{toit}\right) \quad \text{if} \quad h_b \leq h_{toit} \text{ and } d \geq 0.5km \tag{7.10}$$

$$k_a = 54 \quad \text{if} \quad h_b > h_{toit} \tag{7.11}$$

$$k_d = 18 - 15\frac{h_b - h_{toit}}{h_{toit}} \quad \text{if} \quad h_b \leq h_{toit} \tag{7.12}$$

$$k_d = 18 \quad \text{if} \quad h_b > h_{toit} \tag{7.12}$$

$$k_f = -4 + 0.7\left(\frac{f}{925} - 1\right) \quad \text{in medium-sized towns and urban areas} \tag{7.13}$$

$$k_f = -4 + 1.5\left(\frac{f}{925} - 1\right) \quad \text{in dense urban environments} \tag{7.14}$$

The term k_a here represents the increase in attenuation that occurs when the base station antenna is located below the top of the roofs of the adjacent buildings, while terms k_d and k_f represent the dependence of the attenuation due to multiple edge diffraction on the distance and the frequency respectively.

France Telecom R&D Model. In this model, the field strength is calculated by recurrence over each diffraction screen using the Kirchhoff-Huygens integral, in order to obtain the amplitude of the field H_n at the top of the edge n. The numerical resolution is then performed by linearly approximating the amplitude and the phase of this integral over a suitably selected discrete interval. As demonstrated by Walfish and Bertoni, H_n tends to become constant for values of the incidence angle α and values of n high enough. The following equation for Q is deduced by interpolation:

$$Q = \left(\frac{\alpha}{0.03}\left(\frac{b}{\lambda}\right)^{1/2}\right)^{0.9} \tag{7.15}$$

It might be pointed out that the following equation can be derived under the assumption that $\alpha \approx \Delta Hb/d$, where ΔHb is the height of the emitter above roof level:

$$L_{msd} = -20 \log_{10}(Q) \tag{7.16}$$

Losses due to the Last Diffraction and Reflection at Buildings and behind the Mobile (L_{rts})

These losses describe the coupling of the wave which occurs as the wave propagates above buildings towards the mobile located in the street.

COST231 Model.

$$L_{rts} = -16.9 - 10\log_{10} w(m) + 10\log_{10} f(MHz) + 20\log(h_{toit} - h_m) + L_{ori} \tag{7.17}$$

where:

$$L_{ori} = \begin{cases} -10 + 0.354\varphi & 0° \le \varphi < 35° \\ 2.5 + 0.075(\varphi - 35°) & 35° \le \varphi < 55° \\ 4.0 - 0.114(\varphi - 55°) & 55° \le \varphi < 90° \end{cases} \tag{7.18}$$

where w is the street width, f is the frequency and φ is the angle formed by the street axis and the direction of incidence of the wave.

France Telecom R&D Model. At the level of the mobile, the field can be seen as resulting from the superposition of the wave diffracted by the last building before the mobile (path 1) and of the wave reflected by the building located after the mobile (path 2). All other paths can be neglected. Let E_1 be the field associated with path 1 and E_2 be the field associated with path 2. The resulting field E is thus expressed by the following equation:

$$E = \sqrt{E_1^2 + E_2^2} \tag{7.19}$$

Since E_1 and E_2 actually result from the diffraction by the nth last edge, they can be determined using Fresnel equations under the assumption that all necessary conditions are fulfilled. The paths considered here are w_1 for E_1 and $2W\text{-}w_1$ for E_2 respectively. We are thus presented with the two following equations expressing E_1 and E_2 respectively in the form of functions of the free-space attenuation E_o:

$$E_1 = \left(\frac{0.225}{\sqrt{2}}\right) * Eo * \frac{\sqrt{\lambda * w1}}{\Delta Hm} \qquad (7.20)$$

$$E_2 = \left(\frac{0.225}{\sqrt{2}}\right) * Eo * \frac{\sqrt{\lambda * (2W - w1)}}{L_r * \Delta Hm} \qquad (7.21)$$

where ΔHm is the difference between the height of the mobile and the height of the last building, while the parameter L_r represents the inverse of the reflection coefficient for path 2. Assuming that $L_r = 2$, the total value of the field strength as well as the value, expressed in dB, of the losses L_{rts} can be deduced from these two equations:

$$E = \left(\frac{0.225}{\sqrt{2}}\right) * Eo * \frac{\sqrt{\lambda * \left(w1 + \dfrac{2W - w1}{L_r^2}\right)}}{\Delta Hm} \qquad (7.22)$$

and:

$$L_{rst} = -20 \log_{10}\left(\frac{E}{E_0}\right) \qquad (7.23)$$

It might be pointed out that the expression of this model as it was defined within the framework of the COST 231 European group amounts to considering that $w1 = W / 2$, i.e. that the mobile stands in the centre of the street, and that the following relation holds true :

$$L = L_0 + L_{msd} + L_{rts} \qquad (7.24)$$

Losses due to the Diffraction by a Main Edge (L_{deg})

In the case of a single edge represented in Fig. 7.9, the attenuation due to diffraction can be determined using Deygout approximate relations derived from Fresnel rigorous theoretical relations. The parameter h represents the height of the edge with respect to the straight line joining E and R ($h > 0$ if the edge intersects with the direct line-of-sight path between E and R, and $h < 0$ if it is not the case), while the parameter r represents the radius of the Fresnel ellipsoid. The losses L_{deg} expressed in dB are then given by the following set of equations:

$$L_{\text{deg}} = 0 \qquad \text{if } \frac{h}{r} < -0.5 \qquad (7.25)$$

$$L_{\text{deg}} = 6 + \frac{12h}{r} \qquad \text{if } -0.5 < \frac{h}{r} \leq 0.5 \qquad (7.26)$$

$$L_{\text{deg}} = 8 + \frac{8h}{r} \qquad \text{if } 0.5 < \frac{h}{r} \leq 1 \qquad (7.27)$$

$$L_{\text{deg}} = 16 + 20\log_{10}\left(\frac{h}{r}\right) \qquad \text{if } \frac{h}{r} \geq 1 \qquad (7.28)$$

The determination of the profile is a most important step. It consists in extracting the parameters necessary for the calculation of attenuation from an actual profile, like the one represented in Fig. 7.8, which usually is constructed from land use data contained in baseline datasets. For more detail on geographical databases in general, the reader is referred to Appendix N. The parameters extracted from this profile are either geometrical variables, like the average width of the streets, the width of the street where the mobile is located, the average height of the buildings or the orientation of the street where the mobile is located, or qualitative variables characterising the environment, for instance whether the receiver is in line-of-sight or in non line-of-sight from the emitter, whether the considered environment is a suburban or an urban area or the presence of vegetation in the environment.

In order to adapt the model to actual configurations and improve its performances, the construction proceeds in a semi-statistical manner. The total attenuation is therefore expressed with this model in the following form:

$$AFF = L_0 + \alpha L_{msd} + \beta L_{rts} + \gamma L_{\text{deg}} + \sum_i c_i P_i \qquad (7.29)$$

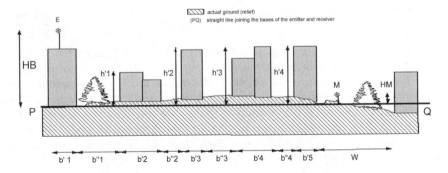

Fig. 7.8. Actual profile

Reception along the measurement path

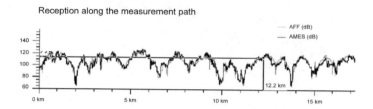

Profile of the measurement path: reception at the point (97.5 km, 231 km)

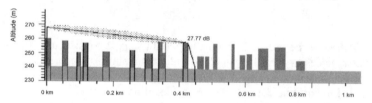

Fig. 7.9. Behaviour of the COST-France Telecom R&D (COST-CNET) model along a measurement path in Mulhouse at the 900 MHz frequency

where the terms P_i stand for the different geometrical variables involved, while α, β, γ and c_i are coefficients obtained by simple linear regression over a given set of measured values.

7.5.2 Lee Model

In this model, the attenuation is given by the following equation, based on measurements realised in different types of environments (free space, open environment, suburban environment, and urban environment) and on the determination of the effective height of the emission antenna (Lee 1982, 1983):

$$L = 10n \log_{10} d - 20 \log_{10} h_b(eff) - P_0 - 10 \log h_m + 29 \qquad (7.30)$$

where d is the length of the link in kilometres, n and P_0 are parameters derived from experimental measurements and depending on the nature on the environment (see Table 7.1), while h_m is the height of the mobile and $h_b(eff)$ is the effective height of the base station antenna. The latter is determined by the projection of the slope of the ground in the near vicinity of the mobile onto the site of the base station. In Fig. 10 is schematically represented the dependence of the effective height of the emitting antenna on the slope of the ground for different locations of the mobile (Saunders 1999).

Table 7.1. Parameters used in the Lee model

Environment	n	P_0
Free space	2	- 45
Open space	4.35	- 49
Suburban	3.84	- 61.7
Urban : Philadelphia	3.68	- 70
Urban : Newark	4.31	- 64
Urban : Tokyo	3.05	- 84
Urban : New York	4.08	- 77

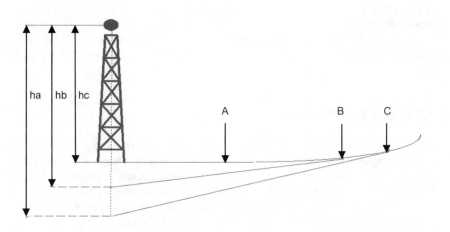

Fig. 7.10. Determination of the effective height of the base station antenna

7.6 Microcell Models

7.6.1 Introduction

Microcell modelling is based on a duality between a line-of-sight calculation and a non line-of-sight calculation. This results from the fact that the distances considered in this context are generally limited, and rarely larger than a few hundreds metres. If the emitter is located in the street and below the roof level, the propagation will guided along the streets, either along the very street where the transmitter is located, hence in line-of-sight, or along adjacent streets if at least one street corner is present along the propagation path, hence in non line-of-sight. These two modes of calculation are present in most two-dimensional analytical models found in the literature, in particular in the two-slope model (Xia 1993): this model assumes a d^2 attenuation ($20\log_{10}d$) in line-of-sight over short distances and a d^4 attenuation ($40\log_{10}d$) in non line-of-sight and in line-of-sight over long distances, where the direct line-of-sight path is combined with the path reflected off the ground. The location of the breakpoint is generally given by the following equation:

$$d = 2\pi\, h_b h_m / \lambda \tag{7.31}$$

where h_b and h_m represent respectively the height of the base station antenna and the height above the ground of the mobile.

The actual value of the slope in non line-of-sight situations is strongly dependent on the angle α between the streets and is therefore very difficult to determine accurately in practice. Only through the use of the *uniform theory of diffraction* (Keller 1962; Kouyoumjian 1974) can a precise analytical formulation of propagation phenomena in urban areas be derived (Bertoni 1994; Berg 1995; Jakoby 1995).

7.6.2 MicroG-CNET Model

The MicroG-CNET model is a 2 x 2-dimensional analytical model extending and generalising the duality described above. In situations where the propagation along streets is the dominant phenomenon, the attenuation is determined in the horizontal plane (Wiart 1993). In contrast, in situations, either in line-of-sight or in non line-of-sight, where the propagation above roofs becomes the dominant phenomenon, the attenuation is determined through a profile calculation in the vertical plane defined by the emitter and the receiver. Such situations occur for instance in the case of long distances or in the presence of woods or suburban environments where the average height of buildings is relatively low compared to the height of the emitter. There are therefore two calculation categories, in the horizontal plane along the streets and in the vertical profile between the emitter and the transmitter.

Horizontal Plane Calculation

An analysis of the propagation of radio waves around a street corner at the 900 MHz and 1800 MHz frequencies was conducted through simulations based on the electromagnetic methods of the *unified theory of diffraction*. This analysis led to simplified analytical expressions of the propagation mechanisms in line-of-sight: the attenuation due to reflections and diffractions was thus expressed in terms of the widths of the starting point and destination streets, referred to as W_1 and W_2 respectively, of the distance D between the emitter and the street corner, of the distance X between the street corner and the receiver, of the angle α between the two streets and of the emission frequency. The total attenuation *Aff* at the level of the receiver beyond the street corner is expressed by the following equation:

$$Aff = AffVis + \max (AffRef, AffDif) \qquad (7.32)$$

where:

- *AffVis* is the line-of-sight attenuation at the distance D, as given by the equation:

$$AffVis = 32.4 + 20 \log_{10} f + 20 \log_{10} D \qquad (7.33)$$

— *AffRef* is the attenuation due to reflection (Wiart 1993) :

$$AffRef = 32.4 + 20 \log_{10} f + 20 \log_{10} D \, (D + X) + SX \qquad (7.34)$$

where:

— S represents the decrease gradient of the field at the street corner, defined by the equation :

$$S = (D / W_1 W_2) \, f (\alpha) \qquad (7.35)$$

— W_1 and W_2 are the street widths in line-of-sight and in non line-of-sight respectively,
— X is the distance between the receiver and the street corner,
— $f(\alpha)$ is a function of the angle α formed by the street.
— *AffDif* is the attenuation due to diffraction, as given by the following equation (Wiart 1993) :

$$AffDif = 32.4 + 20 \log_{10} f + 10 \log_{10} (X (X + D) D) + 2D_a \qquad (7.36)$$

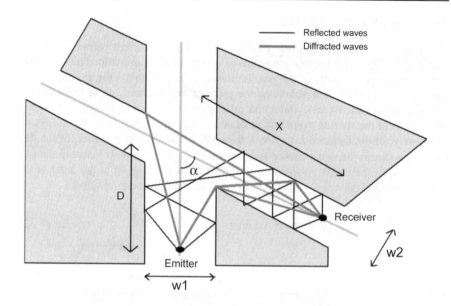

Fig. 7.11. Typical example of microcell propagation at a crossroads

where:

$$D_a = (45/2\pi)\, a \tan (X / W_2) - (23/\pi)\, (\alpha - \pi/2) \tag{7.37}$$

This procedure can be extended recursively to the situation of several successive street corners: the street corner n - 1 assumes the role of the emitter, while the street corner n + 1 assumes the role of the receiver. In order to determine the attenuation between an emitter and a receiver, the MicroG-CNET model searches for all possible paths between the emission and the reception points. The contributions in watts from the different paths are summed, before converting the result into decibels.

Vertical Profile Calculation

For the determination of attenuation in specific environments not typically microcell, for instance in the case of long distances, either in line-of-sight or in non line-of-sight, of large areas, of public gardens or in the presence of woods, a calculation along streets turns out to be unrealistic and results in errors. In such cases, a calculation in the vertical profile needs to be introduced. This mode of calculation relies on the principle that the propagation above the roof level or through vegetation becomes dominant as compared to the propagation along the streets. The model applied in this context consists in:

– the interpretation of an actual profile in order to reduce the problem to a calculation of multiple diffractions above roofs, and to a Walfish-Bertoni-Ikegami model for the calculation of the diffractions and reflections occurring at the level of the mobile and the emitter,
– the adaptation of theoretical formulas commonly used in a small cell context to the microcell environment.

In order to adapt the model to actual configurations and improve its performances, the construction proceeds in a semi-statistical manner. The total attenuation L_0 is therefore expressed with this model in the following way:

$$L_0 = L_d + \alpha L_{msd} + \beta L_{rtse} + \gamma L_{rtsm} + \delta L_{deg} + \sum_{i=0}^{n} \alpha_i P_i \qquad (7.38)$$

This model considers the following parameters for the determination of the attenuation:

– the attenuation associated with distance L_d,
– the attenuation due to diffraction L_{msd},
– the attenuation due to diffraction and reflection at the level of the emitter L_{rtse},
– the attenuation due to diffraction and reflection at the level of the receiver L_{rtsm},
– the attenuation due to diffraction by a main edge and calculated using the Deygout method L_{deg},
– a set of geometrical variables P_i, containing in particular variables associated with the distance in vegetation,
– the coefficients α, β, γ, δ and α_i, obtained by simple linear regression over a set of given measured values.

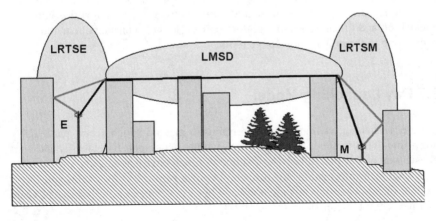

Fig. 7.12. Microcell profile

Reception along the measurement path

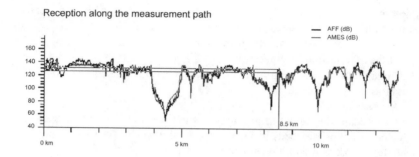

Profile of the measurement path: reception at the point (65 km, 260 km)

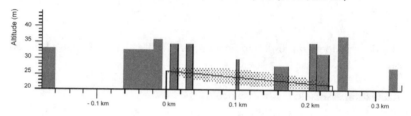

Fig. 7.13. Comparison between the calculated (AFF) and the measured (AMES) values of attenuation expressed in dB along a measurement path in Lille at the 900 MHz frequency. Profile of the measurement path at the x-coordinate of 8503 metres (MicroG-CNET model)

Fig. 7.13 represents the behaviour of the MicroG-CNET model at the 900 MHz frequency along a microcell measurement path in Lille. The main characteristics of an urban microcell environment can be observed in these measurements : the large (nearly 100 dB) amplitude variation in attenuation, the amplitude ranging from 20 to 30 dB of the abrupt variations occurring over short distances (transition from a line-of-sight location to a completely obstructed location after turning an street angle), the presence in the vicinity of the emitter of a region with a very low attenuation and the presence of a region with a relatively constant attenuation corresponding to the propagation above roofs.

7.7 Ray Launching Model

The ray launching technique is a deterministic method which seems particularly promising. This method, based on very accurate geographical databases and on a physical theory, is particularly well suited to urban environments and provides results with rich information, for instance impulse responses. As a consequence, a number of models using this method have been developed in the last recent years (Kurner 1993; Lawton 1994; Liang 1998).

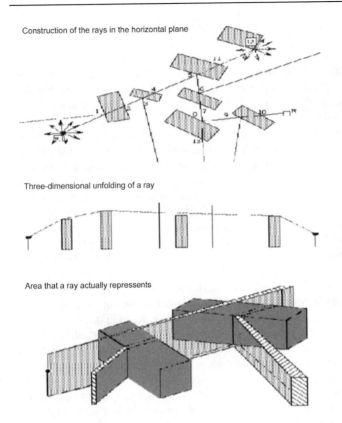

Construction of the rays in the horizontal plane

Three-dimensional unfolding of a ray

Area that a ray actually repressents

Fig. 7.14. Principles of a ray launching modelling (RAYON model)

The RAYON model developed by France Telecom R&D is a model based on a ray launching technique. This model is intended at performing a systematic research of the paths joining the emitter to the receiver where several different physical phenomena are combined, for instance reflection, diffraction, either horizontal or vertical, penetration and propagation in vegetation. All these different phenomena are deduced exclusively from their theoretical values, as given either by the *uniform theory of diffraction* in the case of diffractions or by Fresnel relations and Beckmann's formula in the case of reflections.

The classical ray launching method consists in launching uniformly in all directions a dense set of rays from an emitter into the environment. Each ray is represented by a straight line, which actually denotes the part of the electromagnetic wave present in a cone around this line. In order to limit computation times, certain assumptions were introduced for the simplification of the geometry of the problem. Fig. 7.14 represents the actual procedure (Rossi 1991, 1992).

In the first stage, only the horizontal plane is considered: both the base station E and the obstacles are represented by their vertical projections in the horizontal plane. Rays are then projected into this plane at regular angular intervals from the emitter E. At this stage, a ray no longer corresponds to a cone around a straight line in the three-dimensional space, but to a section of space in the vertical plane. As a ray encounters the edge of a building, represented by point 1 in Fig. 7.14, it may either propagate over this building or be reflected at the surface. Both cases are investigated by defining two branches originating from this point: the first branch is directed along the line-of-sight, while the other is directed along the direction of specular reflection.

This procedure is repeated each time the ray encounters an obstacle until attenuation becomes too high or until the ray reaches an area where ground data are no longer available. When the ray encounters a receiver R, the third dimension is considered by unfolding the ray path according to its curvilinear geographical coordinates: as described in Fig. 7.14, reflection buildings are represented as simple vertical bars. The path followed by the ray corresponds to the shortest path between the base station and the mobile propagating over the diffraction buildings. A rigorous determination of the attenuation along the link can then be performed: the average received power is the sum of the powers of the rays reaching this point.

Two problems arise with this approach: first, the model is heavily dependent on the precision and accuracy of the geographical database and the calculations are extremely sensitive to the shortcomings of the database. In addition, the computation times may increase very rapidly. In order to adjust the model, additional parameters need therefore to be introduced. These parameters are of two types: physical parameters characterising reflection and diffraction, for instance the permittivity and the conductivity of materials used in buildings, which are always difficult to determine, and specific parameters used for the optimisation of the computation times, which may have an influence on the final result. The main parameters to be considered are either of physical nature or of non-physical nature. Examples of physical parameters are the percentage of walls without openings (windows), the average height of the inhomogeneities of the walls, the losses in grazing diffraction or the attenuation due to vegetation. Examples of non-physical parameters are the number of reflections, the allowed number of diffractions, the maximum length of a ray path or the minimum power associated with a ray. Figure 7.15 presents an example of a set of rays joining an emitter to a receiver with a power in a 20 dB range.

7.9 Building Penetration Loss Models

The building penetration loss is defined as the power loss that an electromagnetic wave undergoes as it propagates from outside a building towards one or several places inside this building. This parameter is determined from the comparison between the external field and the field present in different parts of the building where the receiver is located. Penetration loss models, being integrated into the coverage prediction tools, must take into account the environment around the buildings which is under consideration.

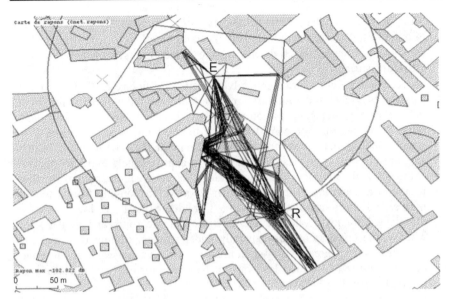

Fig. 7.15. Diffracted and reflected ray paths between an emitter E and a receiver R

The electromagnetic field inside buildings is subject to the influence of different parameters, including the position of the building with respect to the emitter and to other buildings or the architectural characteristics of the building, for instance the materials, the interior layout or the size of the windows. Values obtained for a given building should not, therefore, be generalised to any kind of building when performing coverage predictions. While a database integrating a large enough number of characteristics would help improving the precision of building penetration loss models, the development on a large scale of such a database remains difficult (Toledo 1998).

The value of the building penetration loss is influenced by a number of different physical parameters, whose effects intermingle most of the time. Among these different parameters the following are traditionally distinguished:

– the near environment : a distinction is drawn between districts with high towers more or less separated from each other and more traditional districts with buildings of average height,
– the reception depth in buildings: the amplitude of the field decreases as the mobile moves from the front of the building towards a room located inside it, while the influence of the inhomogeneities decreases as the penetration depth inside a building increases. Waves penetrate more easily through window panes than through brick walls. Accordingly, the paths followed by radio waves will be more or less attenuated, and might even be occulted. The building penetration loss is generally 6 dB lower for glazed walls compared with non glazed

walls (Rappaport 1994). Although the level of attenuation is higher in the back of buildings, it is also much more homogeneous there.

- the incidence angle, which determines the reflection and transmission coefficients at a surface,
- the reception height, more commonly described as 'floor effect'. This parameter induces effects in the form of a reduction of the building penetration loss or of a relative power gain, as compared with the lower floor. The calculation starts therefore from the consideration of the building penetration loss at the ground floor, as determined from a comparison with the external field. In small cell, power gains are typically of the order of approximately 2 to 3 dB per floor at the 900 and 1800 MHz frequencies. However the great diversity of situations results in a scattering of these gains per floor; values ranging from 4 to 7 dB have already been measured in practice (Gahleitner 1994). The lower floors are illuminated by rays reflected and diffracted at the roofs and in the street, whereas the upper floors are generally under stronger illumination, and even sometimes under direct line-of-sight illumination. As a consequence, the values of the building penetration loss vary between the lower floors and the upper floors (Rappaport 1994; Walker 1983).
- the distance between the transmitter and the receiver, when the building where the mobile is located is in line-of-sight of the transmitting antenna. The building penetration loss in this case depends on distance as predicted by the free-space propagation law.
- the height of the emitting antenna,
- the frequency,
- the nature of the materials. The attenuation that electromagnetic waves undergo varies from 4 dB in the case of wood to 10 dB in the case of concrete walls (COST231 1999).

Several different measurement techniques have been developed in order to characterise the building penetration loss associated with the materials of the buildings. In particular, the method based on the use of two reverberation rooms can be mentioned here (Foulonneau 1996).

The most classical models draw on the Motley-Keenan model (Motley 1988) used for the study of propagation inside buildings. The parameters considered in these models for the determination of the building penetration loss include:

- the distance between the emitter and the external wall of the building where the receiver is located,
- the distance between the external wall and the receiver,
- the number of internal walls present along the emitter-receiver profile,
- the floor effect,
- the attenuation due to the external walls of the building,
- the attenuation due to the internal walls of the building.

The loss path L is expressed as the sum of the free-space loss L_0, of the losses due to the obstacles present along the direct line-of-sight path (tiles, walls, doors, windows) and of a constant L_c (Motley 1988). Databases can be used for differentiating between the different obstacles to which specific attenuation values are attached. This model is the most commonly used.

$$L = L_0 + L_c + \sum_{j=1}^{N} N_j L_j + N_f L_f \qquad (7.39)$$

where N_j is the number of walls of type j present along the line-of-sight path, L_j is the losses due to walls of type j, N is the number of types of walls, N_f is the number of tiles present along the line-of-sight path and L_f is the losses due to tiles.

Typical values of transmission losses for different types of materials used for the external walls in the 1-2 GHz frequency band are summarised in Table 7.2 (COST 231 1999).

Table 7.2. Transmission losses for external walls made of different types of materials in the 1-2 GHz frequency band

Materials	Losses [dB]
Porous concrete	6.5
Reinforced glass	8
Concrete (30 centimetres)	9.5
Thick concrete wall (25 centimetres) with large glazed panes	11
Thick concrete wall (25 centimetres) without glazed panes	13
Thick wall (> 20 centimetres)	15
Tile	23

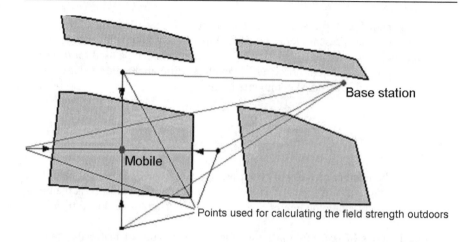

Fig. 7.16. Propagation paths considered in building penetration loss models

The main disadvantage of these models lies in the empirical evaluation of the parameters. These parameters, for instance the value of the wall attenuation, may indeed present significant fluctuations from one building to another, which results in a reduction of the degree of precision of the prediction of the field strength. Furthermore, the internal architecture of buildings is not at the present time integrated by any database. Even though this solution could be considered for certain specific buildings, it is doubtful that it could ever be implemented on a large scale.

Building penetration loss models can be enhanced through the use of microprofiles joint the indoor mobile to external reference points, as represented in Fig. 16. The information relating to the external environment and to the interior of buildings and extracted from the microprofiles allows determining the total attenuation, which consists of two terms: a building penetration loss and an outdoor path loss between the base station and the building (COST 231 1999).

7.9 Indoor Propagation Models

The propagation of radio waves in buildings depends primarily on the nature of the environment. The propagation environment can be characterised as either dense (office type buildings), open (office type buildings, offices with large capacity), broad (buildings with very large rooms, for instance warehouses, airports or railway stations) and corridor (in the case where the emitter and the receiver are located in the same corridor). Indoor propagation is also multipath: the predominant propagation mechanisms are reflection, transmission, diffraction and scattering (Hashemi 1993b; Valenzuela 1997).

7.9.1 Empirical Models

Different empirical models for the evaluation of the path loss exist. The following types of models may be mentioned here: long-distance models, Motley-Keenan models (Keenan 1990) and multi-ray models, also referred to as corridor models.

In the case of the long-distance models, the parameters considered are the frequency and the distance between the transmitter and the receiver. Two long-distance models have been advanced: while the first such model (COST 231 1999) assumes a logarithmic dependence of the attenuation on distance, the second model assumes a linear dependence (COST 231 1999). Both models can be used in non line-of-sight situations.

Motley-Keenan models apply to non line-of-sight situations in dense environments, like an office. Like the building penetration loss, the path loss L is expressed as the sum of the free-space loss L_0, of the losses due to the different obstacles along the direct line-of-sight path (tiles, walls, doors, windows) and of a constant L_c (Motley 1988). Databases can be used for differentiating between the different obstacles to which specific attenuation values are attached. This model is the most commonly used.

$$L = L_0 + L_c + \sum_{j=1}^{N} N_j L_j + N_f L_f \tag{7.40}$$

where N_j is the number of walls of type j present along the line-of-sight path, L_j is the losses due to walls of type j, N is the number of types of walls, N_f is the number of tiles present along the line-of-sight path and L_f is the losses due to tiles.

A few typical values of the losses in the 1-2 GHz frequency band for different types of materials used in the internal walls are summarised here in Table 7.3 (COST 231 1999).

The waveguide effect of the corridors imposes to develop specific models taking into account a diffraction term in non line-of-sight situations (Dersch 1994a; Lafortune 1990).

7.9.2 Deterministic Models

Due to the density of the propagation environment, models relying on the ray launching method are generally three-dimensional models. These models have been developed by several authors (Rappaport 1994; Seidel 1994; Cichon 1994) and have led to encouraging results (Rappaport 1994). They can be used as reference models for developing multipath models. However, due to long computation times, their use remains somewhat limited for engineering application.

The ray tracing method consists in the construction of rays based on the image theory. This method enables to easily take into account the different propagation mechanisms, for instance multiple reflections or diffractions, and has been used by several authors (Mc Kown 1991; COST 231 1999; Valenzuela 1994; Jenvey 1994;

Laurenson 1993). An exhaustive study has been conducted for the construction of ray tracing models (Honcharenko 1992). The performances of these models are close to those of the ray launching method (COST 231 1999).

The two other approaches, namely the finite difference method (Murch 1994; Lauer 1994) and the Tayleig-Gans approximation method (Lu 1993) use relations derived from Maxwell's equations. These two methods use a mesh representation of the propagation environment. The field propagates in the space in the vicinity of the emitter iteration after iteration. The computation times required for the actual implementation of these methods are very long due to the fact that the mesh step must be much smaller than the wavelength (approximately $\lambda/8$). These two methods however are still in an early stage of development, and representative results on their performances are not yet available.

Table 7.3. Transmission losses for different types of materials (internal walls) in the 1-2 GHz frequency band

Material	Losses [dB]
Plasterboard	1.5
Wood	1.5
Glaze	2
Brick wall with limited thickness (< 14 centimetres)	2.5
Brick	2.5
Concrete wall with limited thickness (< 10 centimetres)	6
Double concrete wall (2*20 centimetres)	17
Floor	23

7.10 Broadband Models

7.10.1 Introduction

A broadband model is a model representing or modelling the behaviour of the propagation channel over a given bandwidth and for a given configuration as regards such parameters as the frequency, the bandwidth, the mobility type (fixed, pedestrian, car, etc.), the environment (indoors or outdoors, rural or urban environment), the use of single-antenna or multi-antenna systems or the diversity. These models account for the time dispersion of the impulse response (multiple paths) which is caused by a variety of phenomena involved in the propagation process, like for instance reflection, diffraction or scattering phenomena.

Broadband models can be classified into two main families: simulation models and prediction models. In the first case the propagation channel is represented with its time and space variability characteristics for a given environment from experimental measurements, in order for instance to evaluate the effects induced by modifications of these characteristics on the quality of the numerical signal. Broadband simulation models are primarily intended for being integrated in simulation environments of digital transmission systems using software like COSSAP.

In contrast, broadband prediction models are aimed at predicting a minimum number of characteristics from data pertaining to the propagation environment. These models are intended for engineering application. The function of these models is to predict the impulse response of the propagation channel, or at least the value of parameters directly affecting the quality of transmission, like for instance the number of paths or the delay spread.

Different types of broadband models have been developed. The principles of these models will be summarised thereafter, before presenting some examples of existing models.

7.10.2 Path Loss Models

COST 207 (or GSM) Models

The principle of path loss models is far from being new and was initially defined in the context of the COST 207 European cooperation on digital land mobile radio communications. These models are based on the representation of the impulse response in the form of a limited number of discrete paths for different propagation environments, in rural area (RA), typical urban (TU) bad urban (BU) and hilly terrain (HT) profiles respectively. Each path is defined by an amplitude, a delay and a Doppler spectrum type (Rayleigh, Gauss or Rice). Due to the simplicity of this principle, this type of model can be easily implemented in software and was therefore widely employed during the period of the GSM standardisation. The very simplicity of this principle also brings considerable disadvantages that the users tend to overlook, like for instance the limited numbers of paths, or the fact that

these models are stationary. Further, the discretisation of the temporal space associated with delay times results in a frequency periodicity. At last, these models are
not very practical in frequency hopping mode.

The associated power profile can be represented by a Dirac-type discontinuous
set describing the significant ray paths of the profile within an unlimited frequency
band. The following equation describes the impulse response:

$$h(t) = \sum_i a_i \delta(t - \tau_i)$$
(7.41)

where a_i represents the amplitude of the wave arriving at time τ_i.

The power of the wave arriving at time τ_i is given by the equation:

$$P(\tau_i) = P_0 \int_{-f_d}^{+fd} S^2(\tau_i, f) df$$
(7.42)

where P_0 is the standardised power, f_d is the maximum Doppler frequency, while
$S(\tau_i, f)$ represents the diffusion function of the propagation channel and characterises the Doppler spread of the signal along the path I, caused by the displacements
of the mobile.

The maximum Doppler frequency is equal to v / λ for a mobile moving at speed
v, λ being the wavelength. Four different traditional types of functions (Rayleigh,
flat, Gauss and Rice functions) have been defined in order to model the Doppler
spectrum associated with echoes. The choice of the function relates to the amplitude and delay of the echo, and depends on the diffusion function.

The classical Rayleigh function is based on the Clarke model. The angles of arrival of the incident waves are assumed to be uniformly distributed, and there is no
privileged direction : the associated fast fading follows in this case a Rayleigh law,
while the Doppler spectrum has a U shape and assumes values in the interval $]-f_d$
$,+ f_d[$. The boundaries of this interval correspond to incidence angles, either positive or negative, in the same direction than the displacement of the mobile.

$$S(\tau_i, f) = \frac{A}{\sqrt{1 - \left(\dfrac{f}{f_d}\right)^2}}$$
(7.43)

where A is a constant, $\tau_i \in [0, 0.5 \ \mu s]$, and $f \in]-f_d, +f_d[$.

The following equation defines the flat function:

$$S(\tau_i, f) = 1$$
(7.44)

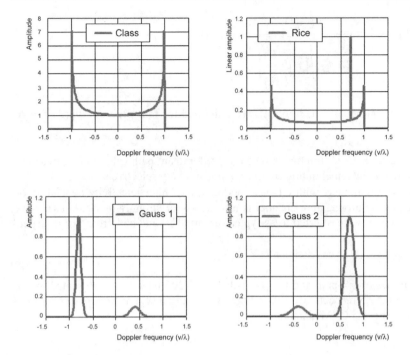

Fig. 7.17. Graphical representation of the Doppler spectra: classical Rayleigh function, Rice function and Gauss function

where $f \in]-f_d, +f_d[$.

The Gauss type function is the sum of two functions $N(m,s)$ where m and s are two functions of the form αf_d and βf_d respectively. Depending on the value of the delay:

$$S(\tau_i, f) = Ae^{\frac{(f+0.8f_d)^2}{2(0.05f_d)^2}} + 0.1Ae^{\frac{(f-0.4f_d)^2}{2(0.1f_d)^2}}, \text{ for } 0.5\mu s < \tau_i < 2\mu s \tag{7.45}$$

$$S(\tau_i, f) = Be^{\frac{(f-0.7f_d)^2}{2(0.1f_d)^2}} + 0.03Be^{\frac{(f+0.4f_d)^2}{2(0.15f_d)^2}}, \text{ for } \tau_i \geq 2 \ \mu s \tag{7.46}$$

These two functions define a solid angle where the incident waves are confined.

The Rice function is based on the consideration of a predominant path, which generally is a direct line-of-sight path, and corresponds to the superposition of a Rayleigh function and a Dirac function:

$$S(\tau_i, f) = \frac{0.41}{2\pi f_d \sqrt{1 - \left(\dfrac{f}{f_d}\right)^2}} + 0.91\delta(f - f_d) \tag{7.47}$$

Six-path and twelve-path models were defined for four different types of environments:

- rural area (RA) environments where a predominant path can be identified. The propagation channel in this environment is not very selective.
- urban environments with either a moderate or a high selectivity depending on whether the propagation channel is moderately or strongly blocked. These two environments are referred to as typical urban (TU) and bad urban (BU) respectively,
- hilly terrain (HT) environments.

Examples of six-path and twelve-path COST 207 models are presented in Tables 7.4 and 7.5 respectively, while Fig. 18 provides a schematic representation of a twelve-path TU GSM channel.

Table 7.4. Six-path model in bad urban (BU) environment

Number of paths	Delay [μs]	Power	Doppler spectrum	DS [μs]
1	0	- 3	Rayleigh	2.4
2	0.4	0	Rayleigh	2.4
3	1	- 3	Gauss 1	2.4
4	1.6	- 5	Gauss 1	2.4
5	5	- 2	Gauss 2	2.4
6	6.6	- 4	Gauss 2	2.4

Table 7.5. Twelve-path model in hilly terrain (HT) environment

Number of paths	Delay [μs]	Power	Doppler spectrum	DS [μs]
1	0	- 10	Rayleigh	5
2	0.2	- 8	Rayleigh	5
3	0.4	- 6	Rayleigh	5
4	0.6	- 4	Gauss 1	5
5	0.8	0	Gauss 1	5
6	2	0	Gauss 1	5
7	2.4	- 4	Gauss 2	5
8	15	- 8	Gauss 2	5
9	15.2	- 9	Gauss 2	5
10	15.8	- 10	Gauss 2	5
11	17.2	- 12	Gauss 2	5
12	20	- 14	Gauss 2	5

ATDMA Models

ATDMA models (RACE 1994) were elaborated from data files of measurements carried out in macrocell, in microcell and in picocell environments within the RACE ATDMA project framework. These measurements were realized using RUSK 5000 propagation channel sounders in outdoor environments and the France Telecom R&D propagation channel sounder in indoor environments. The analysis band was equal to 5.75 MHz in outdoor environments and to 50 MHz in picocell environments. The reference stored data for the modelling in COST 207 format were selected from the distribution of the moments of orders 1 and 2 of the delay dispersion. They were constructed by subdividing the average power profile into six equidistant parts. For each of these regions, the delay was estimated from the filtering power contained in the region. The amplitude associated with delay τ_i was estimated by filtering the power contained in the observation window T_i.

Four propagation scenarios were considered: indoor propagation, penetration, microcell propagation and macrocell propagation.

- in the indoor propagation scenario, the distance between the emitter and the receiver varies from 60 to 150 metres. Three different types of environments were studied: offices, shopping centres and subways.
- the penetration scenario corresponds to the penetration into buildings of waves emitted at a normal incidence and at a small distance from the emitter (100 metres). The indoor environment is an office type environment with windows on the walls.
- the microcell propagation scenario describes an urban environment, and includes both microcell environments, where the emitting antenna is located below the roof level, and small cell environments, where it is located above the roof level.
- the macrocell propagation scenario corresponds to an emission above the roof level and to a wider coverage area.

The measurements were realised using either omnidirectional or directive antennas. Typical files correspond to a configuration where both the emitting and the receiving antennas are omnidirectional antennas, whereas atypical files correspond to configurations where at least one directive antenna is used, either at the emission site or at the reception site.

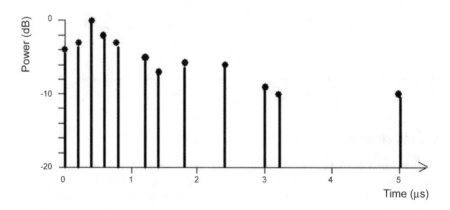

Fig. 7.18. Schematic representation of a twelve-path TU GSM channel

Table 7.6. Examples of ATDMA models in indoor environment

Number of paths	Typical data files		Atypical data files	
	Delay [µs]	Power	Delay [µs]	Power
1	0	0	0	- 11
2	0.077	- 3.3	0.102	- 11.1
3	0.186	- 9.3	0.161	0
4	0.299	- 14.3	0.289	- 20.1
5	0.404	- 20.3	0.377	- 25.6
6	0.513	- 26.8	0.475	- 29.2

A schematic representation of an atypical ATDMA model in indoor environment is presented in Fig. 7.19.

ITU-R Models

The models recommended by the ITU-R were defined at the WP TG8/1 meeting held in Mainz in 1996 on the subject of the Appendix C of the REVAL evaluation guidelines. Three propagation scenarios were considered: office indoor environment, pedestrian and vehicle. The experimental configurations for these three scenarios are successively reported here in Table 7.7. For each propagation scenario, two types of propagation channels were considered, which are referred to as channels A and B respectively.

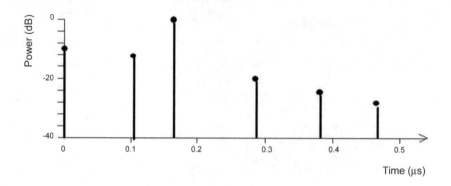

Fig. 7.19. Representation of an atypical ATDMA model in indoor environment

The propagation channel A describes the average behaviour of a propagation channel in situations where it is not very selective. This channel is constructed from models which have been validated by the European Telecommunications Standards Institute (ETSI).

The propagation channel B describes the average behaviour of the propagation channel in frequently encountered critical cases.

Tables 7.8 and 7.9 provide examples of ITU-R models presented in the COST 207 format.

Table 7.7. Propagation scenarios for the ITU-R model

	Radio coverage [km^2]	Distance [km]	Speed of the mobile [km/h]	Type of cell
Indoor office environment	0.01	0.1	3	Picocell in open space environment
Pedestrian mode	4	2	3	Microcell
Vehicle	150	13	120	Macrocell

Table 7.8. Indoor office environment. Channel A

Number of paths	Delay [µs]	Power	Doppler spectrum	DS [µs]
1	0	0	flat	
2	0.5	- 3	flat	
3	0.11	- 10	flat	
				0.035
4	0.17	- 18	flat	
5	0.29	- 26	flat	
6	0.31	- 32	flat	

Table 7.9. Indoor office environment. Channel A

Number of paths	Delay [µs]	Power	Doppler spectrum	DS [µs]
1	0	0	flat	
2	0.1	- 3.6	flat	
3	0.2	- 7.2	flat	
				0.1
4	0.3	- 10.8	flat	
5	0.5	- 18	flat	
6	0.7	- 25.2	flat	

A schematic representation of an ITU-R model in vehicular environment (propagation channel A) is presented here in Fig. 7.20.

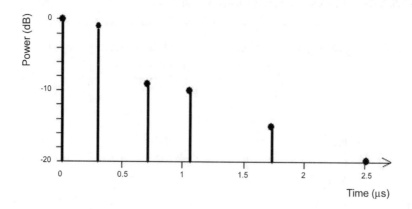

Fig. 7.20. Representation of an ITU-R model (vehicular environment, channel A)

CSELT Models

These models were developed at the CSELT for the study of UMTS systems, and may consist either of 6 or 12 paths. The distinction between indoor building environments and external environments is not taken into account. A criterion between line-of-sight situations and non line-of-sight situations results in the definition of two different propagation scenarios. These models were constructed from measurements carried out either in microcell environment or inside open space buildings. The receiving antenna is omnidirectional, while the emitting antenna is directive in the H plane (70 °) and in the E plane (60 °). The propagation scenarios considered for the models are as defined in Table 7.10.

Table 7.10. Propagation scenarios for the CSELT models

Propagation scenario	Classification	Height of the emitter [m]	Height of the receiver [m]	Distance between the emitter and the receiver [m]	Analysed band [MHz]
Indoor/microcell	Line-of-sight	3	1.6	20 - 150	20
Microcell/indoor	Non line-of-sight	1.9	1.6	20-150	20

Table 7.11. Five-path CSELT model in line-of-sight indoors (hall) and in microcell

Number of paths	Delay [μs]	Power	Doppler spectrum	DS [μs]
1	0	0	Rayleigh	
2	0.07	- 3	Rayleigh	
3	0.14	- 10	Rayleigh	0.102
4	0.28	- 18	Rayleigh	
5	0.40	- 26	Rayleigh	

Table 7.12. Ten-path CSELT model in non line-of-sight indoors (hall) and in microcell

Number of paths	Delay [μs]	Power	Doppler spectrum	DS [μs]
1	0	0	Rayleigh	
2	0.07	- 18	Rayleigh	
3	0.14	- 10	Rayleigh	
4	0.21	- 18	Rayleigh	0.133
5	0.3	- 15	Rayleigh	
6	0.4	-12	Rayleigh	
7	0.45	- 20	Rayleigh	

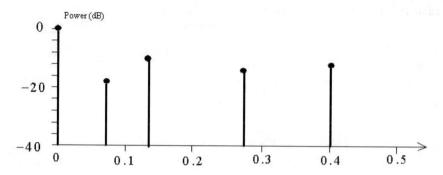

Fig. 7.21. Representation of a CSELT model in non line-of-sight situation

Table 7.12. (cont.)

8	0.5	- 20	Rayleigh
9	0.55	- 20	Rayleigh
10	0.6	- 20	Rayleigh

A schematic representation of a CSELT model in non line-of-sight situation can be seen in Fig. 7.21.

7.10.3 Representation Models

Representation models are models describing a propagation channel considered as representative of a given propagation scenario and defined through the statistical analysis of the selectivity of the propagation channel (Siaud 1996, 1997a, 1997b).

Deterministic Propagation Models

The propagation channel causes modulations of the signal in amplitude, phase and frequency. The obstacles along the path generate echoes on the received signal. These echoes present delays which are dependent on the relative distance of the reflectors from the direct path. The amplitude and phase of the echoes are variable in time and therefore induce a phase modulation as well as an amplitude modulation of the received signal. The frequency modulation is a consequence of the mobility of the propagation environment. Significant echoes are damaging to the quality of transmission of the link if the delays are longer than the symbol duration of the system generating inter-symbol inferences. The propagation channel behaves like a frequency selective filter. The selectivity of the filter with respect to the simulated system depends on the bandwidth of the system under consideration and on the propagation environment.

The propagation channel is therefore simulated in the form of a time-variable numerical filter with a sampling rate equal to the sampling rate of the transmitted modulated signal. The construction of this time-variable transversal filter relies on the principles of the theory of multirate systems (Crochière 1981). The refresh frequency of the coefficients of the model is adjusted from the simulated speed of the receiver and from the amplitude of the variations occurring in the environment during the simulation of the transmission chain (shadowing effect). The modelling of a given propagation scenario is realised from the input file of the model, which corresponds to a measurement file of impulse responses. This measurement file is itself selected on the basis of the statistical analysis of the selectivity parameters of the measured propagation channel.

This model is based on the representation of the propagation channel by a time-variable transversal filter. The temporal variability of the filter depends on the speed of the mobile, on the temporal variability of the environment and on fast fading.

The analysis of the variability of the propagation channel proceeds through the interpolation of the temporal variations of the propagation channel by intercalating computed impulse responses between two successive measured impulse responses. The determination of the coefficients of the filter relies on the assumptions that the propagation channel is invariant during a period T' and that the coefficients of the filter change at the frequency $F' = 1 / T'$ corresponding to the interval between two impulse responses after the interpolation has been performed. The interpolation is performed in order to obtain a continuous temporal evolution of the filter and is not intended at simulating a vehicle speed different from the measurement speed.

A transversal filter models each impulse response. The filter used in this method is a non-recursive discrete filter with finite impulse response. This type of a filter has no poles in the convergence region and presents a limited number of coefficients.

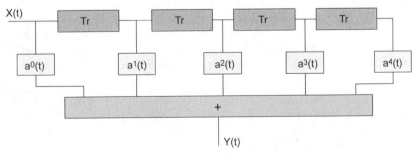

Fig. 7.22. Modelling of the propagation channel

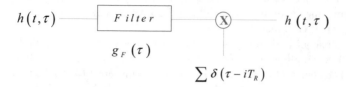

$$h(t,\tau) \quad\text{---}\quad \boxed{\begin{array}{c} Filter \\ \\ g_F(\tau) \end{array}} \quad\text{---}\quad \otimes \quad\text{---}\quad h(t,\tau)$$

$$\sum \delta(\tau - iT_R)$$

Fig. 7.23. Modelling of the propagation channel by a transversal filter

The canonical structure of a transversal filter consists in a set of delay lines, where T_r represents the sampling step of the transmitted signal and where the coefficients r are the complex amplitudes of each path. The order of the filter depends on T_r and on the measurement environment. As represented in Fig. 7.22, it is constant for the data files of the measurements under consideration.

The equipartition of the paths implies that they are not to be considered as corresponding to significant actual paths. Each coefficient $a_i(t)$ is calculated in order to approximate the value that would have been obtained at the ith path of the impulse response measured at time t if the bandwidth used for the channel sounding had been equal to $1/T_r$.

The modelling of each response by a transversal filter with respect to delays may be compared to a real sampling performed during the digitisation process of the measurements. The impulse response of the transversal filter at time t can be expressed in the form:

$$h_F(t,\tau) = h(t,\tau) * g_F(\tau)\delta(\tau - iT_r) = \sum a_i(t)\delta(\tau - iT_r) \qquad (7.48)$$

where:

- $h(t, \tau)$ is the impulse response of the propagation channel measured at time t,
- $g_F(\tau)$ is the impulse response of the filter used in the sampling function.

The interpolation of the fast variations of the propagation channel proceeds in a similar manner. The interpolated impulse response must approximate the impulse response that would have been measured if the acquisition frequency had been equal to $1/T'$. The assumption that the propagation channel is invariant is expressed through the use of a particular sampling function, referred to as the block sampler. The coefficients of the transversal filter at time $t = iT'$ remain unchanged during a period T'. The time-dependent transversal filter can be expressed in the following form:

$$h_I(t,\tau) = \sum h_F(t,\tau)\delta(\tau - kT') * g_I(t) = \sum hF(kT',\tau)g_I(t - kT') \qquad (7.49)$$

$$h_I(t,\tau) = \sum_{k,i} a_i(kT')g_I(t - kT')\delta(\tau - iT_r) \qquad (7.50)$$

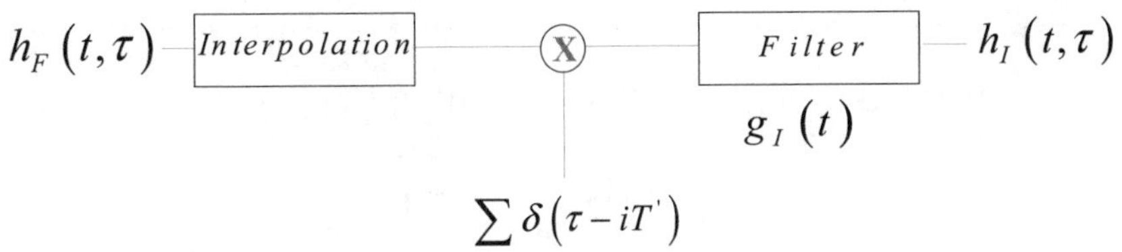

$$h_F(t,\tau) \boxed{Interpolation} \quad \otimes \quad \boxed{Filter} \quad h_I(t,\tau)$$

$$g_I(t)$$

$$\sum \delta(\tau - iT')$$

Fig. 7.24. Modelling of the propagation channel by a transversal filter (interpolation and filtering)

Deterministic Propagation Models with Frequency Hopping

The frequency selectivity of the propagation channel can be achieved by using frequency diversity. Two successive sequences of the communication signal filtered by two different frequency bands of the propagation channel assumed to be independent from one another generate a reduction of the level of fading of the resulting signal. This enhancement may be significant provided that the frequency hopping is higher than the channel correlation bandwidth, i.e. that the propagation channel is selective. In contrast, in the case of a nonselective propagation channel, the frequency hopping rate creates a temporal diversity. If the duration between two frequency hops is higher than the correlation time of the propagation channel, the two successive filters are independent and result in an enhancement of the quality of transmission.

The input files used for these models were taken from measurements realised within the RACE ATDMA project in a 50 MHz bandwidth using France Telecom R&D propagation channel sounder. Four propagation scenarios were considered: indoor propagation (office, shopping centre), penetration, microcell and small cell propagation, macrocell propagation. A few significant results concerning broadband parameters are presented in the following tables: the mean m and the standard deviation σ for the 75 and 90 percent delay windows (W75, W90), for the delay intervals at 6 and 15 dB (I6, I15), for the delay spread (DS) and for the 50 percent correlation bandwidth (BC):

Table 7.13. Indoor propagation (standard office)

	W75 [μs]	W90 [μs]	I6 [μs]	I15 [μs]	DS [μs]	Bc [MHz]
M	0.104	0.155	0.128	0.232	0.046	5.06
σ	0.03	0.044	0.044	0.06	0.01	1.67

Table 7.14. Indoor propagation (standard shopping centre)

	W75 [μs]	W90 [μs]	I6 [μs]	I15 [μs]	DS [μs]	Bc [MHz]
M	0.142	0.219	0.153	0.311	0.069	5.32
σ	0.05	0.07	0.06	0.01	0.02	1.73

Table 7.15. Penetration

	W75 [μs]	W90 [μs]	I6 [μs]	I15 [μs]	DS [μs]	Bc [MHz]
m	0.095	0.157	0.109	0.217	0.074	4.69
σ	0.08	0.154	0.07	0.18	0.05	2.13

Table 7.16. Microcell

	W75 [μs]	W90 [μs]	I6 [μs]	I15 [μs]	DS [μs]	Bc [MHz]
m	0.216	0.338	0.164	0.403	0.129	4.27
σ	0.223	0.311	0.194	0.379	0.1	1.71

Table 7.17. Small cell

	W75 [μs]	W90 [μs]	I6 [μs]	I15 [μs]	DS [μs]	Bc [MHz]
M	0.632	0.953	0.532	0.134	0.351	2.77
σ	0.53	0.66	0.66	1.2	0.24	1.73

Table 7.18. Macrocell

	W75 [µs]	W90 [µs]	I6 [µs]	I15 [µs]	DS [µs]	Bc [MHz]
M	0.236	0.448	0.274	0.854	0.234	4.19
σ	1	0.139	1.32	2.05	0.479	1.79

7.10.4 Ray Models

Ray models are deterministic models which rely on an accurate knowledge of the actual propagation environment and therefore require the use of such geographical databases as baseline databases or indoor databases. For more detail on geographical databases, the reader is referred to Appendix N. These models can be used for the prediction of the different propagation paths in a given configuration. Further, by performing an adjustment to the frequency band under consideration, these models allow to conduct parametric studies by simulation, in order for instance to analyse the influence of the radiation pattern of the antennas or the characteristics of materials, in a much less expensive way than the realisation of multiple measurement campaigns. Two examples of ray tracing in outdoor environment and in indoor environment respectively are presented in Figs. 7.25 and 7.26 (Chaigneaud 2001a, 2001b, 2002).

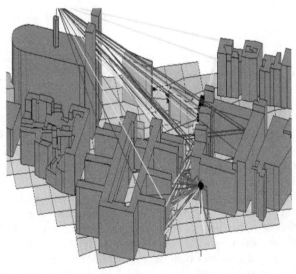

Fig. 7.25. Example of ray tracing in external environment

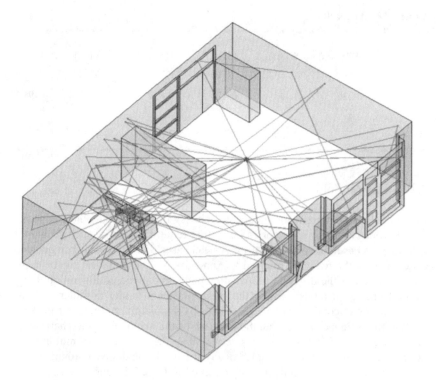

Fig. 7.26. Ray tracing in indoor environment

7.10.5. Geometrical Models

Geometrical models were first suggested within the RACE CODIT project framework and really started being developed in the context of the European METAMORP project and of the COST 259 European cooperation.

The concept underlying geometrical modelling is the relation between the angular and temporal powers profiles on the one hand, and the location within the propagation environment of the reflectors and the scatterers or clusters. As can be seen in Fig. 7.27, the form of the impulse response directly depends on the position of the dominant scatterers with respect to the base station and to the mobile. A propagation path is regarded as being composed of several rays in order to account for the notions of spatial and temporal spreading. In order to obtain realistic enough Doppler spectra, these rays are defined by a radio propagation path containing two successive reflections, the first one at reflectors or at distant scatterers, the second one at a scatterer at a short distance from the mobile. In contrast, as can be seen in Fig. 7.27, the direct line-of-sight path between the emitter and the receiver consists only of rays scattered by reflectors at a short distance from the mobile.

The modelling phase consists in the identification, from measured temporal and angular power profiles, i.e. from space-time representations of the impulse response of the propagation channel, of the location within the propagation environment of the dominant scatterers. The localisation of each cluster with respect to the base station and to the mobile, as well as its attenuation law, which assumes a form $N_i log_{10}$ *(length of the ray)*, can be determined from the information pertaining to the delay, the direction of arrival and the gain associated with this cluster. A representation of the propagation environment in the measurement configuration can thus be obtained.

Three specific radio channel measurement configurations were identified and reproduced using the method outlined above from a broadband multisensor experiment realised in dense urban environment in Paris at the 2 GHz frequency and in a small cell context (antenna located above the average roof level):

− a configuration displaying an average temporal and spatial selectivity compared to all the measurement paths considered in this experiment,
− a configuration displaying a strong temporal selectivity,
− a configuration displaying a strong spatial selectivity.

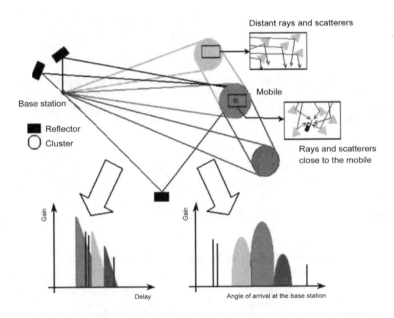

Fig. 7.27. Dependence of the form of the temporal power profile on the location within the propagation environment of the reflectors and scatterers

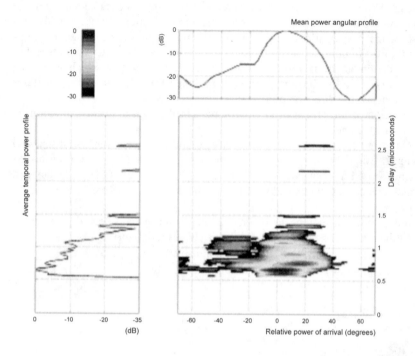

Fig. 7.28. Spatial-temporal representation of the impulse response in a configuration with average temporal and spatial selectivity. *a.* angular power profile, *b.* temporal power profile, *c.* average spatial-temporal power distribution. The origin of the angles corresponds to the pointing axis of the base station antenna

Fig. 7.28 presents an example of the spatial-temporal representation of the complex impulse response measured during a broadband multisensor experiment realised in dense environment at the 2 GHz frequency and in small cell context (antenna located above the roof level) (Laspougeas 2000).

7.10.6 Uses of Broadband Models in Simulation Software

Propagation channel models are intended at the evaluation of the quality of service of a digital transmission where distortions due to propagation play an essential part. This evaluation is conducted with the use of simulation software packages, among which MATLAB, COSSAP and PTOLEMY are the more commonly used, and which consist of several software modules implementing different functions present in the transmission chain, for instance coding and modulation functions, access techniques, distortion correction techniques and of course propagation channel models.

Although the implementation of ray models is relatively simple, these models do not represent in a completely realistic manner the behaviour of the propagation channel. This results from such factors as the fact that the Doppler spectrum and the frequency correlation in these models are not representative enough, the existence of spatial properties which have not been modelled, or the fact that antenna effects were not taken into account.

This becomes even more sensitive in the analysis of radio communication systems using intelligent multisensor antennas. These techniques aim either at improving the evaluation of the link through the use of very directive antennas and the rejection of jammers or at increasing the overall capacity system through the introduction of an additional spatial dimension (*spatial division multiple access*).

The multisensor channel model should therefore allow reproducing the behaviour as well as the correlation of the signals received by the different sensors. For this purpose, several different models have been developed, for the most part in the UHF band (Ertel 1998). In order to implement these models, a method called MASCARAA (Méthode Adaptée a la Simulation du Canal Large Bande avec Antennes Adaptatives, *method adapted to the simulation of adaptive antenna broadband channels*) was developed. This method is intended at reproducing, through the modelling of the spatial and frequency selectivity, the broadband behaviour of the propagation channel in a given configuration as regards such parameters as the frequency, the bandwidth, the sampling speed, the mobility (for instance fixed, pedestrian or vehicle), the environment (indoor, outdoor, urban or rural) or the configuration (single-sensor or multisensor). Since the simulation block is very general, this method can be used for simulating most types of models with a single software module: COST type models, geometrical models based on the statistical spatial distribution of the scatterers derived from experimental measurements, models based on the results of a ray model like for instance the GRIMM model as well as models based on a typical measurement file representative of a given propagation environment.

Figure 7.29 presents examples of temporal and spatial power profiles for a comparison between measurements and simulations.

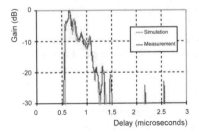 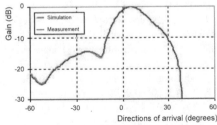

Fig. 7.29. Temporal and spatial power profiles: comparison between measurements and simulations

By way of conclusion, the different elements required for the evaluation of the quality of a radio digital transmission system can be summarised as follows : the availability of a simulation software for the transmission chain, the implementation of a software module simulating as realistically as possible the propagation with its different variability and selectivity characteristics, the determination of the most adequate propagation models to use with this simulation module, the optimisation and the validation of these models through a large enough number of measurements as to cover the different types of propagation environments. This is a long term program, but several elements essential to the completion of this project are already available at France Telecom R&D.

11 References

Barbot JP, Levy AJ, Bic JC (1992) Estimation of fast fading distribution functions. Com. URSI Commission F Open Symposium

Bic JC, Charbonnier A, Duponteil D, Ruelle D, Tabbane S, Taisant JP (1995) Radiocommunications et mobilité. Annales des Télécommunications, 50, 1: 114-141

Berg JE (1995) A recursive method for street microcell path loss calculations. PIMRC'95, Toronto, Canada, pp 140-143

Bertoni HL, Honcharenko W, Maciel LR, Xia HH (1994) UHF propagation prediction for wireless personal communications. Proceedings of the IEEE, vol 82, 9: 1333-1359

Boithias L (1987) Radio Wave Propagation. MacGraw-Hill, New York

Bourdeilles (1997) Modélisation de la propagation radio pour l'ingénierie radio des systèmes de communications avec les mobiles. SEE : Propagation électromagnétique dans l'atmosphère du décamétrique à l'angström, pp 115-120

Braun WR, Dersch U (1991) A physical mobile radio channel. IEEE Transactions on Vehicular Technology vol 40, 2 : 472-482.

Chaigneaud L, Guillet V, Vauzelle R (2001a) 3D ray tracing method for indoor propagation modelling at 60 GHz. European Conference on Wireless Technology, London

Chaigneaud L, Guillet V, Vauzelle R (2001b) A 3D ray tool broadband wireless system. Vehicular Technology Conference, Atlantic City

Chaigneaud L, Guillet V, Vauzelle R (2002) Méthode de tracé de rayon 3D pour la modélisation de la propagation en intérieur à 60 GHz. Propagation électromagnétique dans l'atmosphère du décamétrique à l'angström, Rennes

Cichon DJ, Wiesbeck W (1994) Indoor and outdoor propagation modelling in pico cells. PIMRC'94, Personal Indoor Mobile Radio Communications

Clarke RH (1968) A statistical theory of mobile-radio reception. BSTJ: 957-1000

CNET/CSELT Cooperation (1998) Data transmission on DECT standard. Definition of common propagation models, regeneration scheme and performance evaluation criteria for the aligment of the two radio link simulators

COST 259 (2000) COST 259 Web informations : www.lx.it.pt/cost259

COST 231 (1999) Evolution of land mobile radio (including personal) communications. Final report, Information, Technologies and Sciences, European Commission

Crochiere RE, Rabiner LR (1981) Interpolation and decimation of digital signals - a tutoral review. Proceedings of the IEEE vol 69, 3: 300-331

Failly M (1989) Final Report of COST 207, Digital Land Mobile Radio Communications. CEE Luxemburg

Foulonneau B, Gaudaire F, Gabillet Y (1996) Measurement method of electromagnetic transmission loss of building components using two reverberation chambers. Elect. Letters 7 vol 32, 23: 2130-2131

Gahleitner R, Bonek E (1994) Radio waves penetration into urban buildings in small cell and microcells. Technische Universität Wien, Vienna, Austria, Proceedings Vehicular Technology Conference, Stockholm, pp 887-891

Gfeller FR, Bapst URS (1979) Wireless in-house data communication via diffuse infrared radiation. Proceedings of the IEEE vol 67, 11

Hashemi H (1993) The Indoor Radio Propagation Channel. Proceedings of the IEEE vol 81, 7: 943-968

Hata M (1980) Empirical formula for propagation loss in land mobile radio service. IEEE Transactions on Vehicular Technology vol 29: 317-325

Ikekami F, Yoshida S, Takeuchi T, Umehira M (1984) Propagation factors controlling mean field strength on urban streets. IEEE Transactions on Antennas and Propagation vol 32, 8: 822-829

ITU-R (1996) International Telecommunication Union Study Groups 'Guidelines for evaluation of radio transmission technologies for IMT-2000/FPLMTS'. FPLMTS.REVAL Question ITU-R Document 8/29-E

Jakoby R, Liebenow U (1995) Modelling of radiowave propagation in microcells. Proc. Intern. Conference on Antennas and Propagation. ICAP, Eindhoven, the Netherlands, pp 377-380

Jenvey S (1994) Ray optics modelling for indoor propagation at 1.8 GHz. Proceedings of the IEEE 44th Vehicular Technology Conference, Stockholm, Sweden

Kattenbach R, Fruchting H (1995) Calculation of system and correlation functions for WSSUS channels from wideband measurements. Frequenz 49 3-4: 42-47

Keenan JM, Motley AJ (1990) Radio Coverage in Buildings. British Telecom Technol. J. vol 8, 1

Keller JB (1962) Geometrical theory of diffraction. JOSA vol 52: 116-130

Kouyoumjian RG, Pathak PH (1974) A uniform geometrical theory of diffraction for an edge in a perfectly conducting surface. Proc IEEE vol 62, 11: 1448-1461

Kurner T, Cichon DJ, Wiesbeck W (1993) Concepts and Results for 3D Digital Terrain Based Wave Propagation Models : an overview. IEEE Trans. Selected Areas in Com., vol SAC 11, 7 : 1002-1012

Lagrange X (2000) Les réseaux mobiles. Chapitre 2 : Propagation radioélectrique. In: Sizun H, Bic JC (eds) Réseaux et Télécoms, Information-Commande-Communication, Hermès, Paris

Laspougeas R, Pajusco P, Bic JC (2000) Radio propagation in urban small cells environment at 2 GHz : Experimental spatio-temporal characterization and spatial wideband channel model. Proc. IEEE Vehicular Technology Conference VTC'2000, Boston

Lauer A, Bahr A, Wolff I (1994) FDTD simulations of indoor propagation. Proceedings of the 44th Vehicular Technology Conference, Stockholm, Sweden

Laurenson DI, McLaughlin S, Sheikh AUH (1993) The application of ray tracing and the GTD to indoor channel modelling. IEEE Conf. GLOBECOM'93, Houston, USA

Lavergnat J, Sylvain M (1997) Propagation des ondes radioélectriques. Collection Pédagogique de Télécommunication, Masson, Paris

Lawton MC, McGeehan JP (1994) The application of a deterministic ray launching for the prediction of radiochannel characteristics in small cell environment. IEEE Transactions on Vehicular Technology, vol 43, 4: 955-969

Liang G, Bertoni HL (1998) A new approach to 3D ray tracing for propagation prediction in cities. IEEE Transactions on Antennas and Propagation vol 46, 6

Lu YE (1993) Site precise radio wave propagation simulations by time domain finite difference methods. Proceedings of the 43th Vehicular Technology Conference, Meadowlands, USA

McKown JW, Hamilton RL (1991) Ray tracing as a design tool for radio networks. IEEE Network Magazine

McNamara DA, Pistorius CWI, Malherbe JAG (1990) The Uniform Geometrical Theory of Diffraction. Artech House, London

METAMORP Project (2000) Description of the modeling method. Deliverable C2/1 www.nt.tuwien.ac.at/mobile/projects/METAMORP/en/

Motley AJ, Keenan JM (1988) Personnal communication radio coverage in building at 900 MHz and 1700 MHz. Electronics Letters vol 24, 12

Murch RD, Cheung KW, Fong MS, Sau JHM, Chuang JCL A new approach to indoor propagation prediction. Proceedings of the 44th Vehicular Technology Conference, Stockholm, Sweden

Parsons JD (1992) The mobile radio propagation channel. Pentech Press Publishers

RACE ATDMA Project (1994) Channel models Issue 2. R084/ESG/CC3/DS/029/b1 Gollreiter R (ed)

Rappaport TS, Sandhu S (1994) Radio Wave Propagation for Emerging Wireless Personal Communication Systems. IEEE Antennas and Propagation Magazine vol 36, 5:14-23

Rossi JP, Barbot JP, Levy AJ (1997) Theory and measurement of the angle of arrival and time delay of UHF radiowaves using a ring array. IEEE Transactions on Antennas and Propagation vol 45, 5: 876-884

Rossi JP, Bic JC, Levy AJ, Gabillet Y, Rosen M (1991) A ray launching method for radiomobile propagation in urban area. IEEE Antennas and Propagation Symposium, London, Ontario, vol 3: 1540-1543

Rossi JP, Levy AJ (1992) A ray model for decimetric radio-wave propagation in an urban area. Radio Science vol 27, 6 : 971-979

Saunders SR (1999) Antennas and Propagation for wireless communications systems. Wiley, London

Siaud I (1996) A digital signal processing approach for the mobile radio propagation channel simulation with time and frequency diversity applied to an indoor environment at 2.2 GHz. Personal indoor mobile radio communications conference, PIMRC'96, Taiwan

Siaud I (1997a) A mobile propagation channel model with frequency hopping based on a digita signal processing and statistical analysis of wideband measurements applied in micro and small cells at 2.2 GHz. IEEE Vehicular technology Conference, Phoenix, Arizona vol 2, pp 1084-1088

Siaud I (1997b) Simulation du canal de propagation radiomobile en environnement urbain pour l'étude des performances des systèmes de communication de 3iéme génération avec diversité de fréquence. 3^{ièmes} journées d'étude 'Propagation électromagnétique dans l'atmosphère du décamétrique à l'angström' pp 277-282

Seidel SY, Rappaport TS (1994) Site-specific propagation prediction for wireless in building personal communication system design. IEEE Transactions on Vehicular Technology vol 43, 4

Valenzuela RA (1994) Ray tracing prediction of indoor radio propagation. PIMRC'94, Personal Indoor Mobile Radio Communications

Valenzuela R, Landron O, Jacobs DL (1997) Estimating Local Mean Signal Strength of Indoor Multipath Propagation. IEEE Transactions on Vehicular Technology vol 46, 1: 203-121

Walfish J, Bertoni HL (1988) A theoretical model of UHF propagation in urban environments. IEEE Antennas and Propagation vol 36, 12: 1788-1796

Walker EH (1993) Penetration of Radio Signals into Buildings in the Cellular Radio Environment. The Bell System Technical Journal vol 62, 9: 2719-2730

Wiart J, Marquis A, Juy M (1993) Analytical Microcell Path Loss Model at 2.2 GHz. PIMRC'93, Yokohama

Xia HH, Bertoni HL (1993) Radio propagation characteristics for line-of-sight microcellular and personal communications. IEEE Antennas and Propagation vol 41, 10

Yang H, Lu C (2000) Infrared wireless LAN using multiple optical sources. IEE Proc OptoElectron vol 147, 4

Appendices

A The Sun and the Solar Activity

A.1 Introduction

The Sun is for us the most important star: dispenser of light and warmth, the Sun is the source of energy which maintains life on Earth. The Sun is a spherical planet located at a distance of $1.5 \ 10^{11}$ meters (150 million kilometres) from the Earth, with an apparent diameter of approximately 32' arcsecond, i.e. 695 000 kilometres, which represents about 109 times the diameter of the Earth. The Sun is a gaseous planet formed of hydrogen (81.76 percent), helium (18.17 percent) and, albeit in negligible quantities, carbon, nitrogen, oxygen and different other metals. The Sun has a mass equal to $332 \ 10^3$ times the mass of the Earth, equal to 2.10^{27} tons. The average density of the Sun is relatively low, hardly higher than the density of water: it is equal to only 1.41, to be compared with the average density of the Earth, equal to 5.52. Finally, the gravity at the solar equator is equal to 27.9 times the gravity observed at the surface of the Earth.

Unlike the rotation speed of a solid body, the rotation speed of the Sun depends on the latitude. The rotation period of the Sun thus varies between a 25 day minimum period in equatorial areas and a 34 day maximum period around the poles. The average period of rotation of the Sun is commonly assumed to be equal to 27 days.

Like any star, the Sun has its own motion, moving with respect to the centre of the galaxy at the speed of 20 kilometres per second in the direction of a point of space located in the vicinity of the star Vega in the constellation of Hercules.

All the energy emitted by the Sun results from thermonuclear fusion reactions whereby hydrogen atoms are combined into helium atoms. The energy thus produced inside the Sun is then conveyed towards the external layers through different physical processes.

The temperature at the surface of the Sun is equal to 5780 °K. The Sun radiates in electromagnetic form an energy of $3.9 \ 10^{33}$ erg/s over the entire wavelength spectrum, ranging from X-rays (very short waves) to radio waves (long waves). This energy reaches the Earth after an eight minute travel time.

Most of the energy emitted by the Sun is contained within the visible spectrum and originates primarily from the layer immediately above the solar interior. This 400 kilometre thick layer is referred to as the photosphere and constitutes the visible surface of the Sun. The layers of the solar atmosphere extending above the photosphere are much fainter and therefore they are under normal conditions invisible on the disk of the sky. Only during a total solar eclipse, at the moment when the photosphere disappears behind the lunar disk, can it be seen that a relatively bright and irregular reddish fringe expands outward from the photosphere. This region has a thickness of approximately 8000 kilometres, and is referred to as the chromosphere. Above the chromosphere lies a relatively faint white halo, without definite boundaries, and which may extend at a distance of several solar radii from the solar edge. This region is known as the solar corona and expands itself outward in the form a constant stream of matter known as the solar wind. The solar wind can be directly detected by satellite measurements and generates several different terrestrial phenomena, for instance auroras and magnetic disturbances. The activity of the Sun is primarily due to local modifications of its magnetic field and after a solar flare may result in the production of radio emissions and corpuscular radiations.

A.2 The Photosphere

The photosphere is the luminous surface delimiting the apparent contour of the Sun: this region is the most stable layer of the solar atmosphere and the main source of radiations emitted by the Sun.

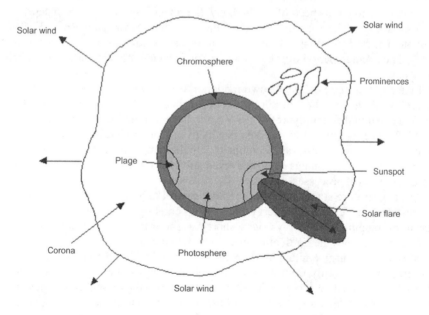

Fig. A.1. Schematic representation of the different features of the Sun

The essential features of the photosphere are granules and sunspots. Historically, the study of sunspots led to the determination of the rotation period of the Sun. Galileo used this method with his newly discovered telescope to deduce that the Sun has a rotational period of approximately one month. These sunspots are associated with very strong magnetic fields and have an important part in the solar activity and in the disturbances affecting the relations between the Sun and the Earth.

A.2.1 Granulation

The photosphere exhibits a mottled appearance described under the name of granulation. This phenomenon is caused by the convection currents moving beneath the photosphere with a velocity of 0.3 kilometre per second. The diameter of granules varies approximately from 700 to 1000 kilometres. These features have a lifetime of about 8 minutes and they present fluctuations in temperature of the order of 100 °K. The convection currents within the photosphere which are the cause for the appearance of granules are intimately related to the solar activity and to the chromosphere.

A.2.2 Sunspots

Observed in hydrogen-alpha (6563 Å), sunspots appear in the form of roughly circular dark surfaces called umbras, surrounded by less dark region known as penumbras. Sunspots are located within called plages or faculae, relatively brighter than the rest of the solar disk. The penumbra of a sunspot appears to be composed of filaments which seem to stream in a spiral pattern. Sunspots are the most evident as well as the earliest demonstration of the existence of the solar activity.

Sunspots appear dark only by contrast with the rest of the solar disk. In reality their temperature is still very high, ranging approximately from 4000 to 5000 °K, while their brightness is still equal to a third of the brightness of the Sun.

The dimensions of sunspots are extremely variable: while some sunspots hardly exceed an arcsecond, corresponding to a few hundreds kilometres, and can hardly be distinguished from the granulation of the photosphere, other sunspots frequently exceed the diameter of the Earth. Sunspots extending across several hundreds of thousands of kilometres in diameter can sometimes be observed: this was the case for instance in 1947, when the spotted surface of the Sun extended over several millions square kilometres.

The observation of sunspots reveals that they drift westward as time passes: they describe at the surface of the Sun a small circle whose poles are the poles of the Sun. This circle determines the solar equator. The apparent rotational period of the Sun is 27 days. However, if the movement of the terrestrial observer is also taken into account, the true rotational period of the Sun turns out to be equal to 25.4 days.

Sunspots are not permanent features, and their lifetime is variable: whereas small sunspots with diameters of a few thousand kilometres remain visible for less than a day, very large sunspots some tens of thousand kilometres across may last for months, and even, in exceptional cases, for more than one year.

The number of sunspots strongly varies from one year to another. The permanent observations of the Sun conducted since the eighteenth century have shown that the number of sunspots follows an eleven-year cycle. It might be noted however that the periods of this cycle are not constant and that they vary between 9 and 13 years: the extreme values of 7.3 and 15 years are sometimes even mentioned. At the beginning of the cycle sunspots tend to appear about 30 degrees north and south of the solar equator. As time passes, the region where sunspots are forming moves towards the equator: the last sunspots to be seen during a cycle appear at the average latitude of 5 degrees north and south.

The observation of the Zeeman effect, i.e. the splitting of the absorption lines, in the spectrum of sunspots, reveals the existence of very intense magnetic fields which may reach the order of 1000 to 2000 Gauss and occasionally 4000 Gauss. These values are to be compared with the dipolar field of the Sun, equal to 20 Gauss in calm regions, and to the dipolar field of the Earth, which is approximately equal to 0.5 Gauss.

Sunspots generally form by groups of opposite magnetic polarity in opposed hemispheres. Further, in a given hemisphere, the polarity of the leading sunspot, i.e. the first sunspot crossing the solar meridian line changes sign from a solar cycle to the following. This phenomenon demonstrates that the true solar cycle is 22 year long.

Sunspots are associated with swirls forming inside the Sun around the latitudes of 30 degrees north and south of the solar equator before moving in the direction the equator: these swirls produce two groups of sunspots with opposite polarities. From one eleven-year cycle to the following, the rotation direction of the swirls changes, which reverses the polarities of the sunspots.

A.2.3 The Magnetic Fields of Sunspots

The existence of intense magnetic fields inside sunspots was discovered as early as 1908 by G.E. Hale by applying the principle of Zeeman splitting to the study of the solar spectrum. The amplitude of these fields varies from a few hundreds Gauss for small spots to 3000 Gauss approximately for the largest ones.

Inside a given sunspot, the amplitude of the magnetic field reaches its maximum at the centre of the umbra, where the field is approximately vertical. The amplitude of the field then decreases as the distance from the centre of the sunspot increases, while the field tends to be more and more inclined until it becomes almost horizontal in the outer penumbra. The geometry of the field inside the sunspot still remains misunderstood.

The magnetic fields of sunspots and the weak fields in their vicinity rigorously follow remarkable polarity laws : in a given hemisphere and during a given solar activity cycle, the first sunspot in the direction of the solar rotation and the region surrounding it have a defined polarity, while the tail sunspot has the opposite polarity. This polarity arrangement is reversed from one solar hemisphere to another and from a solar activity cycle to the following.

A.3 The Chromosphere

The chromosphere is the layer of the solar atmosphere extending between the photosphere and the solar corona. Its thickness is about 8000 kilometres. Like the photosphere the chromosphere has a negative density gradient. The temperature in this region is increasing with height, from 4300 °K to approximately 10^6 °K at the base of the solar corona.

The chromosphere is most commonly observed in monochromatic light in the red part of the spectrum, at the 6563 Å wavelength (hydrogen-alpha line). The chromosphere presents different types of features such as plages, prominences or filaments, and fibrils.

A.3.1 Plages

Plages, also referred to as faculae or flocculates depending on whether they are observed under Hα, Calcium or white light illumination, are very wide, irregular and bright surfaces which correspond to gaseous emissions occurring near sunspots and moving under the influence of the magnetic fields generated by sunspots. The dimensions of plages are very variable but some of them may reach the order of 100 000 kilometres. Their brightness is determined by the intensity of the local magnetic field.

A.3.2. Prominences and Filaments

The names prominence and filament are essentially describing the same feature. The difference in name indicates where the feature is located in reference to the solar disk. These features appear in the form of immense clouds of relatively cold gas suspended above the surface of the Sun under the influence of the magnetic field. In the vertical plane with respect to the surface of the Sun, these features appear bright on the disk of the sky and are referred to as prominences. In the horizontal plane they are called filaments and appear in the form of very lengthened and dark areas against the brighter chromosphere.

The filaments of the chromosphere may reach a length of the order of $3 \ 10^8$ kilometres and a height of the order of 10^5 kilometres above the photosphere. While stable, or quiescent, prominences may last for several months, others appear by expelling a flow of particles towards the interplanetary medium. The appearance of prominences is a normal event in the evolution of the active regions.

A.3.3 Fibrils

Fibrils, also referred to under the name of spicules, are extremely thin and elongated features appearing in the chromosphere illuminated under Hα light. Fibrils tend to be aligned along the magnetic field lines, and can frequently be seen near active regions.

A.4 The Solar Corona

The solar corona is the outer atmosphere of the Sun, and appears in the form of a huge white light aureole extending around the chromosphere, with streamers radiating out in all directions. In periods of maximum solar activity, the solar corona is nearly circular, while in periods of minimum solar activity it is smaller and extends farther out towards the equatorial region.

The brightness of the solar corona is less than one millionth the brightness of the Sun itself. The solar corona cannot therefore be observed, except during eclipses or using a Lyot coronagraph, which is a device allowing to create an artificial eclipse of the Sun.

The solar corona is a practically isothermal hot gas with an average temperature of approximately $1.6 \ 10^6$ °K. It can therefore be regarded as a plasma. Due to its high temperature, the solar corona emits radio radiations in the X-ray and ultraviolet range from which it can be studied. It might be noted here that the wavelength of the emitted rays is inversely proportional to the distance from the surface of the layer where it originated. This allows monitoring solar emissions as they rise within the solar atmosphere: they will be first detected by short waves, and then by waves with increasingly high wavelengths. The ejected matter then flows out into the interplanetary space and eventually arrives in the vicinity of the Earth, where it induces different types of disturbances, including auroras, magnetic storms and ionospheric storms.

Besides these sporadic emissions, which can be extremely dangerous to astronauts, a stream of charged particles, known as the solar wind, is constantly blown off the Sun.

A.5 The Solar Wind

The temperature in the solar corona is so high that this region cannot remain in hydrostatic equilibrium, and therefore blows off a solar wind of high-speed charged particles. The solar wind streams escape primarily through coronal holes, which are regions of the solar corona where the temperature and the pressure are attenuated: these regions correspond to open magnetic field lines and have a lifetime of several solar rotations.

The existence of the solar wind was demonstrated by space probes such as

Skylab. The solar wind may carry several million tons of solar matter per second. Although it presents very frequent variations, its average speed can be estimated as approximately equal to 400 kilometres per second.

The solar wind has a density at 1 AU (or astronomical unit, equal to the distance between the Sun and the Earth) of a few particles, between 1 and 10 particles per cm^3 and a temperature equal to approximately 10^5 K. The ionic composition of the solar wind is identical to the ionic composition of the solar corona, and consists of H^+, He^{++} particles, accompanied by low energy electrons (a few eV).

The individual ionised particles of the solar wind travel in approximately four days from the Sun to the Earth. The kinetic energy density of these particles is significantly higher than the energy density of the interplanetary magnetic field: the solar wind causes therefore a displacement of the magnetic field lines. Due to the rotation of the Sun, the magnetic field lines are drawn into spirals.

A.6 The Magnetosphere

The existence of the Earth's magnetic field results in the consequence that the Earth acts as an obstacle to the solar wind: when the solar wind encounters the dipolar magnetic field of the Earth, it will therefore be slowed down and to a large extent deflected around it. The region of the circumterrestrial space subjected to the influence of the Earth's magnetic field is known as the magnetosphere. Fig. 2 provides a schematic representation of this region. The magnetosphere deflects the flow of particles emitted by the Sun, and is itself modified by the collisions occurring with these particles.

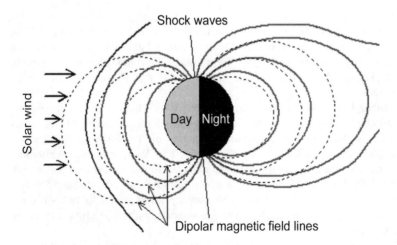

Fig. A.2. Schematic representation of the magnetosphere

The form of the magnetosphere is determined by the hydromagnetic equilibrium between the terrestrial magnetic field and the pressure induced by the solar wind. The interactions between the solar wind and the magnetosphere are at the origin of different aspects of the relations between the Earth and the Sun.

A.7 Solar Activity

The three regions of the atmosphere of the Sun, the photosphere, the chromosphere and the corona, might be locally disturbed by transitory formations called active centres. The active centres of the Sun are induced by the evolution of the magnetic field lines which result in temporary disturbances of the existing energy exchanges. They appear either in the form of sunspots inside the photosphere, in the form of plages, faculae, filaments, protuberances and fibrils inside the chromosphere, or in the form of condensations and jets inside the corona. The more general term of solar activity encompasses all these different phenomena. Two main indices are used for characterising solar activity: the Wolf number and the solar radio flux at the 2800 MHz frequency (10.7 cm wavelength).

A.7.1 Wolf Number

From 1749 to the present, the solar activity has been generally characterised by a solar number R defined from the number of visible spots observed at the surface of the Sun. This number is given by the following equation:

$$R = k(10g + f) \tag{A.1}$$

where k is correction factor, varying between 0.6 and 1 depending on the observation station, and accounting for the experimental characteristics and for the conditions of observation, while g is the number of groups of spots, including groups consisting of a single isolated sunspot, and f is the number of visible individual spots.

The number R is generally referred to as the Wolf number, or as the relative sunspot number. This number provides a good characterisation of the solar activity, and is therefore an important parameter in the study of the various terrestrial phenomena influenced by the latter. The Wolf number nonetheless presents the disadvantage of being discontinuous, as shown by the fact that it may pass directly from 0 to 11k. Further, a given number can be attributed to two different configurations: for instance, the number 22 is given to two different groups of sunspots consisting each of a single sunspot and to one group consisting of twelve sunspots.

The Wolf number ranges from 0 in periods of minimal solar activity to approximately 250 in periods of maximal solar activity. The highest ever observed value was 278, and was reached on January 22, 1959. The monthly number of sunspots (average over one month of the daily Wolf number) has never exceeded the value 220.

Fig. A.3 represents the variations observed between 1749 and 2001 of the Wolf number. The eleven-year cycle followed by solar activity can clearly be seen in this figure.

The analysis of the Wolf number from 1749 to the present day has revealed the existence of solar cycles, the best known being the eleven-year cycle. The successive periods of solar cycles present significant variations, both in duration and in amplitude. According to Balcook, the existence of solar cycles could be due to the differential rotation deforming the magnetic field lines of the Sun. Under the effect of this differential rotation, the relative displacements of two sunspots lying at a small distance from each other but with different latitudes and opposite polarities would induce an increase or a reduction of the gradient of the magnetic field, which may account for the existence of solar cycles.

As can be observed in periods of minimal solar activity, the fluctuations of the Wolf number follow a 27 day period, corresponding to the synodical rotation period of the Sun.

Several different indices are derived from the Wolf number R, among which we may for instance mention the indices R_1, IR_5 and R_{12}. These three indices are defined as follows:

- R_1 is the average value over a month m of the daily Wolf number, as defined by the following equation :

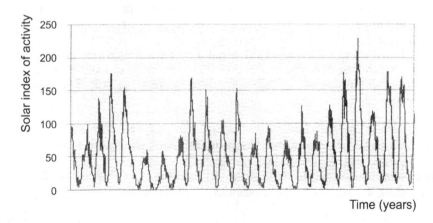

Fig. A.3. Variation of the solar activity index between 1749 and 2001

$$R_1(m) = \frac{1}{N} \sum_{j=1}^{N} R(j) \tag{A.2}$$

where N is the number of days in month m.
− IR_5 is defined by the equation :

$$IR_s(m) = \frac{1}{5}\left(+R(m-3)+R(m-2)+R(m-1)+R(m)+R(m+1)\right) \tag{A.3}$$

− R_{12} is defined by the equation :

$$R_{12}(m) = \frac{1}{12}\left(\sum_{m-5}^{m+5} R_k + \frac{1}{2}\left(R_{m-6} + R_{m+6}\right)\right) \tag{A.4}$$

where R_k is the average value over month k of the Wolf number.

A.7.2 Radio Flux at the 10.7 cm Wavelength

The solar activity may also be characterised by the daily solar radio flux measured at the 10.7 cm wavelength, or 2800 MHz frequency. This index was first introduced in 1947 at the ALGON radio observatory of the *National Research Council of Canada* located in Ottawa.

The numerical values of the daily solar radio flux are expressed in 10^{-22}Wm^{-2}Hz^{-1} and correspond to a single calibration performed at local midday approximately (17 00 UT). This index presents the same characteristics as regards for instance periods or burst than the Wolf.

Since the distance between the Sun and the Earth varies during the year, the observed radio flux observed has to be adjusted to one astronomical unit. The adjusted flux is noted Sa. The relative error is then equal to 2 %, but this adjustment allows comparing absolute values.

A.7.3 Correlation of the Wolf Number and the Solar Radio Flux

The correlation of the Wolf frequency R_{12} and the solar flux at the 10.7 cm wavelength Φ_{12} is given by the standard recommended equation:

$$\Phi_{12} = 63.7 + 0.728 R_{12} + 8.9 * 10^{-4} R_{12}^2 \tag{A.5}$$

A.8 Solar Flares

Sudden disturbances of the solar activity are more particularly caused by solar flares. Solar flares appear in the form sudden bursts of brightness of certain active regions of the chromosphere. This phenomenon may last from few minutes to a few hours and causes disturbances in the solar corona.

A solar flare has typically first an explosive phase during which a very intense emission is detached from the continuum chromosphere and rapidly spreads out over a large surface. The sources of solar flares are always located near a sunspot.

The time duration of a solar flare increases with the surface and with the spectral width of the reinforced Hα emission. Exceptionally, a solar flare may extend across the thousandth part of the solar disk. It then subsides with a slow decrease during two to three hours. A considerable energy is released during a solar flare, in the form of both radiation and kinetic energy. The kinetic energy accelerates the speed of the emitted particles, with the result that a greater number of particles are therefore released from the Sun's atmosphere.

Solar flares are classified either with respect to their maximum surface over the solar disk, their relative intensity and their duration or with respect to the intensity of the emitted X-rays. The relative intensity of a solar flare can be determined from the brightness of the different emission features, generally the Hα emission features, and is noted by the letters F (faint), N (normal) or B (bright).

Solar flares can also be classified according to the intensity of the emitted X-rays in the 1-8 Å band:

- C-class events are solar flares not associated with emissions of X-rays.
- N-class events are solar flares associated with emissions of low intensity X-rays and generally accompanied by sudden ionospheric disturbances (SID), and more specifically by sudden short-wave fadeouts. Further detail on these events can be found in Appendix F devoted to the ionospheric and geomagnetic disturbances associated to solar events.
- X class events are accompanied by emissions of very hard X-rays or by radio bursts in the 10 cm wavelength range. These events are generally accompanied by important sudden ionospheric disturbances.

This classification of solar flares with respect to the intensity of the emitted X-radiation was introduced on January 1 1969 at the *Space Environment Services Center* (SESC) in Boulder. It presents two main advantages over the traditional optical classification:

- it provides a more accurate expression of the geophysical significance of solar events,
- it allows an objective classification of all geophysical events with respect to the location on the Sun of the disturbances.

Different types of emissions are associated with solar flares: solar radio emissions, corpuscular radiations and emissions of shock waves.

A.8.1 Solar Radio Emissions

The solar radio emission has three different components:

- the emission of the quiet sun,
- the slowly variable radio emission due to the stable thermal radiation of plages and coronal condensations,
- bursts: their durations and intensities, as well as their appearances on measurements and their relations to optically observable phenomena are highly variable.

Solar radio emissions are associated with solar flares. With the advent of artificial satellites, their spectrum has been intensely studied in the last recent years. These emissions are classified into X-rays, ultraviolet rays, centimetric, decimetric, metric and decametric waves.

In the X-ray and ultraviolet ray ranges, the 1- 8Å band is the most important: while emissions are particularly weak during quiet periods, they may reach a high level at times of solar flare occurrences.

In the centimetric and decametric ranges, different standard types of radio bursts are distinguished according to their duration and to the characteristics of their dynamic spectrum:

- Type I: very short emissions lasting a few tenths of second with a narrow bandwidth of the order of a few megahertz which are superimposed on continuous emissions and form 'noise storms'. These storms are the more frequently observed solar emissions and are associated with the presence on the solar disk of the most active regions,
- Type II: relatively rare emissions observed at the onset of most chromospheric eruptions. These emissions are generally referred to under the term of slow drift bursts. They initially appear at high frequencies and gradually drift towards lower frequencies at a rate from 0.1 to 1 MHz/s, which means that these emissions originate in regions moving outward from the Sun. The duration of these emissions is of the order of ten minutes.
- Type III: short, intense bursts drifting from high frequencies towards lower frequencies at a much higher rate (10 to 20 MHz/s). These very frequent emissions occur in groups during all chromospheric eruptions regardless their importance,
- Type IV : stable emissions which may last several hours after the most intense chromospheric eruptions and which can be observed in all frequency ranges,
- Type V: emissions over a broad frequency range of the same type as above but much shorter (approximately 1 minute).

A.8.2 Corpuscular Radiation

The strongest solar eruptions and eruptive prominences may accelerate the speed of shock waves and solar particles to the point where they are no longer attracted to the Sun and therefore flow out the solar wind. The particles thus emitted are for the most part protons: hence the name of solar proton radiation. The energy spectrum of these protons extends from 10 Mev to a few Bev.

The interval between a solar emission and the arrival of protons on Earth in the 10 to 60 Mev energy spectrum ranges from approximately 3 to 10 hours for events occurring in the western part of the Sun and from 15 to 60 hours for events occurring in its eastern part. The difference results essentially from the configuration of the magnetic field lines in the solar corona and in the solar wind which tends to prevent particles emitted in the eastern part of the Sun from reaching the Earth.

This interval is much shorter for protons with energy higher than 100 Mev, and is equal to approximately one half hour for the western part of the Sun and one hour and half for the eastern part.

A.8.4 Emissions of Shock Waves

The stronger solar flares often generate magneto-hydrodynamic shock waves which propagate outward from the Sun at speeds ranging from 500 to 1500 kilometres per second. These shock waves reach the Earth 24 to 50 hours after the beginning of the solar flare. The travel time of shock waves is very difficult to predict since it depends on a large number of different parameters, including the speed of the shock wave and its deceleration, or the speed of the solar wind.

The penetration of a shock wave in the Earth's atmosphere causes a compression of the magnetosphere as well as a number of different types of terrestrial phenomena, like for instance auroras or magnetic and ionospheric disturbances.

References

Delabeau F (1973) L'environnement de la Terre. Presses Universitaires de France, Paris
Mangis SJ (1975) Introduction to Solar-Terrestrial Phenomena and the Space Environment Service Center. NOAA Technical Report ERL 315 SEL 32
Mc Intosh PS Dryer M (1972) Solar Activity Observations and Predictions. Massachussets Intitute of Technology, Cambridge, Mass.
Michard R (1974) Le soleil. Presse Universitaires de France, Paris
Zirin H (1966) The Solar Atmosphere, Blaisdell Pub. Co. Walham, Mass.

B Microphysical Properties of Hydrometeors

As condensation intensifies, the diameter of the droplets from which clouds are formed increases, either by coalescence, i.e. by agglomeration, or by the absorption of the steam around them. When their fall speed increases, precipitation occurs, either in the form of drizzle, if the diameter of the droplets lies between 0.1 and 0.5 millimetres, or in the form of rain, if the droplets are of larger dimensions. If the temperature falls beyond 0° C, hydrometeors are present in solid form (snow and hailstones).

B.1 Rain and Drizzle

The shape of raindrops has been the subject of a number of studies based on photographic measurements. These studies have revealed that raindrops of radius higher than 1 millimetres have an oblate, albeit flattened at the poles, spheroid shape (flattened ellipsoidal of revolution) with a flattened base (Pruppacher 1970, 1971). Fig. B.1 shows the shape of thirteen modelled water drops whose equivalent radii (i.e. the radii of spheres with same volume) range from 0.25 to 3.25 millimetres (Oguchi 1983).

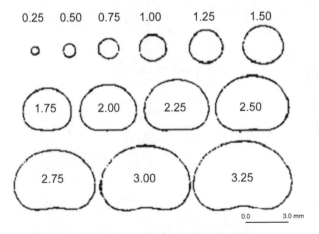

Fig. B.1. Modelling of the shapes of raindrops. The numerical values are the equivalent radii of spheres with the same volume

The fall speed of raindrops increases with their size, while the gradient speed gradually decreases down to the point where it becomes null : their terminal speed reaches its maximum value at approximately 9 ms^{-1} (Gunn 1958). The existence of this limit is a consequence of the deformation that the raindrops are submitted to: indeed, if raindrops were perfectly spherical, their speed would exceed 9 ms^{-1} in the case of raindrops with radius larger than 2.5 millimetres. The speed of raindrops depends on the atmospheric pressure P, on the water content H and on the temperature T. Different expressions for the speed of drops speed exist in the literature : they are summarised here in Table B.1. D is the diameter of the raindrops expressed in millimetres. Fig. B.2 presents a graphical comparison between these different expressions.

Table B.1. Different expressions of the fall speed of raindrops

Authors	Fall speed (ms^{-1})
Gunn and Kinzer	$9.65-10.3\ e^{-0.6D}$
Spilhaus	$4.49\ D^{0.5}$
Sekhon et al.	$4.25\ D^{0.6}$
Liu et al.	$3.35\ D^{0.8}$
Atlas et al.	$3.78\ D^{0.67}$

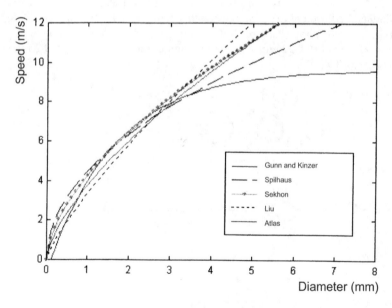

Fig. B.2. Comparison between the different laws expressing the speed of raindrops as a function of their diameter

The size distribution of raindrops is of primordial import in the study of the attenuation of electromagnetic waves, as it is actually directly involved in the equation of the attenuation coefficient. The size distribution of raindrops is written in the form $N(D)dD$ and represents the number of raindrops with an equivalent diameter comprised between D and $D+dD$ per unit of volume (m^3).

This parameter has been the subject of a number of studies, leading in general to results dependent upon the type of rain and the rain regime of the region considered. Different laws can be found in the literature, the best known being the decreasing exponential laws of Marshall-Palmer (Marshall 1948) and Joss *et al.* (Joss 1968) :

$$N(D) = N_0 e^{-\Lambda D} \tag{B.1}$$

where N_0 and Λ are experimentally determined constants. The values for these constants depend on the nature of the rain under consideration.

The distribution law advanced by Joss *et al* differs from the preceding to the extent that it explicitly takes into account the fundamental differences in the size distribution of raindrops which exist between three different types of rain: convective rain, continuous rain and drizzle. Table B.2 presents the parameters of the Marshall-Palmer and Joss *et al.* distributions. The parameter R represents here the rainfall rate in mm/h, i.e. the total height of water precipitated during a given time.

Table B.2. Numerical values of the parameters of the Marshall-Palmer and Joss *et al.* distributions for various types of rain

Authors		$N_0 \, (m^{-3}mm^{-1})$	$\Lambda \, (mm^{-1})$
Marshall-Palmer		8000	$4.1R^{-0.21}$
	drizzle	30000	$5.7R^{-0.21}$
Joss et al.	continuous rain	7000	$4.1R^{-0.21}$
	convective rain	1400	$3R^{-0.21}$

Distribution models of a more complex decreasing exponential form have been advanced by Best, Khergian et al. and by Fujiwara (Best 1950). Gamma and log-normal distribution models, based on theoretical studies and experimental observations, have also been suggested for the distribution of the sizes of raindrops. Table B.3 summarises the contributions by different authors to the modelling of the size distribution of raindrops for different types of climates :

Table B.3. Contributions to the modelling of the size distribution of raindrops for different types of climates

Distribution model	Authors
Deceasing exponential distribution	Marshall-Palmer, 1948
	Joss et al., 1968
	Best, 1940
Gamma distribution	Khergian et al., 1952
	Fujiwara, 1960
	Sekhon et Srivastava, 1971
	Moupfouma et al, 1982
	Ihara et al, 1984
	Gibbins et al., 1992
	Gloaguen, 1994
	Montanari, 1997
Lognormal distribution	Ajayi et Olsen, 1985
	Maciel et al., 1990
	Thimothy et al., 1995
	Montanari, 1997
	Gibbins et al., 1998

Despite being among the most ancient, the Laws-Parson distribution and the decreasing exponential Marshall-Palmer and Joss et al. distributions still remain the most commonly used for the calculation of rain attenuation. These models were established from experimental observations carried out in temperate areas.

A number of devices have been developed for the measurement of the size of raindrops, of their falling speed and of their density. An example of such a device is the dual-beam disdrometer developed by the *Centre of Studies on Terrestrial and Planetary Environments* of the University of Versailles Saint Quentin, which allows to measure raindrops with diameter beyond 0.2 millimetre. More information can be found at the website of this organisation at www.cetp.ipsl.fr.

B.2 Snow

Snow generally falls in the form of flakes or ice crystal aggregates. Although the diameter of these flakes generally lies between 2 and 5 millimetres, it may in some cases reaches the order of 15 millimetres. The shape of snowflakes is extremely complicated. Photographic measurements performed by Magono et al. have revealed that the relation between their maximal horizontal and vertical dimensions is extremely variable (Magono 1965). It has nevertheless been shown that for flakes with diameter lower than 10 millimetres the average value of this ratio is close to unit, while for flakes of a large diameter, this ratio is around 0.9 millimetres. Nishitsuji *et al.* have developed a classification of the different types of snow according to their water contents (Nishitsuji 1971).

Magono et al. have conducted rigorous measurements of the terminal fall speed of snowflakes (Magono 1965), advancing a semi-empirical model valid for large densities of snowflakes :

$$v(D) = 3.94\sqrt{\frac{D(\rho_s - \rho_a)}{2}}$$ (B.2)

where v *(D)* is the terminal speed in m.s^{-1}, ρ_s is the density of snowflakes in g.cm^{-3}, ρ_a is the density of the air in g.cm^{-3} and D is the diameter of the snowflakes in millimetres.

The most commonly used distribution for the sizes of snowflakes is the Gunn distribution (Gunn 1958). This decreasing exponential distribution model was developed on the basis of a large number of experimental measurements, and is based on the consideration of the two following parameters, expressed in mm^{-1}m^{-3} and in cm^{-1} respectively :

$$N_0 = 3.8 \times 10^3 R^{-0.87}$$ (B.3)

$$\Lambda = 25.5 R^{-0.48}$$ (B.4)

where R (mm/h) represents the precipitation rate of snow after the fusion of the snowflakes.

Sekhon and Srivastava have advanced an alternative decreasing law of exponential type. It seems however that this model is mostly designed to snowflakes with relatively small water contents. The parameters involved in this model are :

$$N_0 = 2.5 \times 10^3 R^{-0.94}$$ (B.5)

$$\Lambda = 22.9 R^{-045}$$ (B.6)

B.3 Hail

Hail is the precipitation of ice particles which takes place inside cumulonimbus clouds at negative temperatures. This phenomenon has a very localised character and is therefore neither very common nor very alarming. For this reason, only a limited number of studies have been devoted to this phenomenon. Hail can be considered as ice containing air bubbles. Hailstones are the largest observable hydrometeors; the equivalent diameter of hailstones is on the average lower than a centimetre, even though it may reach magnitudes between 2 and 8 centimetres.

Hailstones occur in a variety of different forms: spherical, spheroid, conical as well as a number of other irregular forms (Masson 1971). While the spherical form is certainly the most commonly used for the description of small hailstones, it is no longer realistic for the largest ones. The scattering properties of spheres of ice in fusion at millimetre-length wavelengths have been inquired into by Oguchi using the theory of diffusion by two concentric spheres with different refractive indexes as developed by Arden and Kerker (Oguchi 1966; Arden 1951). The mathematical model is represented in Fig. B.3.

The subject of the terminal fall speed of hailstones has been more particularly studied by Bringi, Willis and Douglas (Bringi 1977; Willis 1964; Douglas 1963).

Models for the distribution of the sizes of hailstones have been put forth by Douglas and Smith (Douglas 1964; Smith 1976):

– Douglas :

$$N(D) = 4960e^{-1236D} \tag{B.7}$$

– Smith (weak hail, $R = 10$ mm/h) :

$$N(D) = 1.1 \times 10^5 e^{-2000D} \tag{B.8}$$

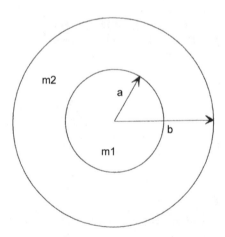

Fig. B.3. Model of a sphere of ice in fusion with radius b consisting of sphere of ice with radius a inside a concentric sphere of water with radius b - a. The refractive indexes of these two spheres are n_1 and n_2 respectively

– Smith (strong hail, $R = 100$ mm/h):

$$N(D) = 5.8 \times 10^4 e^{-1080D} \qquad (B.9)$$

where R represents the precipitation rate of hailstones after fusion.

B.4 Fog

Fog is due to the concentration of fine water droplets in suspension inside the atmosphere, resulting in the formation of a cloud limiting the visibility down to less than one kilometre. The size distribution of these droplets for different types of fogs has been studied by Eldridge among other subjects (Elridge 1957, 1966). The best known distribution model for the sizes of particles is the Gamma distribution model modified by Deirmendjian (Deirmendjian 1969) :

$$N(r) = a.r^\alpha \exp\left(-b.r^\gamma\right) \qquad (B.10)$$

where $N(r)$ is the number of particles per unit of volume with radius between r and $r+dr$, r is the radius of the droplet and a, α, b, γ are positive real constants, a being an integer.

The table presented below summarises the numerical values of the a, α, b and γ constants for different types of fogs (Kim 2001; Chu 1968; Deirmendjian 1969; Shettle 1989).

Table B.4. Numerical values of the coefficients of the Gamma distribution model modified by Deirmendjian for different types of fogs

	typical ray (μm)	a	b	α	γ
thick fog	10	0.027	0.3	3	1
moderate fog	2	607.5	3	6	1
Chu-Hogg model	1	341	4	2	0.5
maritime light fog	0.05	$5.3\ e^4$	8.9	1	0.5
continental light fog	0.07	$5.0\ e^6$	15.1	2	0.5

References

Arden AL Kerker M (1951) Scattering of electromagnetic waves from two concentric spheres. J. Appl. Phys. vol. 22: 1242-1246

Best AC (1950) Empirical formulae for the terminal velocity of water drops fall in through the atmosphere. Quart. J. Met. Soc. vol 76:302-311

Bringi VN, Seliga TA (1977) Scattering from axisymetric dielectrics or perfect conductors immbedded in an axisymetric dielectric. IEEE Trans. A.P. vol AP-25, 4:575-580

Chu TS, Hogg GDC (1968) Effects of precipitation on propagation at 0.63, 3.5 and 10.6 microns. Bell Syst. Tech. J. 47:723-759

Cohn GI (1972) Magnetohydrodynamic wave phenomena in sea water. In: AGARD Conference Proceedings 77 Electromagnetics of the Sea, pp. 25-1- 25-20

Deirmendjian D (1969) Electromagnetic Scattering on Spherical Polydispersions. Elsevier, New York

Douglas RH (1963) Hail size distributions of Alberta hail samples, Mc Gill Univ., Montreal, Stormy Wea. Gp. Sci. Rep. MW-36: 55-71

Douglas RH (1964) Hail size distributions. World Conf. on Radiometeorology, 11th Weather Radar, Boulder, Colorado, pp 146-149

Eldridge RG (1957) Measurement of cloud drop-size distribution, J. Meteor. 14:55-59

Eldridge RG (1967) Haze and fog aerosol distribution. Journal of the Atmospheric Sciences, vol 23: 605- 613

Gunn KLS, Marshall JS (1958) The distribution with size of aggregate snowflakes. J. Meteorol. vol 15: 452-461

Joss J, Thams JC, Waldvogel A (1968) The variation of raindrop size distribution at Locarno" Proc. of the International Conference on Cloud Physics, Toronto

Kim II, McArthur B, Korevaar E (2001) Comparison of laser beam propagation at 785 nm and 1550 nm in fog and haze for optical wireless communications. SPIE-Proceeding-series, 4214, pp 26-37

Marshall JS, Palmer WMcK (1948) The distribution of raindrops with size. Journal Meteorol. vol 5 : 165-166

Moupfouma F, Martin L (1985) L'acquisition de données de propagation et de radiométéorologie par le canal du système de collecte par satellite Argos. NT/LAB/MER/201, CNET

Oguchi T (1983) Electromagnetic wave propagation and scattering in rain and other hydrometeors. Proc. IEEE vol 71, 9

Mason BJ (1971) The Physics of Clouds, second Edition, Clarendon Press, Oxford

Pruppacher HR, Beard KV (1970) A wind tunnel investigation of the internal circulation and shape of water drops falling at terminal velocity in air. Quart. J. R. Met. Soc. vol 96: 247-256

Pruppacher HR, Pitter (1971) A semi-empirical determination of the shape of cloud and rain drops. J. Atmos.Sci. 28:86-94

Renaudin M (1991) Météorologie. Cepadues Editions

Shettle EP (1989) Models of aerosols, clouds and precipitation for atmospheric propagation studies. In: Atmospheric Propagation in the UV, visible, IR and MM-Wave region Related Systems Aspects AGARD Conf. Proc. 454, pp. 15-1 - 15-13

Smith PL, Musil DJ, Webber SF, Spahn JF, Johnson GN, Sand WR (1976) Raindrop and hailstone size distributions inside hailstones. Proc. Cloud Phys. Conf., Amer. Meteor. Soc. Boston, Mass., Boulder, Colorado, pp. 252-257

Stock (1993) La météo de A à Z. La météorologie nationale, Paris

Willis JT, Browning KA, Atlas D (1964) Radar observations of ice spheres in free fall. J. Atmos. Sci. 21, 103 : 348-420

C The Frequency Spectrum

C.1 Introduction

The spectrum management at international level is the responsibility of the International Union of Telecommunications Radiocommunications Sector (ITU-R) located in Geneva. It regularly publishes a frequency allocation table dividing the frequency spectrum into separate bands for each service (fixed, mobile, amateur, broadcasting, radio-astronomy, etc) depending on its geographical localisation. This frequency allocation table is integrated into the Radiocommunications Regulations (RR).

In this appendix the different frequency bands will be successively described: for each band, the atmospheric and terrestrial influence, as well as the system considerations and the associated services are indicated (Hall 1989). The frequency allocation by the IMT-2000 is then given, before listing the frequency bands used in satellite communications. At last, additional information concerning the P, L, S, X, K, Q, V and W bands is provided.

C.2 Frequency Bands

C.2.1 ELF Waves

Wavelength and frequency

- frequencies lower than 3 kHz
- wavelength (λ) higher than 100 kilometres

Atmospheric influence

- the ionosphere forms the higher limit of the propagation waveguide

Terrestrial influence:

− the surface of the ground forms the lower limit of the propagation waveguide

System consideration

− requires the use of extremely long antennas,
− very low information rate,
− sensitivity to the atmospheric noise (storm discharges in the distance and electric discharges in the vicinity)

Associated services

− short and long distance submarine communications,
− underground communications,
− sensitive communications,
− underground remote sensing.

C.2.2 VLF Waves

Wavelength and frequency

− frequencies from 3 to 30 kHz,
− wavelength (λ) : from 10 to 100 kilometres

Atmospheric influence

− the ionospheric D layer is the higher limit of the propagation waveguide

Terrestrial influence

− the surface of the ground is the lower limit of the propagation waveguide

System consideration

− necessity of using extremely long antennas
− difficulty to design and to build directional antennas,
− very low information rate.

Associated services

− world maritime telegraphy,
− long distance fixed services,

- navigation systems (OMEGA),
- standard time diffusion,
- underground and underwater communications,
- geological monitoring.

C.2.3 LF Waves

Wavelength and frequency

- frequencies from 30 to 300kHz,
- wavelength (λ): 1-10 km (kilometric waves).

Atmospheric influence

- sky wave in the propagation waveguide formed by the ground (lower limit) and the ionospheric D layer (higher limit) at frequencies lower than 100 kHz,
- sky wave and ground wave at frequencies higher than 100 kHz.

Terrestrial influence

- effects of the Earth's curvature on the propagation of the ground wave.

System consideration

- necessity of using very long antennas
- difficulty to design and to build directional antennas.

Associated services

- long distance maritime communications,
- long distance fixed services,
- radio diffusion,
- radio navigation.

C.2.4 MF waves (300-3000 kHz)

Wavelength and frequency

- frequencies from 300 to 3000 kHz,
- wavelength (λ) : from 100 - 1000 m (hectometric waves)

Atmospheric influence

- ground wave for short distances and lower frequencies,
- sky wave for long distances and higher frequencies. The night field is higher even for short distances.

Terrestrial influence:

- reflection of the wave at the ground.

System consideration:

- possibility of using directional antennas (antennas with multiple elements, L or T inverted antennas).

Associated services:

- radio diffusion,
- radio navigation,
- mobile communications (land, maritime, aeronautical),
- fixed services.

C.2.5 HF waves

Wavelength and frequency

- frequencies from 3 to 30 MHz
- wavelength (λ): 10-100 m (decametric waves)

Atmospheric influence

- ground wave for short distances more particularly above the sea.
- sky wave for long distances (beyond the first hop : 1E or 1F)

Terrestrial influence

- reflection and scattering.

System consideration

- use of log-periodical antenna arrays (either vertical or horizontal),
- use of vertical and horizontal dipole antennas.

Associated services

- point-to-point fixed services,
- mobile services (aeronautical, terrestrial, maritime),
- long distance radio diffusion.

C.2.6 VHF Waves

Wavelength and frequency

- frequencies from 30 to 300 MHz,
- wavelength (λ): 1-10 m (metric waves)

Atmospheric influence

- refraction and reflection due to irregularities of the refractive index, enabling forward-scatter links,
- ionospheric scatter (more particularly in the sporadic E layer), enabling forward-scatter links,
- Faraday rotation and ionospheric scintillations along Earth-space links.

Terrestrial influence

- screening effects by hills and mountains,
- diffraction inside valleys,
- reflection at the surface of flat lakes. The surface of the seas and the ground enhance multipath propagation.

System consideration

- use of multi-element dipole antennas (Yagi aerials),
- use of helicoidal and slot antennas.

Associated services

- radio diffusion (up to 100 kilometres),
- communications with mobiles (terrestrial, aeronautical, maritime),
- mobile phones,
- aeronautical radionavigation beacons.

C.2.7 UHF waves

Wavelength and frequency

- frequencies from 300 to 3000 MHz
- wavelength (λ) : 10 - 100 cm (decimetre waves)

Atmospheric influence

- atmospheric refraction,
- reflection at atmospheric layers at lower frequencies,
- ducting effects at higher frequencies,
- fluctuations of the refractive index,
- forward scattering beyond the horizon at frequencies higher than 500 MHz.

Terrestrial influence

- screening effects by hills and buildings

System consideration

- use of multi-element dipole antennas (Yagi aerials),
- possibility of using broadband systems,
- use of antennas with reflectors at the highest frequencies.

Associated services

- television,
- aeronautical navigation,
- radars,
- point-to-point fixed services,
- communications with mobiles,
- telemetry,
- cellular communications.

C.2.8 SHF Waves

Wavelength and frequency

- frequencies from 3 to 30 GHz,
- wavelength (λ) : 1 - 10 cm (centimetre waves)

Atmospheric influence

- attenuation by hydrometeors (rain, snow, hail),
- refraction,
- duct propagation,
- scintillations due to fluctuations of the refractive index.

Terrestrial influence

- diffraction by buildings,
- screening effects (buildings...),
- reflection and scattering by the environment (buildings, trees, etc).

System consideration

- use of parabolic aerials or horns.
- use of waveguides,
- possibility of using several channels at each carrier frequency.

Associated services

- fixed services (voice, multichannel television),
- radars,
- fixed Earth-satellite links,
- mobile services,
- satellite telemetry.

C.2.9 EHF Waves

Wavelength and frequency

- frequencies from 30 to 300 GHz
- wavelength (λ): 1 - 10 mm (millimetre waves)

Atmospheric influence

- important attenuation caused by hydrometeors (rain, snow, hail).
- attenuation by clouds, fog, dust, smoke, etc,
- scattering by hydrometeors (rain, snow, hail),
- scintillations due to fluctuations of the refractive index fluctuations of the index,

− attenuation by oxygen and water vapour.

Terrestrial influence

− screening effects (buildings, trees, etc).

System consideration

− use of small parabolic antennas.

Associated services

− short distance line-of-sight communications (either fixed or mobile),
− satellite teledetection.

C.2.10 Sub-UHF waves

Wavelength and frequency

− frequencies from 300 to 3000 GHz
− wavelength (λ) : 0.1 - 1 mm (sub-millimetre waves)

Atmospheric influence

− important attenuation by meteors (rain, snow, hail, clouds, dust, smoke),
− appearance of localised refractive index gradients (mirage),
− scintillations due to fluctuations of the refractive index,
− attenuation by atmospheric gases (oxygen, water vapour).

Terrestrial influence

− screening effects (trees).

System consideration

− use of antennas equipped with reflectors or with lens.

Associated services

− short line-of-sight communications,
− teledetection.

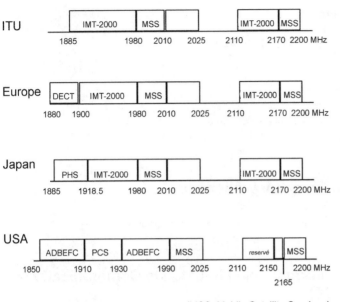

(MSS: Mobile Satellite Services)

Fig. C.1. IMT-2000 frequency allocation

C.2.11 Infrared and Light Waves

Wavelength and frequency

- frequencies from 3 to 430 THz (infrared waves) and from 430 to 860 THz (light waves)
- wavelength (λ) : 0.1mm - 0.4μm (infrared and light waves)

Atmospheric influence

- important attenuation by meteors (rain, snow, hail, clouds, dust, smoke),
- appearance of localised refractive index gradients (mirages),
- scintillations due to fluctuations of the refractive index,
- attenuation by atmospheric gases (carbon dioxin).

Terrestrial influence

- screening effects by small-sized objects.

System consideration

- use of antennas equipped with reflectors or with lens,
- use of lasers.

Associated services

- communications along very short paths,
- indoor communications.

C.3 IMT-2000 Frequency Allocation

These frequencies are allocated to third-generation mobiles (see Fig. C.1)

C.4 Frequency Bands used in Satellite Communications (COST255 2000)

Table C.1. Frequencies used in satellite communications

Frequency band	Frequencies (uplink/downlink)	Examples of service
L	1.6/1.5 GHz	Mobile communications (INMARSAT, IRIDIUM)
S	2 /2.4 GHz	Mobile communications
C	6/4 GHz	PSTN, Video (INTELSAT, ANIK, US Domestic, PALAPA)
X	8/7 GHz	Military communications (SKYNET)
Ku	14/11 and 14/12 GHz	PSTN, video (INTELSAT, EUTELSAT)
	17/12 GHz	(ASTRA)
Ka	30/20 GHz	Multimedia (TELEDESIC)
	44/20 GHz	Military communications (MILSTAR, SKYNET IV)
V	48/47 GHz	Multimedia (High Altitude Platform)
	40/50 GHz	Multimedia (PANAMSAT)

C.5 P, L, S, X, K, Q, V and W band (Freeman)

Table C.2. P Band

Sub-band	Frequency (GHz)	Wavelength (cm)
	0.225	133.3
	0.390	76.9

Table C.3. L Band

Sub-band	Frequency (GHz)	Wavelength (cm)
	0.390	76.9
P	0.465	64.5
C	0.510	58.8
L	0.725	41.4
Y	0.780	38.4
T	0.900	33.3
S	0.950	31.6
X	1.150	26.1
K	1.350	22.2
F	1.450	20.7
Z	1.550	19.3

Table C.4. S Band

Sub-band	Frequency (GHz)	Wavelength (cm)
	1.55	19.3
e	1.65	18.3
f	1.85	16.2
t	2.00	15.0
c	2.40	12.5
q	2.60	11.5
y	2.70	11.1
g	2.90	10.3
s	3.10	9.67
a	3.40	8.32
w	3.70	8.10
h	3.90	7.69
z[a]	4.20	7.14
d	5.20	5.77

Table C.5. X Band

Sub-band	Frequency (GHz)	Wavelength (cm)
	5.20	5.77
a	5.50	5.45
q	5.75	5.22
y_a	6.20	4.84
d	6.25	4.80
b	6.90	4.35
r	7.00	4.29
c	8.50	3.53
l	9.00	3.33
s	9.60	3.13
x	10.00	3.00
f	10.25	2.93
k	10.90	2.75

Table C.6. K Band

Sub-band	Frequency (GHz)	Wavelength (cm)
	10.90	2.75
p	12.25	2.45
s	13.25	2.26
e	14.25	2.10
c	15.35	1.95
ub	17.25	1.74
t	20.50	1.46
qb	24.50	1.22
r	26.50	1.13
m	28.50	1.05
n	30.70	0.977
l	33.00	0.909
a	36.00	0.834

Table C.7. Q Band

Sub-band	Frequency (GHz)	Wavelength (cm)
	36.0	0.834
A	38.0	0.790
B	40.0	0.750
C	42.0	0.715
D	44.0	0.682
E	46.0	0.652

Table C.8. V Band

Sub-band	Frequency (GHz)	Wavelength (cm)
	56.0	0.652
a	48.0	0.625
b	50.0	0.600
c	52.0	0.577
d	54.0	0.556
e	56.0	0.536

Table C.9. W Band

Sub-band	Frequency (GHz)	Wavelength (cm)
	56.0	0.536
	100.0	0.300

References

COST 255 (2002) Radiowave propagation modelling for SatCom services at Ku-band and above. Final report, European Space Agency, SP1252, 2002.

Hall MPM (1983) Overview of Radiowave Propagation. Chapter 1, Radiowave Propagation, edited by MPM Hall and LW Barclay, IEEE Electromagnetics Waves series 30, Peter Peregrinus Ltd. On Behalf of the Institution of Electrical engineers

Freeman RL (1998) Telecommunication Transmission Handbook. 4rd edition, Wiley, New York

D Cross-Polarisation induced by the Atmosphere

D.1 Introduction

In order to increase the line capacity of a given link without increasing the bandwidth of the signal, one generally resorts to orthogonal polarisations, either rectilinear or circular. However, when a wave propagates through the atmosphere, a part of the energy emitted with a given polarisation becomes orthogonally polarised, causing thereby interferences between the two communication channels. This phenomenon is essentially due to the presence of asymmetrical raindrops and ice particles, and even to the clear atmosphere itself.

Depolarisation is the phenomenon whereby all or part of a radio wave transmitted with a given polarisation has no longer any defined polarisation after the propagation.

The phenomenon of transpolarisation is characterised by the appearance, during the course of propagation, of a polarisation component (cross-polar component) orthogonal to the initial polarisation (copolar component). This phenomenon can originate inside rain or inside hydrometeors present in the troposphere, in multipath propagation and in tropospheric or ionospheric scintillation.

Two parameters allow the characterisation of the phenomenon of cross-polarisation: the cross-polarisation discrimination or decoupling ratio (*XPD*) and the cross-polarisation isolation (*XPI*).

D.1.1 Cross-Polarisation Discrimination

For an electromagnetic wave emitted with a given polarisation, the cross-polarisation discrimination or decoupling ratio (*XPD*) is the ratio, at the reception point, of the power received in initial polarisation to the power received in orthogonal polarisation. This parameter expresses the degree to which a wave emitted with a given polarisation becomes orthogonally polarised after having propagated. The value of this parameter, expressed in decibels, depends on the nature of the propagation medium, and is given by the following equation:

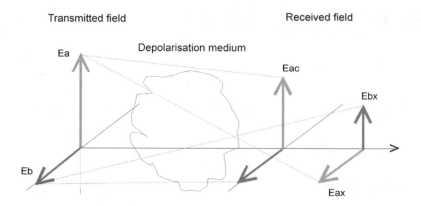

Fig. D.1. Schematic representation of the phenomenon of cross-polarisation

$$XPD = 20\log_{10} (Eac/Eax) \qquad (D.1)$$

The cross-polarisation discrimination is intimately related to the copolar attenuation of the wave. Several cross-polarisation discrimination models have been developed and will be described in the next section.

D.1.2 Cross-Polarisation Isolation

Let us consider two components with orthogonal polarisations *Ea* and *Eb* emitted in the same plane. The cross-polarisation isolation is defined as the ratio of the copolar power *Eac* (or *Ebc*) to the cross-polar power *Ebx* (or *Eax*). These two ratios play an important part in the design of systems. It might be noted that the *Eac/Ebx* ratio value is not necessarily equal to that of the *Ebc/Eax* ratio. The cross-polarisation isolation expresses the degree of interference at the reception of two signals simultaneously transmitted in orthogonal polarisation. It depends on the nature of the propagation medium and is expressed in decibels by the following equation:

$$XPI = 20 \log_{10}(Eac/Ebx) \qquad (D.2)$$

D.2 Cross-Polarisation Discrimination Models

D.2.1. Cross-Polarisation caused by Rain

Flock model

This model, developed by the NASA (Flock 1983) is very simple to implement and turns out to be a satisfactory approximation for frequencies higher than 10 GHz. In this model, the cross-polarisation due to rain is directly connected to the copolar attenuation *CPA* through the following equation:

$$XPD = a - b\log_{10}(CPA) \tag{D.3}$$

where *CPA* is the copolar attenuation in dB, while a and b are constants: $a = 5.8$ and $b = 13.4$.

ITU-R Model

This model, more specifically valid in the 8-35 GHz frequency band, is based on a distinction, for a given time percentage, between the cross-polarisation caused by rain and the cross-polarisation caused by ice.

The cross-polarisation discrimination (*XPD*) induced by rain for a given time percentage p is a function of the copolar attenuation, of the frequency, of the polarisation, of the elevation angle and of the obliquity angle of the raindrops. It is given by the following equation:

$$XPD_{1,p} = U - V\log_{10}(A_p) - C_\tau - C_\beta + C_\sigma \tag{D.4}$$

where:

- $U = 30\log(f)$, f being the frequency in GHz,
- $V = 12.8 f^{0.19}$ for f contained in the 8-20 GHz frequency range (Fukuchi 1984),
- $V = 22.6$, for f contained in the 20-35 GHz frequency range (Fukuchi 1984),
- A_p is the copolar rain attenuation exceeded for the time percentage p (dB),
- $C_\tau = 10 * \log_{10}\left[1 - 0.484 * (1 + \cos(4\tau))\right]$ where τ is the tilt angle between the horizontal plane and the rectilinear polarisation ($\tau = 45°$ in circular polarisation),

– $C_\beta = 40 * \log_{10}(\beta)$ where β is the elevation angle of the wave in degrees. The model is valid provided that β is lower or equal to 60°,

– $C_\sigma = 0.0052 * \sigma_p^2$ is a parameter associated with the fluctuations of the canting angles of the raindrops during a given rain event as well as between different such events. The values 0°, 5° and 10° are generally assumed for σ_p if p is equal to 1, 0.1 or 0.01 per cent respectively.

The cross-polarisation discrimination *XPD* due to ice for a given time percentage *p* is determined by the following empirical formula:

$$XPD_{2,p} = WPD_{1,p} * 0.5 * \left[0.3 + 0.1\log_{10}(p)\right] \tag{D.5}$$

while the cross-polarisation discrimination *XPD* due to rain and ice for a given time percentage *p* is expressed by the formula:

$$XPD_p = XPD_{1,p} - XPD_{2,p} \tag{D.6}$$

Based upon a model first suggested by Nowland (Nowland 1977) this model was tested within the COST255 framework on the basis of ITALSAT measurements, and was further extended to the V frequency band, i.e. to frequencies ranging between 40 and 50 GHz (COST 255 2002).

$$U = 26\log(f) \quad V = 20 \tag{D.7}$$

Dissanayake-Haworth-Watson Analytical Model

This model can be applied in the 9-30 GHz frequency range. It determines the relation existing between the cross-polarisation discrimination and the copolar attenuation, and is based on approximations applied to fields with low amplitude and to the phase in the transmission matrix of rain (Dissanayake 1980):

$$XPD_{1,p} = U' - V'\log_{10}(A) - C_\beta - C'_\tau + C'_\sigma \tag{D.8}$$

where:

– $U' = 84.8 - 88.8 * x * f^y + (50.32 * x * f^y - 21.9) * \log_{10}(f)$ with $x = 0.759$ and $y = 0.08$,

– $V' = 20$,

- $C_\beta = 40 * \log_{10} [\cos(\beta)]$, where β is the elevation angle of the wave measured in degrees,
- $C_\tau' = 20 \log_{10} [\sin(2\tau)]$, where τ is the tilt angle between the horizontal plane and the rectilinear polarisation,
- $C_\sigma = 17.37 * \sigma^2$, where σ is the standard deviation of the canting angles of the raindrops expressed in radians.
-

Chu Model

In this model, the relation between the cross-polarisation discrimination and the copolar attenuation is given by the following equation (Chu 1982):

$$XPD_{1,p} = U - 20 \log_{10}(A_p) - C_\tau - C_\beta + C_\sigma - C_x + 11.5 \qquad \text{(D.9)}$$

where:

- $U = 20 \log(f)$, f being the frequency in GHz,
- $C_\tau = 10 * \log_{10} \{0.5 * [1 - \cos(4\tau) * \exp(-0.0024 * \sigma_\varphi^2)]\}$ where σ_φ is the standard deviation of the canting angles of the raindrops during a thunderstorm and is equal to 3 degrees,
- $C_\beta = 40 * \log_{10} [\cos(\beta)]$, where β is the elevation angle of the wave measured in degrees,
- $C_x = 0.075 * \cos^2(\beta) * \cos(2\tau) * A_p$: a difference appears here between horizontal and vertical polarisations.
- $C_\sigma = 0.0052 * \sigma_\varphi^2$ is a parameter associated with the fluctuations of the canting angles of the raindrops during a given rain event or between different such events.

Stutzman-Runyon Model

In this model, the relation between the cross-polarisation discrimination and the copolar attenuation is given by the following equation (Stuzman 1984):

$$XPD_{1,p} = U - 19 \log_{10}(A_p) - C_\tau - C_\beta + C_\sigma + 9.5 - \log_{10}(r) \qquad \text{(D.10)}$$

where:

- $U = 17.3 \log(f)$, f being the frequency in GHz,

- $C_\tau = 10*\log_{10}\left\{0.5*\left[1-\cos(4\tau)*\exp(-0.0024*\sigma_\varphi^2)\right]\right\}$ where σ_φ is the standard deviation of the canting angles of the raindrops during a thunderstorm and is equal to 3 degrees.
- $C_\beta = 42*\log_{10}\left[\cos(\beta)\right]$, where β is the elevation angle of the wave measured in degrees,
- $C_\sigma = 0.0052\sigma_p^2$ is a parameter connected to the fluctuations of the raindrops canting angles during the rain events or between different such events,
- r is the relative ratio of non-spherical raindrops.

Nowland-Sharofsky-Olsen Model

Referred to as the NOS model (Nowland, Sharofsky, Olsen), this model connects the cross-polarisation discrimination to the copolar attenuation through the following equation (Nowland 1977):

$$XPD_{1,p} = U - V*\log_{10}(A_p) - C_\tau - C_\beta + C_\sigma + 4.1 + (V-20)*\log_{10}(L) \quad \text{(D.11)}$$

where:

- $U = 26*\log_{10}(f)$, f being the frequency in GHz,
- $V = \begin{cases} 20 & ; & 8 \le f \le 15\,\text{GHZ} \\ 23 & ; & 15 \le f \le 35\,\text{GHZ} \end{cases}$
- $C_\tau = 20*\log_{10}|\sin(2\tau)|$, τ is the polarisation tilt angle in the case of a rectilinear polarisation,
- $C_\beta = 40*\log_{10}\left[\cos(\beta)\right]$, β is the elevation angle of the wave measured in degrees,
- $C_\sigma = 0.0052*\sigma_p^2$ is a parameter associated with the fluctuations of the canting angles of the raindrops during a given rain event or between different such events.
- L is the length of the path through rain.

Van de Kamp Model

The relation between the polarisation discrimination and the copolar attenuation is given by the following equation (Van de Kamp 1999):

$$XPD_{1,p} = U - V*\log_{10}(A_p) - C_\tau - C_\beta - C_x + S \quad \text{(D.12)}$$

where:

- $U = 20\log(f)$, f being the frequency in GHz,
- $V = 16.3$,
- $C_\tau = 20 * \log_{10} |\sin(2\tau)|$, where τ is the tilt angle between the horizontal plane and the rectilinear polarisation,
- $C_\beta = 41 * \log_{10} [\cos(\beta)]$, where β is the elevation angle of the wave measured in degrees,
- $C_x = 0.075 * \cos^2(\beta) * \cos(2\tau) * A_p$; a difference appears here between the horizontal and the vertical polarisations,
- S is a constant equal to 8.

This model applies if the following conditions are fulfilled:

- $11 \text{ GHz} < f < 30 \text{ GHz}$,
- $3° < \tau < 50°$,
- $4° < \tau \pm 90° < 86°$,
- $1 \text{ dB} < A_p < 25 \text{ dB}$.

D.2.2. Cross-Polarisation caused by Ice Particles

The cross-polarisation discrimination (*XPD*) caused by ice for a given time percentage p can be derived from the discrimination induced by rain by applying a corrective factor, as suggested by the following equation (COST 255 2002):

$$XPD_{2,p} = XPD_{1,p} * 0.5 * [0.3 + 0.1 * \log_{10}(p)] \qquad (D.13)$$

The corrective parameter introduced here was empirically determined on the basis of comparisons drawn with regard to experimental measurements conducted at various frequencies, climatic polarisations and areas.

D.2.3 Cross-Polarisation caused by Dust and Sandstorms

The sand and dust particles transported by equatorial storms are not spherical (Mewan 1983; Ghobrial 1987). In the absence of turbulence or wind shears, the hydrodynamic forces tend to direct these particles in such a direction that their principal axis will be contained within the vertical plane. This anisotropy may induce a degradation of the cross-polarisation discrimination factor both in linear and in circular polarisations (Bashir 1986).

The value of the cross-polarisation discrimination in circular polarisation is a function of the visibility, as described by the following equation (Ghobrial 1987):

$$XPD = 91.6 - 20*\log_{10}(f*d) + 21{,}4* \log_{10} V \qquad (D.13)$$

where f is the frequency expressed in GHz, d is the length of the path in kilometres and V is the visibility in kilometres.

The values of the cross-polarisation discrimination XPD are reduced from 1.7 dB if the dust contains 4 percent of surface moisture (CCIR 1990).

D.2.4 Cross-Polarisation in Clear Atmosphere

The significant deteriorations of the cross-polarisation discrimination XPD which can be observed in clear atmosphere are generally caused by the deep fading of the copolar signal in propagation by multiple paths. However, it has been shown that for a number of different links and under low fade depths a good correlation could be observed between the variations of the cross-polarisation discrimination and the copolar signal. Several different possible mechanisms of cross-polarisation have been advanced in order to explain the reduction of the cross-polarisation discrimination (Olsen 1981). These mechanisms can be classified into two categories, depending on whether the cross-polarisation diagram of the antennas plays any part at all (CCIR 1990)

In the first class of mechanisms, a diffusing or reflective medium produces a cross-polarised diffused wave. The causes for such a phenomenon may be:

- the presence of heterogeneous turbulences along the path, causing a depolarisation of the copolar signal,
- the depolarisation of some indirect component of the copolar signal by reflection off an inclined layer,
- the depolarisation of some indirect component of the copolar signal by diffusion or by reflection off the ground or the water surface along the path.

In the second class of mechanisms, the cross-polarisation is primarily caused by the additional effects induced by the cross-polarisation diagrams of the antennas. Among these effects, the following may be mentioned:

- the coupling of some indirect component of the reflected or refracted signal, either by an atmospheric layer or by the ground,
- the bending of the direct path by refraction of the wave at the receiving antenna and at a certain angle with respect to the principal axis.

Three different methods for the evaluation of the statistical distribution of the cross-polarisation discrimination have been found in the literature:

- the first method is based on a log-normal distribution of the conditional distribution of the field of the cross-polar signal (Martin 1977; Rooryck 1977).

$$\text{Median (XPD)} = - CPA + C_1 \quad \text{for } CPA > 15 \text{ dB} \tag{D.14}$$

- the second method is based on a Nakagami-Rice distribution (Flax 1977; Motti 1977). The root-mean-square value of the cross-polarisation attenuation is given by the equation :

$$\text{RMS (XPD)} = - CPA + C_2 \tag{D.15}$$

The similarities and the differences between these two methods have been examined by Olsen (Olsen 1981).

- the third method requires that the unconditional distribution of the reduction of the cross-polarisation discrimination XPD follow a Rayleigh distribution for values higher than 15 dB (Morita 1979). In this case, the cross-polarisation attenuation is given by the equation:

$$XPD = - CPA + C_3 \tag{D.16}$$

If the C_3 coefficient is expressed by the equation $C_3 = XPD_0 + Q$, where XPD_0 is the static value of XPD in the absence of fading, then the Q coefficient is found to be inversely proportional to the slope of the cross-polarisation diagram of the antenna within the vertical plane.

The C_1, C_2 and C_3 coefficients used in these three methods are essentially empirical parameters.

References

COST Action 255 (2002) Radiowave propagation modelling for SatCom services at Ku-band and above. Cost255 final report, European Space Agency

Bashir SO, McEwan NJ (1986) Microwave Propagation in Dust Storms, a Review. IEE Proc. Pt H. vol 133: 241-247

Ghobrial SI, Sharief SM (1987) Microwave attenuation and cross-polarisation in dust storms. IEEE Trans. Ant. Prop. vol AP-35 4: 418-425

Lin SH (1977) Impact of microwave depolarisation during multipath fading on digital radio performance. BSTJ vol 56: 654-674

Martin L (1977) La réutilisation des fréquences en polarisation croisée avec diversité d'espace sur 50 km à 7 GHz, Ann. des Télécomm. vol 32 11-12: 552-559

McEwan, NJ Bashir SO (1983) Microwave propagation in sand and dust storms; the theoretical basis of particle alignment. IEE Conf. Publ.219 Pt. 2 pp 40-44

Morita K, Sakagami S, Murata S, Mukai T, Ohtani N (1979) A method for estimating cross-polarisation discrimination ratio during multipath fading. Trans Inst. Electron. Comm. Engrs. Japan vol 62-B: 998-1005

Motti TO (1977) Dual-polarized channel outages during multipath fading. BSTJ vol 56: 675-701

Olsen R (1981) Cross-polarization during clair-air conditions on terrestrial links: a review. Radio Sci. vol 16 5: 631-647

Rooryck M, Martin L (1977) Disponibilité des liaisons hertziennes numériques utilisant des polarisations croisées à la même fréquence. Ann. Des Télécomm. vol 32 : 560-564

E Fresnel Equations

E.1 Introduction

Let us consider here, as represented in Fig. E.1, an incident plane wave (E_i, H_i, k_i) striking a plane surface separating two media (1) and (2). The permittivity and the permeability of these two media are referred to as ε_1, μ_1 and ε_2, μ_2 respectively. These media are considered as purely dielectric i.e. waves propagate without loss: the conductivities σ_1 and σ_2 are null. It is further assumed that the separation surface, extending within the x,y plane, is of very large dimensions and that the irregularities are quite small compared to the wavelength. The x,z plane is the incidence plane and contains the vector k_i. The incidence angle, i.e. the angle between vector k_i and the normal to the separation surface is equal to θ_i.

From the resolution of Maxwell's equations and from the continuity of the tangential components of the electric and magnetic fields as well as that of the continuity of the normal components of the electric and magnetic inductions at the separation surface both a reflected wave (E_r, H_r, k_r) and a transmitted wave (E_t, H_t, k_t) can be determined. The continuity conditions imply that:

$$\theta_i = \theta_r \ \textit{(reflection law),} \tag{E.1}$$

and:

$$n_1 \sin(\theta_i) = n_2 \sin(\theta_t) \ \textit{(refraction or Descartes-Snell law)} \tag{E.2}$$

The reflection and transmission coefficients are deduced from the two following equations:

$$R = \frac{E_r}{E_i} \ \text{and} \ T = \frac{E_t}{E_i} \tag{E.3}$$

The relative amplitudes of the different waves (incident, reflected and transmitted) result from the continuity conditions but are dependent upon the polarisation. Two cases are generally distinguished: horizontal polarisation and vertical polarisation, the separation plane being assumed to be horizontal. The general case can be regarded as a linear combination of the two previous cases.

E.2 Horizontal Polarisation

The electric field is horizontal, i.e. parallel to the separation plane. The continuity conditions lead to the following equations:

$$E_i + E_r = E_t \tag{E.4}$$

$$(H_i - H_r)\cos\theta_i = H_t \cos\theta_t \tag{E.5}$$

$$\mu_1(H_i + H_r)\sin\theta_i = \mu_2 H_t \sin\theta_t \tag{E.6}$$

From these equations it turns out that the reflection $(R_{//})$ and transmission $(T_{//})$ coefficients depend on the impedance Z_1 and Z_2 of the media as well as on the incidence and refraction angles:

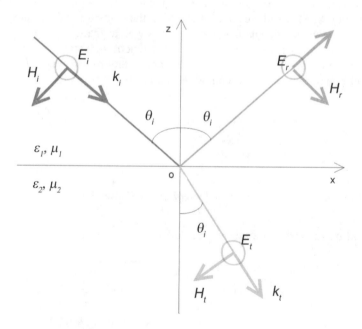

Fig. E.1. Schematic representation of the reflection and transmission of an electromagnetic wave in horizontal polarisation

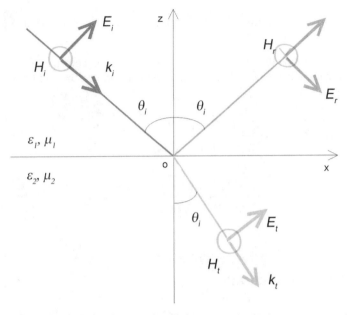

Fig. E.2. Schematic representation of the reflection and transmission of an electromagnetic wave in vertical polarisation

$$R_{//} = \frac{E_{r//}}{E_{i//}} = \frac{Z_2 \cos\theta_t - Z_1 \cos\theta_i}{Z_2 \cos\theta_t + Z_1 \cos\theta_i} \tag{E.7}$$

$$T_{//} = \frac{E_{t//}}{E_{i//}} = \frac{2Z_2 \cos\theta_t}{Z_2 \cos\theta_t + Z_1 \cos\theta_i} \tag{E.8}$$

E.3 Vertical Polarisation

In vertical polarisation, the magnetic field is horizontal, i.e. parallel to the separation plane. The electric field itself is not vertical as the term would suggest. Here again, the continuity conditions lead to the following system of equations:

$$(E_i - R_r)\cos\theta_i = E_t \cos\theta_t, \; E_i + E_r = E_t \tag{E.9}$$

$$H_i + H_r = H_t \tag{E.10}$$

$$\varepsilon_1 \left(E_i + E_r \right) \sin \theta_i = \varepsilon_2 E_t \sin \theta_t \tag{E.11}$$

It results from these equations that the reflection and transmission coefficients $R_\perp$ and $T_\perp$ depend on the impedances Z_1 and Z_2 of the two media and on the incidence and refraction angles:

$$R_\perp = \frac{E_{r\perp}}{E_{i\perp}} = \frac{Z_2 \cos\theta_i - Z_1 \cos\theta_t}{Z_2 \cos\theta_i + Z_1 \cos\theta_t} \tag{E.12}$$

$$T_\perp = \frac{E_{t\perp}}{E_{i\perp}} = \frac{2 Z_2 \cos\theta_i}{Z_2 \cos\theta_i + Z_1 \cos\theta_t} \tag{E.13}$$

E.4 Unspecified Polarisation

In the general case where polarisation is not specified, the different fields (incident, reflected and transmitted fields) are given by the following system of equations:

$$E_i = E_{i//} u_{//} + E_{i\perp} u_\perp \tag{E.14}$$

$$E_r = R E_i \tag{E.15}$$

$$E_t = T E_i \tag{E.16}$$

where $u_{//}$ and $u_\perp$ are the unit vectors parallel and perpendicular to the separation plane between the two media respectively.

$$E_i = \begin{bmatrix} E_{i//} \\ E_{i\perp} \end{bmatrix}, \quad E_r = \begin{bmatrix} E_{r//} \\ E_{r\perp} \end{bmatrix} \text{ and } E_t = \begin{bmatrix} E_{t//} \\ E_{t\perp} \end{bmatrix} \tag{E.17}$$

$$R = \begin{bmatrix} R_{//} & O \\ O & R_\perp \end{bmatrix} \text{ and } T = \begin{bmatrix} T_{//} & O \\ O & T_\perp \end{bmatrix} \tag{E.18}$$

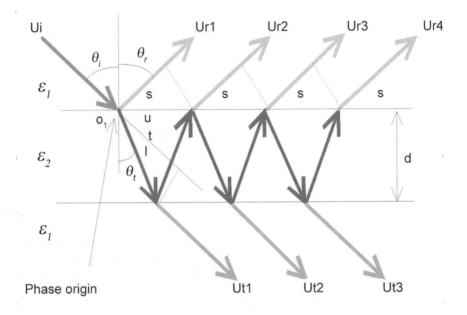

Fig. E.3. Schematic representation of multiple reflections and transmissions

E.5 Multiple Reflections and Transmissions

The case where multiple reflections and transmissions arise is schematically represented in Fig. E.3.

In the case of multiple reflections, the paths to be considered are parallel to one another. For a dielectric material, the multiple reflection coefficient tends towards the value determined by the following equation:

$$R = \left(\frac{R_1 \left(1 - P_a P_d^{\,2}\right)}{1 - R_1 P_a P_d^{\,2}} \right) \tag{E.19}$$

where R_1 is the reflection coefficient of the wave at the first interface, $P_a = e^{jk_1 s}$ and $P_d = e^{-jk_2 l}$.

In the case of multiple transmissions, the rays transmitted through the plate with parallel faces are themselves parallel to one another. For a dielectric material, the multiple transmission coefficient tends towards the value determined by the following equation:

$$T = \left(\frac{P_d\left(1-R_1^2\right)}{1-R_1^2 P_a P_d^2} P_t \right)$$

(E.20)

where:

- R_1 is the reflection coefficient of the wave at the first interface,
- $P_a = e^{jk_0 s}$,
- $P_d = e^{-jk_r l}$,
- $P_t = e^{jk_1 t}$.

T is calculated from the point Q_1, i.e. from the origin of the phases of the reflected field. The term with phase P_t is added to the value thus determined, in order to account for the delay corresponding to the time of propagation in the first medium.

The origin of the phases of the reflected and transmitted fields is therefore located at the same point Q_1.

F Ionospheric and Geomagnetic Disturbances associated with Solar Events

F.1 Introduction

Solar flares appear in the form of sudden, rapid and intense variations of brightness of certain active regions in the chromosphere, more specifically in the X-ray and ultraviolet ray ranges. This phenomenon is accompanied by emissions of particles over a broad energy level range:

- cosmic rays of solar origin with an energy close to 1 Gev,
- sub-relativist protons with an energy ranging from 1 to 1000 Mev, which spiral along the field lines of the terrestrial magnetic field in polar regions 80 minutes to 4 hours after a solar eruption,
- clouds of ions and electrons with lower energy, which envelop the Earth between 20 and 40 hours after a solar flare.

All these events interact with the Earth's atmosphere and result in several different terrestrial phenomena:

- the simultaneous increase of X radiation and ultraviolet radiation causes a modification of the electronic density of the ionosphere, which is referred to as a sudden ionospheric disturbance (SID).
- cosmic rays are recorded on Earth using neutron monitors. Any modification in their intensity is referred to as a ground level event (GLE).
- sub-relativist particles, for the most part protons, penetrate in the polar region along the terrestrial magnetic field lines and increase the absorption of the waves propagating in this region. Such phenomena are referred to under the term polar cap absorption (PCA).
- the slower electronic and ionic particles generate at the same time magnetic storms as well as complex modifications of the ionisation in the ionosphere, more particularly in the F layer, which are referred to as ionospheric storms.
- at high altitudes, electronic and ionic particles also generate luminous phenomena called polar auroras.

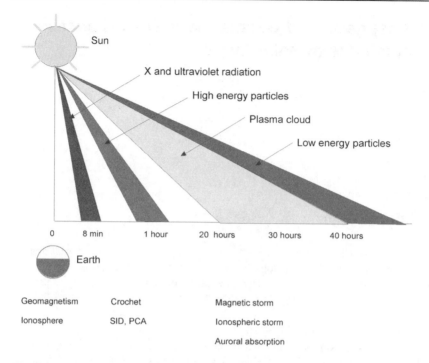

Fig. F.1. Schematic representation of the sequence of events occurring after a solar flare

The sequence of events occurring after a solar flare is schematically represented in Fig. F.1. All these different phenomena may disturb radio communications over a broad frequency band (ELF, VLF, LF, MF, HF and VHF frequencies). This appendix will therefore be devoted to a survey of these phenomena.

F.2 Sudden Ionospheric Disturbances

In the last recent years, the variations of X and ultraviolet solar radiation have been intensively recorded using satellites. Frequencies in the 1-10 Å range are both the most sensitive to solar flares and the most penetrating in the ionosphere, until altitudes lower than 100 kilometres.

For this very reason, these altitudes are also the most affected by solar flares. Although the E and F layers are also submitted to the influence of solar flares, the effects induced in these layers are less pronounced and their observation is therefore more difficult. These variations of the ionisation inevitably results in significant modifications affecting signals propagated in the ionosphere illuminated by the Sun over a broad frequency range. The term sudden ionospheric disturbance has been given to this sudden, total and more or less prolonged fading of the field of short and medium wavelength waves propagated by ionospheric reflection. This phenomenon is so sudden and unforeseen that radio operators immediately tend to question the systems themselves. This phenomenon however is general and affects all ionospheric reflections occurring in the illuminated hemisphere. In certain rare cases, disturbances have been reported in the dark hemisphere. In the VLF and LF frequency range, this phenomenon causes an increase of the electromagnetic field.

The maximum fading of the signal is reached two or three minutes after the beginning of the solar flare, and lasts for a more or less long period. The electromagnetic field then increases, before gradually returning to its normal diurnal level after a period ranging from a few tens of minutes to one hour and a half. Fig. F.2 is an example of a sudden ionospheric disturbance. This phenomenon may last for a shorter duration, and the weakest such phenomena are difficult to distinguish. The sensitivity to disturbance of a given link depends on the frequency (the lowest frequencies being always the most affected), on the local noise (of industrial, cosmic or atmospheric origin) and on the normal fading more specifically caused by the pulsations of the elementary fields propagated along several simultaneous paths.

The techniques used for the observation of solar flares require the reception of natural signals emitted by atmospheric storms at the ELF, VLF and LF frequencies, or by the galaxy at the HF and VHF frequencies, as well as signals, either impulse or continuous, emitted by terrestrial transmitters at the VLF, LF, MF, HF and VHF frequencies. Any unusual effect observed with respect to these signals is referred to as a sudden ionospheric disturbance (SID). These phenomena are classified into different categories:

– abrupt increases or decreases of atmospherics the VLF and LF frequencies, referred to as sudden enhancements or decreases of atmospherics (SEA or SDA) respectively,
– abrupt decreases of the intensity of the received signals at the MF and HF frequencies, known as sudden-short wave fade-outs (S - SWF),
– abrupt decreases of the intensity of the cosmic noise, generally observed at frequencies close to 20 MHz, referred to as sudden cosmic noise absorptions (SCNA),
– abrupt variations in the intensity of the signals received in the LF and VLF frequencies, referred to as sudden enhancements of signal (SES).
– abrupt variations of the phase or frequency of signals, known as sudden frequency deviations (SFD). The effects of solar flares can also be studied by considering the variations of the observable minimal frequency f_{min} and the variations of the critical frequencies in the E and F layers, more particularly f_oF2.

F.2.1 Sudden Enhancement of Atmospherics

Radio atmospherics are defined as the impulsive electromagnetic signals launched by individual lightning discharges. Although the emitted spectrum in relatively large, ranging from 10 to 50 kHz, the 27.5 kHz frequency (11 km wavelength) has been found to be the most sensitive and presents very interesting characteristics in terms of amplitude and propagation.

The observed number of atmospherics depends essentially on the frequency of the wave, on the hour, on the season, and in the first place on the localisation of the storm centres and on the conditions of propagation in the D layer of the ionosphere. As might be recalled here, this region of the ionosphere, extending between 80 and 85 kilometres of altitude, is both reflective and absorbing.

After a bright and strong solar chromospheric flare, very short wavelength X-rays are emitted, causing an increase of the ionisation of the D layer. In the case of signals at the 11 kHz frequency, reflection becomes metallic, thereby enabling the propagation of atmospherics. An increase in the number of atmospherics can therefore be observed at the time of a solar flare: this phenomenon is referred to as a sudden enhancement of atmospherics (SEA). It is thus apparent that sudden enhancements of atmospherics do not result from an increase in the number of emitted atmospherics, but rather from an improvement of their conditions of propagation.

At the same time, as the D layer becomes much more absorbing than it generally is, the high frequency waves used in radio communications are no longer capable of reaching the F layer, leading thereby to the abrupt interruption of communications.

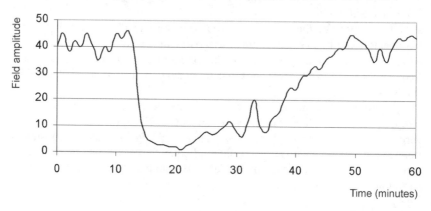

Fig. F.2. Temporal variation of a radio HF field during a typical sudden ionospheric disturbance

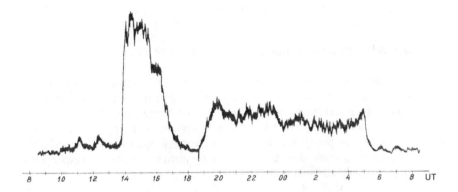

Fig. F.3. Temporal variation of the number of atmospherics observed in Poitiers after a solar flare at the 27.7 kHz frequency

The sources of atmospherics extend over a large area and are not well localised. As a consequence, the atmospheric noise is but an integrated effect which does not allow studying the physic of the medium. Recordings of the atmospheric noise nonetheless constitute an excellent and not very expensive means of detecting solar flares and sudden ionospheric disturbances.

Fig. F.3 presents an example of a recording showing the variation of the number of atmospherics observed after a solar flare at the 27.5 kHz frequency.

At lower frequencies, ranging from 0.8 to 10 kHz, the number of atmospherics decreases: this phenomenon is referred to as a sudden decrease of atmospherics (SDA). The maximum effect can be observed at frequencies close to 5 kHz.

F.2.2 Sudden Enhancements of Signals

The occurrence of sudden enhancements of signals can be demonstrated from the recordings of the field of VLF and LF radio waves propagating between two known stations. At VLF frequencies as well as for long distances at LF frequencies, a reinforcement of the intensity of the signal can be observed. In general however, the effect turns out to be more complex at LF frequencies.

At the 164 kHz frequency, solar flares may induce different types of effects:

- positive effect: a sudden increase in the intensity of the signal, followed by a gradual return to normal intensity,
- negative effect: a sudden decrease in the intensity of the signal, followed by a gradual return to normal intensity,
- complex effect: a decrease in the intensity of the signal during a few minutes, followed by a sudden signal reinforcement.

These three phenomena are successively represented in Fig. F.4. These recordings are an excellent means of detecting solar flares.

F.2.3 Sudden Phase Anomalies at Very Low Frequencies

In order to measure the phase variations of a stable VLF signal propagated in the ionosphere, a reference phase has to be considered at the reception station. For short distances, i.e. distances shorter than 1000 kilometres, the reference phase is the phase of a wave transmitted by way of telephone. For larger distances, it is necessary to have at the reception an extremely stable local oscillator, preferably with the same frequency than the emission frequency or with a constant phase shift compared to this frequency.

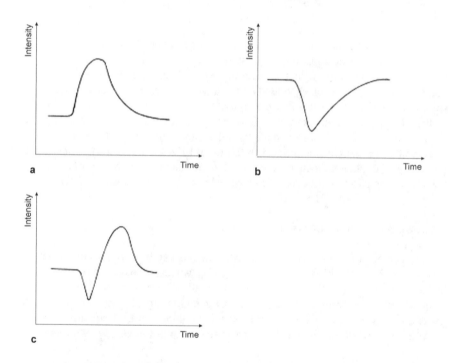

Fig. F.4. Temporal variation of the amplitude of the field at the 164 kHz frequency after a solar flare. *a.* positive effect, *b.* negative effect, *c.* complex effect

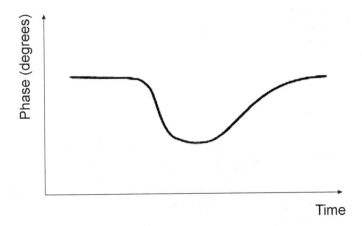

Fig. F.5. Variations of the phase of a VLF signal during a solar flare

The X-rays which are being emitted after a solar flare induce an ionisation of the D layer and cause the reflection height to decrease from 2 to 10 kilometres. This leads to a sudden phase advance which can be measured in degrees. Fig. 5 represents an example of such a phenomena.

This method, which applies to VLF waves propagating between two stations, allows investigating the ionosphere during a solar flare at the reflection point of the waves rather than along the entire path.

F.2.4 Short-Wave Fadeouts

At short frequencies, solar flares may cause the sudden interruption or the sudden fading of the field. This phenomenon, referred to as short-wave fadeout (SWF), is also known as the Möller-Dellinger effect.

Short-wave fadeouts can be easily observed on recordings. It should be noted however that these fadeouts are not to be confused with different other phenomena, among which may be mentioned phenomena of abnormal fading due to modifications of the propagation modes, interferences between several different signals, disturbances affecting ionospheric conditions and caused by dynamic or geomagnetic phenomena.

As illustrated in Fig. F.6, three different types of short-wave fadeouts can be distinguished:

- S.SWF: depending on the recording time-constant, the sudden fadeouts last between 1 and 5 minutes. The return to normal is then gradual.
- slow S-SWF: the fadeouts last between 5 and 15 minutes. The decrease is slower than in the case of a S.SWF and the return is gradual.
- G-SWF: gradual short-wave fadeouts, which are irregular both in their initial phase and in their recovery phase.

F.2.5 Sudden Cosmic Noise Absorptions

Sudden cosmic noise absorptions (SNCA) were discovered in 1954 by Shain and Mitra. These phenomena are observed at frequencies close to 20 MHz, i.e. at frequencies higher than the ordinary critical frequency of the F_2 layer.

These decreases in the intensity of the cosmic noise are due to the absorption of the signal in the D layer which after a solar flare has become strongly ionised. The return to the normal level is gradual.

Four different types of sudden cosmic noise absorptions are commonly distinguished:

- typical SCNA: sudden decreases in the intensity of the cosmic noise, followed by a gradual return to normal intensity,
- SCNA of U type: slow decreases in the intensity of the cosmic noise,
- SCNA accompanied at the onset by a solar flare
- SCNA accompanied by more than one solar flare.

These different types of sudden cosmic noise absorptions are represented in Fig. F.7. All these events have a duration of approximately 30 minutes.

F.2.6 Sudden Increases of the Minimal Observable Frequency

The effect of absorption can also be demonstrated through the variations of the minimal frequency f_{min} observed during a solar flare with a traditional ionosonde. The minimal frequency f_{min} is defined as the lowest frequency observable by an ionosonde. The different factors which may influence the value of fmin are:

- the amplification of the signal by the recording system and the variation of its sensitivity with frequency,
- ionospheric absorption and its variations with frequency,
- the level of local noise.

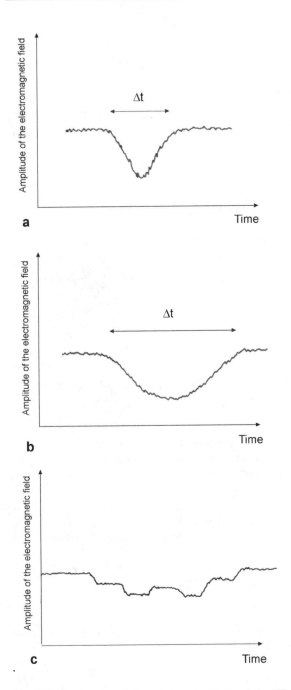

Fig. F.6. Temporal variation of the amplitude of the HF field during a solar flare: *a*. S-SWF, b. slow S-SWF, c. G-SWF

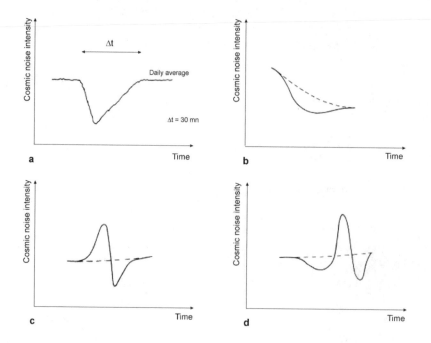

Fig. F.7. Temporal variations of the intensity of the cosmic noise during a solar flare. *a.* typical SCNA, *b.* SCNA of U type, *c.* complex SCNA, *d.* complex SCNA

Provided that certain precautions are taken, this parameter can be used as an indicator for the evaluation of the absorption in the ionosphere and may even provides a reasonable measurement in certain conditions. The study of this parameter has led for instance to:

– the distinction between strong and weak absorption days,
– the interpolation between absorption values determined using other measurement methods,
– the study of variations from day to day and the identification of abnormal conditions such as the winter abnormal conditions,
– the study of the occurrence of polar blackouts.

The interest of this parameter for the study of these phenomena remains however limited due to the radio noise and its variations.

F.2.7 Sudden Frequency Deviations

At high frequencies, the frequency of a stable wave reflected by the ionosphere is submitted to variations caused by the modifications affecting the phase path followed by the wave:

$$\Delta f = -\frac{c}{f} \int_s \frac{\partial \mu}{\partial t} \cos(\alpha)\partial s \qquad (F.1)$$

where:

- Δf represents the variation of frequency,
- f is the frequency of emission,
- c is the speed of the light in vacuum,
- μ is the refractive index,
- α is the angle formed by the normal to the wave plane and the direction of propagation,
- s is the path followed by the wave.

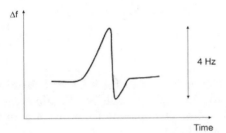

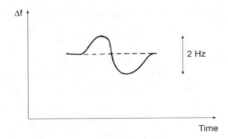

Fig. F.8. Different types of sudden frequency deviations observed at the 10 MHz frequency during a solar flare

The variations affecting the refractive index are essentially due to variations in the electronic density along the path followed by the wave. After a solar flare, due to the emission of X-ray and ultraviolet rays, the E and F_1 layers suddenly become ionised: this results in a modification of the frequency of the waves from a few Hertz (between 0 and 7 Hz). This phenomenon appears through a sudden positive effect lasting a few minutes and followed by a sudden negative effect, as represented in Fig. F.8.

Unlike other phenomena, like for instance sudden enhancements of atmospherics, sudden frequency deviations are relatively brief phenomena, which last only a few minutes and are characterised by fast oscillations between a positive effect and a negative effect. These small frequency variations are recorded using a device called a dopplometer which compares the instantaneous received frequency to the frequency of a stable local oscillator set at the emission frequency of the signal.

Measurements of the sudden frequency deviations at two or more different frequencies can be used for distinguishing between reductions of the phase path due to a decrease of the refractive index in the E layer and reductions of the phase path due to a modification of the reflection height caused by an excess of the ionisation in the F1 layer.

It might be further noted that, due to the high degree of absorption in the D layer, this layer is involved only to a very limited extent in this type of disturbances.

F.3 Polar Cap Absorption

The term polar cap absorption (PCA) was first introduced for describing a particular class of events observed at magnetically high latitudes using riometers, but is employed at present in a broader sense and apply to the different events generated by the penetration in the ionosphere of high energy solar particles. These particles spiral back and forth along the magnetic field lines in polar regions.

The particles observed during such a phenomenon consist for the most part of protons with energy ranging from 1 to 100 Mev, α particles and high energy electrons. They generally appear between 20 minutes and 20 hours after a solar flare, depending on the configuration of the interplanetary magnetic field.

The arrival of these particles appears through an increase of the ionisation in the lowest layers of the ionosphere, causing an absorption of high frequency and very high frequency radio waves. The penetration of solar particles in auroral regions is made easier by the configuration of the terrestrial magnetic field lines. For more detail on this subject, the reader is referred to Appendix A.

Polar cap absorption events may be classified in two categories: simple polar cap absorption events and complex polar cap absorption events.

- a simple event is an isolated event generated by a solar flare and an associated geomagnetic storm. Simple polar cap absorption events are distinguished according to their time variation: events of type F are rapid flux increases, while events of type S have time constants greater than 10 hours.
- a complex event is an event where at least a second solar flare or a second geomagnetic storm occurs. In this case the temporal variations of the signal generally turn to be extremely complex.

F.4 Geomagnetic Disturbances

The terrestrial magnetic field is generated by the metal mass contained inside the Earth and by the system of currents flowing within the ionospheric area.

The interaction of the solar wind (primarily composed of protons, of He^{2+} ions and of electrons emitted by the Sun) with the terrestrial magnetic field confines the latter to the interior of a cavity called the magnetosphere. This region of the interplanetary space deviates the flow of particles and is modified by the collisions with these particles.

The continuous recordings of magnetic parameters at permanent magnetic stations show that different types of geomagnetic disturbances exist: while some of these disturbances are regular and have a daily occurrence, others are irregular and may be on certain days superimposed on regular variations.

Regular geomagnetic disturbances repeat from day to day and can be more particularly observed during quiet days. Solar and lunar diurnal variations should be distinguished here. Whereas the amplitude of the solar diurnal variation S reaches its highest value during the daytime and increases with the solar activity, the diurnal lunar variation L is associated with gravitation forces and has an amplitude which on average is no higher than a tenth of the solar diurnal variation.

Irregular geomagnetic disturbances are due to modifications of the ionospheric current system under the influence of some solar event. Accordingly, these variations may be of different types:

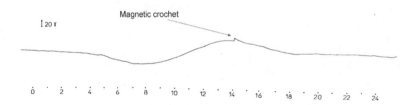

Fig. F.9. Variations of the horizontal component of the terrestrial magnetic field in the presence of a crochet

- magnetic crochets, associated during daytime to solar chromospheric flares. Their amplitude and duration are variable. An example of the variations of the horizontal component of the terrestrial magnetic field during such a phenomenon is presented in Fig. F.9.
- magnetic bays, associated with the ionospheric currents flowing in the auroral region, or auroral electrojets. Their amplitude and duration are higher than in the case of a crochet. An example of such a phenomenon can be seen in Fig. F.10.
- sudden impulses (SI), which are essentially abrupt intensifications of the intensity of the magnetic field not followed by a magnetic storm.
- magnetic pulsations, also known as ultra low frequency waves: these variations are very low oscillations ranging from a few fractions of γ to a few tens of γ at high latitudes. They appear in the form of quasi-sinusoidal oscillations with a period T which may range from a few fractions of second to 500 seconds. Two different types of magnetic pulsations are distinguished:

1. irregular pulsations, or Pi, occurring primarily during the night and accompanying magnetic bays,
2. regular or continuous pulsations, or Pc, subdivided in different classes Pc1 to Pc5 depending on their period T, which may range from 0.2 to 500 seconds. They appear primarily during daytime and may last several hours. Their study is of primary importance for the understanding of the possible interactions that may occur between charged particles and hydromagnetic waves.

- magnetic storms: these storms are of two kinds: gradual and sudden magnetic storms. Gradual magnetic storms recur every 27 days and appear in periods of low solar activity. They are believed to originate inside Bartels M solar areas. Sudden magnetic storms are preceded by chromospheric eruptions, and are caused by the abrupt contraction of the magnetosphere under the influence of the solar wind reinforced by a solar flare. A standard magnetic storm has different phases:

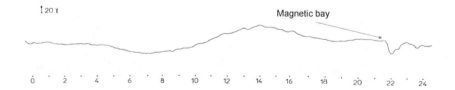

Fig. F.10. Variations of the horizontal component of the terrestrial magnetic field in the presence of a magnetic bay

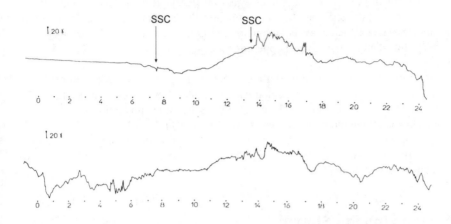

Fig. F.11. Variations of the horizontal component of the magnetic field during a sudden magnetic storm

1. an initial phase, lasting from a few minutes to a few hours, during which the horizontal component H of the field decreases very rapidly,
2. a main phase, lasting from one to three days, during which H is below its initial value,
3. a post-disturbance phase during which H increases,
4. an increase in the diurnal variation.

Fig. F.11 represents the variations of the horizontal component of the magnetic field that were observed during a sudden magnetic storm.

The forecast of the sudden variations of the magnetic field leads to a number of different practical applications:

- at high latitudes, the currents generated during an ionospheric storm may destroy certain protection systems and disturb the electrical current distribution along certain lines,
- as an example, the technique of airborne magnetic prospecting requires magnetically quiet conditions,
- magnetic disturbances may affect the orientation controls of certain communications satellites oriented in space with respect to the magnetic field. Since the orientation of the magnetic field is known for quiet periods, some errors may be introduced if the maneuvering operations of the satellite are performed during a period known to be agitated.
- during a magnetic storm, the magnetopause moves closer to the Earth: in the case of a strong storm, a satellite may thus be directly in the solar wind where the magnetic field is completely different from the Earth's magnetic field, and

where the density of the electric charges is different as well. Electric discharges may occur between the different parts of the satellite, causing local radio interferences which may be wrongly interpreted as command messages and may thus disturb maneuvering operations.

For more detail on the subject of the effects of magnetic storms on systems, the reader is referred to results presented by Lanzerotti or by Allen (Lanzerotti 1999; Allen 1989) or to the AGU website where a schematic illustration of effects of space on spatial technologies is presented (AGU 2001).

Magnetic storms are usually accompanied by ionospheric storms which among other effects cause disturbances of radio communications. We shall hereafter examine these phenomena.

F.5 Ionospheric Storms

Ionospheric storms are disturbances affecting the characteristics of the ionosphere, such as the electronic density or the height of the F2 layer, over a few day period. These storms are generally preceded and accompanied by geomagnetic storms, as illustrated in Fig. F.12.

Ionospheric storms are generally characterised by an increase in the virtual height of the F_2 layer and by a decrease of its maximum electronic density, which may result in the interruption of radio communications along ionospheric paths.

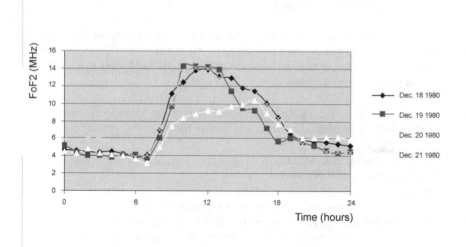

Fig. F.12. Temporal evolution of the critical frequency F_oF_2 observed during an ionospheric storm

Two different types of ionospheric storms can be distinguished: recurring ionospheric storms and sudden ionospheric storms.   Recurring storms are prevalent in periods of minimum solar activity. These storms have a 27 day period recurrence, corresponding to the apparent period of rotation of the Sun, and they are preceded and accompanied by gradual magnetic storms.

Sudden ionospheric storms are associated to solar chromospheric flares, and are preceded and accompanied by sudden magnetic storms. In order to study recurring as well as sudden ionospheric storms, one generally considers the moment of commencement of the gradual or sudden magnetic storms accompanying these two types of storms respectively. This moment is referred to under the name of sudden storm commencement (SSC).

If a magnetic storm starts before 15:00 local time, a very rapid increase of both the maximum density of the F_2 layer and the total electronic content occur around 18:00 (positive phase), followed of an abrupt decrease of the values of these two parameters. During the two following days these parameters have values significantly lower than the values observed before the storm.

If the storm starts after 15 H standard time, a simultaneous decrease of these two parameters can be observed (negative phase). This decrease continues the following day and the return to normal conditions occurs on the third day.

These variations result from simultaneous effects associating the presence of electric fields, of neutral winds and of an energy deposit in auroral regions. In the first phase of the storm, the electric fields and the neutral winds tend to raise the layer, thereby causing a reduction in the recombination rate and consequently a decrease in the electronic density (positive phase). The effects of the energy deposit in auroral regions resulting from the precipitation of solar particles appear through an increase of the scale height of the neutral atmosphere and through a modification of the wind regime. This causes a redistribution of the lightest components (H_2, O, etc) and accordingly a modification in the composition of the atmosphere. Other effects, for instance the heating of the ionospheric medium either by thermal conduction from the magnetosphere or caused by the precipitation of low energy electrons, occurring only during the night, have also been reported.

The variations in the electronic density observed during ionospheric storms increase the absorption of radio waves and the fading effects, leading to the interruption of certain communications, as the operational frequency becomes higher than the maximum useful frequency.

F.6 Auroras

Auroras are the illumination of the sky occurring at night in high magnetic latitudes. This phenomenon occurs through the appearance in the sky of a coloured blue-green arc with pink or red edges and with a height of several hundreds of kilometres. Auroras may be either short, with a duration of only a few minutes, or last all night.

The observation of auroras reveals the existence at all local times of an auroral oval, statistically located between 60 and 75° of latitude at the level of the ionosphere. The northern edge of the oval is often marked by a series of narrow arcs of extremely variable luminous intensity, referred to as discrete auroras. A uniform broad band of light, characteristic of a diffuse aurora, extend over most of the auroral oval and determines its equatorial edge.

The maximum luminosity of most visible auroras is at altitudes extending between 100 and 130 kilometres. These phenomena are generated by the precipitation through auroral horns of particles, for the most part electrons, into the Earth's atmosphere. This precipitation of particles is caused by the increase of magnetospheric hot plasma which appears after a solar flare and by the resulting flow of plasma which occurs, due to influence the convection electric field, from the geomagnetic tail towards the surface of the Earth. The convection electric field is itself directed overall from dawn to dusk and is generated by the interactions occurring between the solar wind and the terrestrial magnetic field. This field acts as a generator in the electric circuit formed by the currents which flow in the magnetosphere, in the ionosphere and along the terrestrial magnetic field lines.

Auroras can be observed visually and cinematographically but also using spectroscopes and even radio receivers in certain cases. Images of auroras combined with measurements taken by a numerical ionospheric sounder carried on airplanes can be used for determining the orientation, the structure and the drift of auroras inside the F layer. Simultaneous measurements of the precipitations of particles realised by DMSP satellites (*Satellite Defense Meteorological Program*) have demonstrated that these arcs are generated by flows of low energy electrons with energy lower than 500 ev. Simultaneous measurements of the amplitude of transionospheric signals show that these auroras are accompanied by irregularities of the order of one kilometre.

F.7 Conclusion

The recording of sudden ionospheric disturbances meets essentially two different objectives:

- the detection of solar flares using simple radio means,
- the study of the variations in the ionospheric density and the determination of the physical mechanisms involved in these variations.

However, these two objectives are not always mutually compatible: for instance, techniques allowing the detection of solar flares may not lend themselves to the study of the physics of the ionosphere. As an example, the monitoring of sudden decreases in the intensity of some received signal (SWF) is an excellent means of detecting of solar flares, but does not allow a quantitative determination of ionospheric variations. On the other hand, although the variations of f_oE and f_oF_2 can be immediately interpreted in terms of variations of the ionisation, these variations are difficult to observe and are therefore not well suited to the detection of solar flares.

If the objective is primarily the study of the physics of the ionosphere, then the region of the ionosphere to be investigated determines the choice of the method. While most methods can be employed for the study of the ionisation in the D layer, only the sudden frequency deviations, the variations of the critical frequency f_oF2 and to a more limited extent the sudden cosmic noise absorptions allow to study the F layer. As a consequence, several different detection methods of solar flares have to be employed if these two objectives are to be address. The routine methods meeting simultaneously these two objectives rely on the consideration of sudden cosmic noise absorptions, sudden enhancements of atmospheric and sudden frequency deviations.

The study of polar cap absorption events is more particularly interesting for the investigation of the phenomena occurring at high magnetic latitudes and generated by the penetration in the atmosphere of high energy solar particles.

The study of magnetic storms allows investigating the ionospheric currents modified by the contraction of the magnetosphere under the influence of the solar wind. The study of these phenomena should also contribute to the development of methods for the forecast of ionospheric storms.

The study of ionospheric storms has an important place in the field of the short-term forecasts of the conditions of propagation of radio waves at high frequencies.

At last, the occurrence of auroras can be used as a means for detecting precipitations of solar particles through auroral horns. The study of auroras further allows determining the energy of these particles.

All these different phenomena, sudden ionospheric disturbances, polar cap absorption events, magnetic and ionospheric storms, auroras, generate significant disturbances affecting radio communications over a broad frequency band (ELF, VLF, LF, MF, HF and VHF). These disturbances are monitored and studied by different organisations in order to improve forecasts and minimise the effects induced by these disturbances on radio systems, for example the Ecole Nationale des Télécommunications de Bretagne, the European Space Research and Technology Centre, the NASA, Space Weather Canada or the Space Environment Center.

References

Lanzerotti LJ, Thompson DJ, MacLennan CG (1999) Engineering issues in space weather. Modern Radio Science, Oxford Science Publications

Allen J, Frank L, Sauer H, Reiff P (1989) Effects of the March 1989 Solar Activity. EOS 70 46

American Geophysical Union (AGU) A flowchart by G.R. Davenport, illustrating the various impacts of space weather on space technologies. www.agu.org/eos_elect/95183e.html

Ecole Nationale des Télécommunications de Bretagne www-iono.enst-bretagne.fr

European Space Research and Technology Centre. Space weather page of the European Space Agency (ESA) www.estec.esa.nl/wmwww/spweather/spweathstudies.htm

National Space Science Data Center. Space weather and related pages of the National Space Science Data Center. www.nssdc.gsfc.nasa.gov/space

Space Weather Canada www.spaceweather.com

Space Environment Center. Space weather forecast, nowcast, alerts, intuitive diagrams. www.sel.noaa.gov

G Investigation Methods of the Ionosphere

The study of the ionosphere proceeds primarily through the use of vertical radio sounding, either bottomside or topside, of backscatter or incoherent scatter soundings, of riometers, and of low frequency and very low frequency receivers. These different methods will be successively described in this appendix.

G.1 Vertical or Oblique Bottomside Sounding

This is the most classical method of study of the ionosphere using radio systems located at the surface of the Earth. This method is based on the emission along the zenith direction of radio impulses with variable frequency and on the reception of the echoes reflected by the ionosphere after a more or less important delay: this delay corresponds to the propagation time of the wave and determines the virtual reflection height. The variation with frequency of the virtual reflection height is known as an ionogram.

In order to minimise the amount of emitted power, present day systems employ frequency modulation techniques and a digital coding of the signal. They may function either at a fixed or at a variable frequency, in frequency scanning and in vertical or oblique incidence. These systems generally possess the following features: presentation of ionograms, of the received levels, of the Doppler frequencies, of the diffusion function (virtual heights/Doppler shifts), of the noise spectrum, of the angle of arrival and of the correlation function of signals received at different antennas (Le Roux 2002).

Fig. G.1 represents a vertical ionogram obtained in Dakar, Senegal using the SCIPION sounder developed at the ENST-Bretagne. The colour code used in this figure indicates the relative amplitudes of the echoes. The diffusion function at the 3.63 MHz frequency exhibits a curved shape which is characteristic of a delay coupled with a Doppler effect (Le Roux 2002).

The average power supplied to the antenna is generally no higher than a few watts. The average noise level measured at each sounded frequency is presented under the ionogram, while under the diffusion function can be seen the noise spectrum in a + or - 4.5 kHz frequency band (600 Hz resolution) around the frequency at which the diffusion function is presented.

The frequency bands commonly used range approximately from 1 to 20 MHz. The consideration of such ionograms allows determining the virtual reflection heights as well as the ordinary and extraordinary critical frequencies of different ionospheric layers. These frequencies are the highest frequencies at which waves can reach the maximum electronic density of the different layers of the ionosphere and be reflected by these layers. In this context, the most important parameter is the critical frequency f_oF2 of the F2 layer. Although this technique is limited to the study of the region of the ionosphere extending beneath the maximum ionisation, it is perfectly suited for the study of ground-to-ground links.

G.2 Vertical Topside Sounding

In this method, the sounding systems are carried on satellites or on sounding rockets. The topside soundings of the ionosphere performed with this method allow the determination of the virtual depth from the ionosonde, as well as certain density levels or particular phenomena occurring between the ionosonde and the electron density maximum of the F layer.

Topside sounders can be used for realising ionograms at points located at a short distance along their trajectory to one another. The geographical area explored by a satellite is considerable. Ionisation contours have for instance be drawn from 50 degrees north to 45 degrees south, leading to the demonstration of their strong dependence on the configuration of the geomagnetic field lines. The observation of topside ionograms also allowed the discovery of phenomena of the resonance of electrons at the gyrofrequency and at its harmonics, as well as at hybrid frequencies of interference between the gyrofrequency and the plasma frequency. These phenomena were previously unseen

The combination of these two types of measurements, i.e. of the bottomside and topside soundings of the ionosphere, allows the determination of an ionisation profile between the ground and the satellite.

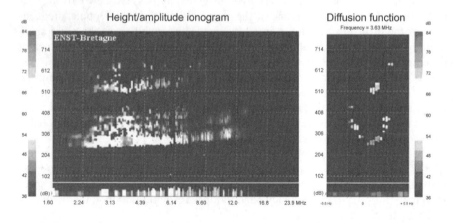

Fig. G.1. Example of an ionogram obtained in Dakar on February 11 2001 at 01:00 UT (the diffusion function is realised at the 3.63 MHz frequency)

G.3 Backscatter Sounding

A backscatter sounder is a device consisting of a transmitter and a receiver located at the same point. Its principle can be summarised as follows: after having been refracted in the ionosphere, an emitted wave strikes the surface of the ground at a different point than its emission point. The energy is then scattered around this point in all directions, including the direction of arrival of the wave. The fraction of energy propagated along this direction returns along the reverse path to the emission point of the wave. By considering only the group time and the amplitude of the signal, different types of diagrams can be obtained with respect to the different variable parameters, like the frequency, the azimuth, the elevation or the time. The temporal soundings obtained at a constant frequency and with the antenna at a fixed orientation provide the evolution of the group path expressed as a function of time. These soundings are more specifically suited for the study of the temporal variations of electronic profiles.

A particular application of the method of backscatter sounding is the forward scatter radars used for the detection of planes or boats beyond the horizon or for the remote determination of the height above the ocean of waves.

G.4 Incoherent Scatter Sounding

The basic principles of incoherent scattering can be summarised as follows: under the action of an electromagnetic field, a charged particle performs an oscillatory movement and scatters energy. Within a gas, each such scatterer induces a phase difference due to its relative location. If the medium is homogeneous, the phase distribution is also homogeneous and the total diffused energy is null. Conversely, if the medium is non-homogeneous, the fluctuations of thermal origin result in a non-homogeneous phase distribution and in a non-null total diffused energy.

The most interesting feature of this technique is the possibility of simultaneously measuring from the ground several different parameters over a large altitude range below and above the maximum of electron density Examples of these parameters are the electronic density, the ion and electron temperatures or the speed of the ions along the terrestrial magnetic field lines.

The determination of the ion and electron temperatures and of the frequencies of collisions, especially in polar regions, is very interesting with respect to the characterisation of plasma. The variations of the frequency of collisions provide information pertaining at the same time to the arrival of solar high-energy particles in the ionosphere and to the absorption of radio waves propagating in this medium. This technique constitutes therefore an excellent means of detecting polar cap absorptions (PCA) events. The determination of the frequency of collisions is also interesting for the calculation of different types of conductivities (Pedersen, Hall, parallel to magnetic field) which modulate the polar current system as well as for the estimation of the density of neutral particles. At last, the electron and ion temperatures allow the determination of the quantity of energy penetrating in the regions under consideration.

G.5 Riometers

A riometer, or relative ionospheric opacity meter, is a radio receiver used for monitoring the cosmic radio noise at the frequency of approximately 30 MHz. This frequency is a compromise between the low frequencies, which are more absorbed and at which interferences between the radio transmissions and the industrial noise are a serious drawback, and the high frequencies at which absorption is limited and proportional to f^{-2} but at which measurements are also more difficult. Furthermore, the 30 MHz frequency is unaffected by the normal absorption occurring in the D layer.

The cosmic radio noise can be assumed to be constant at ground level. Accordingly, the variations in the intensity of the received signal are essentially due to the radio opacity of the ionosphere, which primarily depends on the solar activity, and more particularly on the emission of X-rays (SCNA) and protons of solar origin with an energy ranging from 5 to 50 Mev. The measurement of the cosmic radio noise is however an integrated measurement over the lowest part of the ionosphere, and the quantitative analysis is therefore limited.

G.6 Low Frequency and Very Low Frequency Receivers

Like riometers, low frequency and very low frequency receivers are sensitive detectors of solar protons. The phase lead at VLF frequencies is almost linearly proportional to the decrease of the reflection height which itself directly depends on the flow of protons of solar origin Phase measurements at VLF frequencies have therefore practically the same sensitivity to flows of solar protons than measurements performed using riometers. Since the amplitude of the phase variation over long distance paths at VLF frequencies depends on the integrated properties of the wave path in the ground-ionosphere waveguide, a quantitative study is therefore difficult. This method nevertheless allows the precise determination of the beginning of the event.

At low frequencies, the most important fluctuations may last for several days. One of the most distinctive features of polar cap absorption events is the absence of night fading during these events. This phenomenon expresses the presence of a homogeneous ionisation layer whose reflective properties do not depend on the presence of the solar rays.

References

Le Roux YM, Menard J, Jolivet JP, Toquin C, Bourdillon A (2002) Potentialités expérimentales du sondeur de canal ionosphérique SCIPION. 4^{ièmes} journées d'études Propagation électromagnétique dans l'atmosphère du décamétrique à l'angström, Rennes

H The Terrestrial Magnetic Field and the Magnetic Indexes

H.1 The Terrestrial Magnetic Field

The terrestrial magnetic field is generated by the metal mass contained inside the Earth and by the current system which flows in the ionospheric area. The value in Gauss of the magnetic induction Fx, Fy and Fz along the geographical North and East directions and along the downward vertical direction respectively (see Fig. H.1) is given by the following system of equations (ITU-R P.1239):

$$F_x = \sum_{n=1}^{6} \sum_{m=0}^{n} x_n^m \left[g_n^m \cos m\theta + h_n^m \sin m\theta \right] R^{n+2} \tag{H.1}$$

$$F_y = \sum_{n=1}^{6} \sum_{m=0}^{n} y_n^m \left[g_n^m \sin m\theta - h_n^m \cos m\theta \right] R^{n+2} \tag{H.2}$$

$$F_z = \sum_{n=1}^{6} \sum_{m=0}^{n} z_n^m \left[g_n^m \cos m\theta + h_n^m \sin m\theta \right] R^{n+2} \tag{H.3}$$

The variables x, y and z are defined by the equations:

$$x_n^m = \frac{d}{d\varphi} \left(P_{n,m} \left(\cos \varphi \right) \right) \tag{H.4}$$

$$y_n^m = m \frac{P_{n,m} \left(\cos \varphi \right)}{\sin \varphi} \tag{H.5}$$

$$z_n^m = -(n+1) P_{n,m} \left(\cos \varphi \right) \tag{H.6}$$

where:

- φ is the North colatitude, defined by $\varphi = 90° - \lambda$, where λ represents the geographical latitude expressed in degrees and ranging between -90° and 90°,

- $P_{n,m} (\cos \varphi)$ is the associated Legendre function defined by the following equation:

$$P_{n,m}(\cos\varphi) = \sin^m \varphi \left[\begin{array}{l} \cos^{n-m}\varphi - \dfrac{(n-m)(n-m-1)}{2(2n-1)}\cos^{n-m-4}\varphi + \\[2ex] \dfrac{(n-m)(n-m-1)(n-m-2)(n-m-3)}{(2)(4)(2n-1)(2n-3)}\cos^{n-m-4}\varphi + ... \end{array} \right] \quad \text{(H.7)}$$

- $g^{m,n}$ and $h^{m,n}$ are numerical coefficients for the field,

- R is a proportionality factor dependent on the height and is defined by the equation:

$$R = \frac{6371.2}{6371.2 + h_r} \quad \text{(H.8)}$$

where h_r is the height at which the field is evaluated, and is assumed to be equal to 300 kilometres.

The equation for the total magnetic field F is:

$$F = \sqrt{F_x^2 + F_y^2 + F_z^2} \,, \quad \text{(H.9)}$$

while the magnetic declination I and the gyrofrequency f_H (in MHz) are respectively given by the following two equations:

$$I = tg^{-1}\left(\frac{F_z}{\sqrt{F_x^2 + F_y^2}} \right) \quad \text{(H.10)}$$

$$f_H = 2{,}8\, F \quad \text{(H.11)}$$

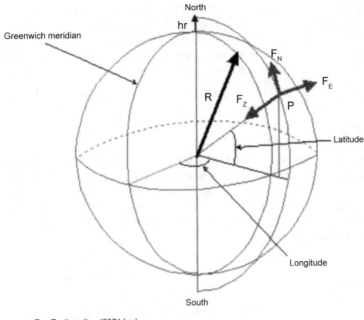

R = Earth radius (6371 km)

Fig. H.1. Schematic representation of the terrestrial magnetic field

H.2 Magnetic Indexes

H.2.1 K Index

The local K magnetic index was first introduced by J. Bartels in 1938 and is determined from the variation in amplitude, observed in the magnetograms of a given observatory, of the horizontal component of the magnetic field within a three hour period. This index takes values 0 to 9 according to a quasi-logarithmic scale specific to the observation station.

H.2.2 K_p Index

The global or planetary K_p index was introduced as a magnetic index by Bartels in 1949. This index is determined from the balanced average of the K indexes derived in thirteen selected magnetic observatories. Like the K index, its values range from 0, characterising magnetically very calm three-hourly intervals, to 9, characterising magnetically severely disturbed magnetically intervals. In contrast to the K values however, the variation range of the K_p is further subdivided into a scale of thirds using the three indices -, $_0$ and $_+$. This leads to 28 different values 0_0, 0_+, 1_-,1_0, 1_+, ... 8_-, 9_0, 9_+, characterising the amplitude of the variations of the geomagnetic field. Since the introduction of this index, magnetic storms are characterised by the K_p index (Chapman 1962).

H.2.3 A_P Index

The K_p index, being expressed according to a logarithmic scale, is not very practical for averaging and exploitation purposes. For this reason, Bartels was led to introduce an equivalent linear index varying from 0 nTesla, corresponding to $K_p = 0$, to 400 nTesla, corresponding to $K_P = 9_0$. Derived from eight three-hourly intervals, this index simultaneously characterises the amplitude and the duration of the main phase of magnetic storms.

H.2.4 AA Index

The magnetic AA index, is determined in the same way than the K_p and A_p indices, with the difference however that it is derived from only two nearly antipodal geomagnetic observatories, the first located in the northern hemisphere, in England, the second located in the southern hemisphere, in Australia. These indices are available for the period between 1868 and 1932.

H.2.5 Dst Index

The preceding magnetic indices characterise magnetic variations occurring at high and medium latitudes, which are more specifically associated with auroral electrojets. Since 1957, the magnetic variations due to the circular current are described by the Dst index. This index is derived from the values of the components of the magnetic field measured at four stations located near the magnetic equator. Since the magnetic field induced by the circular current is directly opposed to the main geomagnetic field, significant magnetic disturbances are therefore characterised by important negative variations of the Dst index.

H.2.6 AE Index

The magnetic AE index characterises magnetic disturbances occurring at very high latitudes. This index is derived from data recorded at twelve geomagnetic observatories located at auroral latitudes.

Readers interested in the determination of the different magnetic indexes are referred to the book by PN Mayaud *Derivation, Meaning and Uses of Geomagnetic Indices* (Mayaud 1980).

The *International Service of Geomagnetic Indices* (ISGI) is the organisation in charge of the elaboration and the dissemination of geomagnetic indexes, and of lists of remarkable magnetic events, with data provided by different geomagnetic observatories. From its foundation until 1987 it was located at the *Koninklijk Nederlands Meteorologish Institut* in De Bilt in the Netherlands. Since then, it has been transferred in France, first at the *Institut de Physique du Globe de Paris* and from 1990 to the present at the *Centre d'Etudes des Environnements Terrestres et Planétaires* located in Saint Maur. It works in close collaboration with several other institutes, the *GeoForschungsZentrum* in Postdam in Germany, the Ebre observatory of Roquétes in Spain, the *World Data Center-C2 for Geomagnetism* in Kyoto in Japan, the *World Data Center for Geomagnetism* in Copenhagen in Denmark, the *Arctic and Antarctic Research Institute* in Saint Petersburg in Russia.

References

Arctic and Antarctic Research Institute. http://www.aari.nw.ru/, 2001

International Service of Geomagnetic Indices (KP, AA, Dst and other indices).
    http://cetp.ipsl.fr/~isgi/homepag1.htm, 2001

Chapman S, Bartels J (1962) Geomagnetism, vols 1 and 2 Clarendon Press, Oxford

The Danish Meteorological Institute. http://web.dmi.dk/fsweb/projects/wdccl/pcn/pcn.html
    Magnetometer data from Tromso on line, http://geo.phys.uit.no/geomag.html, 2001

GeoForschungsZentrum (Kp and Ap indices)
    http//:www.gfz-potsdam.de/pb2/pb23/GeoMag/ niemegk/kp_index/, 2001

Mayaud PN (1980) Derivation, Meaning and Uses of Geomagnetic Indices. American Geophysical Union, Washington

National Oceanic and Atmospheric Administration. www.ngdc.noaa.gov/stp/GEOMAG/aastar.html, 2001

Observari de l'Ebre. Rapid variations of geomagnetic activity
    http://readysoft.es/observebre/

Schelegel K (2001) The strongest geomagnetic storms of the last century. Radio Science Bulletin 298

Space Physics Interactive Data Ressource (SPIDR) data bank of the NOAA with solar, interplanetary, ionospheric and geomagnetic data. http://spidr.ngdc.gov/spidr/, 2001

World Data Centre-C2 for Geomagnetism (Dst, AE indices)
    http://swdcdb.kugi.kyoto-u.ac.jp/, 2001

I Rain Attenuation

I.1 Introduction

Rain attenuation is a most important factor in the design of telecommunication systems, in particular for satellite telecommunication systems. The specific attenuation of the signal induced by rain depends on the temperature, on the size distribution of raindrops, on their final fall speed as well as on their size. Due to the variability of the rain in space and time, the value of the total rain attenuation can be determined only from the knowledge of the characteristics at each point of the path of the raindrops.

While measurements of rain attenuation have been realised, either with satellite beacons or with radiometers, these measurements are temporally and spatially scattered and are severely limited as regards frequency. They cannot therefore be readily generalised at the entire planet. This has led to the development, as an alternative, of several different models based on physical processes and on meteorological data, more particularly on the cumulative distribution of rain intensities, for the evaluation of the margins to be applied in the deployment of telecommunication systems.

I.2 Statistical Models for Rain Attenuation

The literature in this field is quite extensive, and a number of statistical models for the attenuation due to rain are available. A survey of these different models has been realised within the COST 255 framework (COST 255 2002). A few such models shall be described hereafter.

I.2.1 Assis-Einloft Model

For a precipitation rate R_P, the value in dB of the attenuation due to rain along the path is given by the following systems of equations:

$$A_s = [kR_p^{\alpha}D_i + kR_0^{\alpha}(L - D_i)]/\cos\theta \qquad \text{for } Hr \leq 33 \text{ km*tg } \theta \qquad (I.1)$$

$$A_s = [kR_p^{\alpha}D_i + kR_0^{\alpha}(33 - D_i)]/\cos\theta \quad \text{for } Hr > 33 \text{ km* tg } \theta \qquad (I.2)$$

where:

- θ is the elevation angle,
- D_i is the internal diameter of a cylindrical rain cell, as given, for a rain intensity R_P, by the following equation:

$$D_i = 2.2\left(\frac{100}{R_p}\right)^{0.4} \qquad (I.3)$$

– R_0 is the residual rain rate in mm/h in the external rain cell with diameter no larger than 33 kilometres:

$$R_0 = 10\left(1 - e^{0.0105R_p}\right) \qquad (I.4)$$

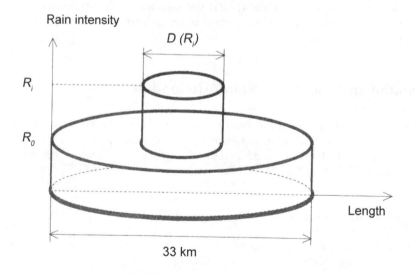

Fig. I.1. Structure of a rain cell

- L is the horizontal projection of the slant path length in kilometres:

$$L = L_s \cos\theta \tag{I.5}$$

- L_s is the length in kilometres of the slant path below the rain height. In the case of an elevation angle $\theta \geq 5°$, L_s is equal to:

$$L_s = \frac{(H_r - H_s)}{\sin\theta} \tag{I.6}$$

while for $\theta < 5°$:

$$L_s = \frac{2(H_r - H_s)}{\left[\sin^2\theta + \dfrac{2(H_r - H_s)}{R_e}\right]^{1/2} + \sin\theta} \tag{I.7}$$

- R_e is the effective Earth radius, equal to 8500 kilometres,
- H_r is the altitude of effective rain in kilometres. It is expressed as a function of latitude λ by the following equations:

$$H_r = \begin{cases} 5 - 0.075(\lambda - 23) & for \quad \lambda > 23 \\ 5 & for \quad -21 \leq \lambda \leq 23 \\ 5 + 0.1(\lambda + 21) & for \quad -71 \leq \lambda \leq -21 \\ 0 & for \quad \lambda < -71 \end{cases} \tag{I.8}$$

The time percentage that the attenuation A_s will be exceeded is thus given by the equation:

$$p(A_s) = p(R_p)L / D_i \tag{I.9}$$

I.2.2 Brazilian Model

In this model the attenuation in dB exceeded at a time percentage p is given by the following equation:

$$A_s = \gamma_p \cdot L_{eff} \tag{I.10}$$

where:

- γ_p is the specific attenuation (dB/km), given by the equation :

$$\gamma_p = k \cdot R_p^{\alpha} \qquad (I.11)$$

- L_{eff} is the effective path length in kilometres, satisfying the equation :

$$L_{eff} = L_s \cdot r_p \qquad (I.12)$$

- R_P is the path length equivalent factor, given by the equation:

$$r_p = \frac{1}{1 + \dfrac{L_s \cos \theta}{L_{R0}}} \qquad (I.13)$$

- L_s is the length of the slant path in kilometres, and is expressed as a function of $h_R,$ θ and H_s:

$$L_s = \frac{(h_R - H_s)}{\sin \theta} \qquad \theta \geq 5° \qquad (I.14)$$

$$L_s = \frac{2(h_R - H_s)}{\left[\sin^2 \theta + \dfrac{2(h_R - H_s)}{R_e} \right]^{1/2} + \sin \theta} \qquad \theta < 5° \qquad (I.15)$$

- R_e is the effective Earth radius, equal to 8500 kilometres,
- h_R is the altitude of the equivalent rain cell, as given by the equation:

$$h_R(R_p, p) = (3.849 + 0.334 \log p)\left[1 + \exp(-0.2R_p)\right] \qquad (I.16)$$

- θ is the elevation angle,
- H_s is the altitude of the base station,
- L_{R0} is the length of the equivalent rain cell, and is a function of the time percentage p and of the precipitation rate R_p exceeded during the time percentage p:

$$L_{R0}(R_p, p) = 200\left[1 + R_p^{(0.425 - 0.089 \log p)}\right]^{-1} \qquad (I.17)$$

I.2.3 Bryant Model

The Bryant model is based on the concept of an effective rain cell and on the variability of the rain height (Bryant 1999). The following equation yields the attenuation along the slant-path:

$$A_s = 1.57 D_m k_n \gamma_p \frac{L_s}{\xi L + D} \qquad [dB] \qquad (I.18)$$

where:

- L_s is the length of the slant path,
- γ_p is the specific attenuation in dB/km,
- $D_m = (2/\pi)D$,
- k_n is the number of cells as given by the formula : $\exp(0.007 R_p)$,
- ξ is given by the following equations:

$$\xi = \begin{cases} \dfrac{1}{\sqrt{2}} \exp(\sin\theta) & \theta \le 55° \\ 1.1 \tan\theta & \theta > 55° \end{cases} \qquad (I.19)$$

- L is the horizontal projection of the slant path, and is given by the equation:

$$L = H_r / \tan\theta \qquad (I.20)$$

- θ is the elevation angle,
- H_r is the rain height in kilometres:

$$H_r = 4.5 + 0.0005 R_p^{1.65} \qquad (I.21)$$

The following equation yields the parameter PR, representing the distribution of the rain attenuation:

$$PR = 1 + \frac{2}{\pi} \frac{L}{D} \qquad (I.22)$$

where D is the diameter of a rain cell, given by the equation:

$$D = 540 \cdot R_p^{-12} \tag{I.23}$$

Several other models can be found in the literature, among which we may mention here the Crane model, the Dah model, the Excell model, the Flavin model, the Garcia model, the ITU-R model, the Karasawa model, the Leitao-Watson model, the Matricciani model, the Misme-Waldeufe model, the *simple attenuation mode* model, the Sviatogor model, etc. For more detail concerning these various models the reader is referred to the final report by the COST255 *Radiowave propagation modelling for SatCom services At Ku-band and above* (COST255 2002).

References

Bryant GF, Adimula I, Riva C, Brussard G (1999) Rain Attenuation Statistics from Rain Cell Diameters and Heights. International Journal of Satellite Communications

COST255 (2002) Radiowave propagation modelling for Satcom services at Ku-band and above. Final report, European Space Agency

J Vegetation Attenuation

J.1 Introduction

In telecommunication applications, the disturbances induced by vegetation on the propagation of waves appear primarily in the form of absorption and diffraction phenomena, resulting in an attenuation of the received signal. The most important physical process involved here is the forward transmission, and the physical quantity given by vegetation attenuation models will be the transmittance or specific attenuation.

J.2 Vegetation Attenuation Models

J.2.1 Exponential Decay Model (EXD Model)

Let P_r and P_e be the received and emitted power respectively and d be the distance between the transmitter and the receiver. The exponential decay model or EXD model expresses the received power in the form of an exponential decreasing function of the distance d travelled inside the vegetal cover:

$$P_r = P_e e^{-\alpha d} \qquad (J.1)$$

where α is the extinction coefficient, which is the sum of the absorption and scattering coefficients α_a and α_d.

The transmittance τ, expressed in natural units, is written in the following form:

$$\tau = e^{-\alpha d} \qquad (J.2)$$

The value of the loss factor or attenuation along the path is expressed in dB:

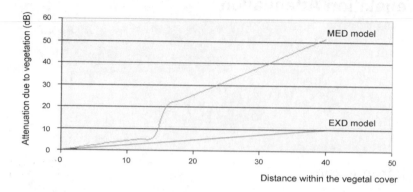

Fig. J.1. Comparison between the attenuation values obtained with the EXD and MED models at the 900 MHz frequency with respect to the distance travelled inside the vegetal cover

$$L = -10\log(\tau) = \alpha d \qquad \text{(J.3)}$$

where α is the specific attenuation in dB/m. The extinction coefficient or specific attenuation is independent of the distance. Depending on the frequency, the extinction coefficient shall assume the following form in the 0.1-3.2 GHz frequency range:

$$\alpha = 0.26 f^{0.77} \qquad \text{(10.4)}$$

where f is the frequency expressed in GHz.

This extremely simple model does not take into account data pertaining to the characteristics of the medium, for instance the heights of the trunks or the density of the branches. Accordingly, the results are far from being extremely precise, and errors may reach 20, 30 dB and even more in some cases.

J.2.2 Modified Exponential Decay Model (MED Model)

This model was developed as an alternative to the EXD model in order to reduce errors arising from the fact the specific attenuation α is not actually dependent on the frequency. The equations for the specific attenuation provided by this model are:

$$\alpha = 1.33 f^{0.284} d^{-0.412} \qquad 14 \le d \le 400 \tag{J.5}$$

$$\alpha = 0.45 f^{0.284} \qquad 0 \le d \le 14$$

In the case where $14 \le d \le 400$, this model allows deriving an expression for the received power as a decreasing function of the form:

$$P_r = P_e e^{-\alpha(f) d^{1-k}} \tag{J.6}$$

where k is equal to 0,412.

As an example, Fig. J.1 presents a comparison between results obtained with the EXD and MED models at the 900 MHz frequency.

J.2.3 Rice Model

In this model, the value of the specific attenuation is given in dB/m by the following Rice-type equation:

$$\alpha = a_1 \sigma + a_2 \exp\left(\frac{-a_3}{f_{GHz}}\right) \log\left(1 + \frac{f_{GHz}}{a_4}\right) \tag{J.7}$$

where σ is the conductivity of the forest environment, while $a_1 = 1637$, $a_2 = 0.2717$, $a3 = 0.09$, $a_4 = 0.1$ and $f > 1$ GHz.

J.2.4 ITU-R Model

The following equation provided by the Recommendation ITU-R P.833-2 gives the additional attenuation due to vegetation in the case where either the emitter or the receiver is located in a wooded area:

$$A_{vegetation} = A_m \left[1 - \exp\left(\frac{-d\gamma}{A_m}\right) \right] \tag{J.8}$$

where d is the length in metres of the path inside the wooded area, γ is the specific attenuation in dB/m for very short paths inside the vegetal medium and A_m is the maximum attenuation in dB at the end of the link in a vegetation with given type and depth characteristics.

The relation between A_m and the frequency is given by the following equation:

$$A_m = 0,18 f^{0,752} \tag{J.9}$$

where f is the frequency in MHz.

J.2.5 Al-Nuami - Hammoudeh Model

This model more specifically applies to the evaluation of the attenuation due to vegetation in an apple orchard and at the 11.2 GHz frequency (Al-Nuami 1994). The following equation gives the attenuation:

$$L = 11.21 d^{0.43} \tag{J.10}$$

J.2.6 Stephens Model

Experimental measurements were conducted at the 11.2 GHz frequency for curtains of trees (Stephens 1995). In this model, the additional attenuation due to vegetation is expressed for the parameter α as a function of the penetration depth in vegetation, with two different slope values:

$$A_{vegetation} = -\left(R_\infty d + k \left(1 - \exp\left(-\frac{(R_0 - R_\infty)}{k} \right) \right) \right) \tag{J.11}$$

where R_0 is the initial specific attenuation rate, R_∞ is the final specific attenuation rate, d is the penetration depth in vegetation, i.e. the distance travelled inside the vegetation, and k is a normalisation coefficient.

For a curtain of trees and at the 11.2 GHz frequency, the following values were obtained with this model:

- $R_0 = 5.67$ dB/m,
- $R_\infty = 0.33$ dB/m,
- $k = 19.10$.

This model was subsequently improved on the basis of measurements carried out at frequencies ranging from 9.6 to 57.6 GHz, in order to take into account the absence or presence of leaves as well as the configuration of the site, characterised by the level of illumination in vegetation (see Recommendation ITU-R P.833-2).

J.2.7 Radiative Transfer Models

Radiative Transfer Theory

The theory of radiative energy transfer is concerned with the study of energy transfers occurring within a medium where scatterers are present. The medium shall be described through the following two matrices:

- the extinction matrix, describing the attenuation of the electromagnetic intensity due to the absorption and the scattering by the particles composing the medium,
- the phase matrix, characterising the coupling between the incident and scattered intensities at each point of the medium.

Let I (r,s) be the incident intensity arriving inside an elementary cylindrical volume with a basis of unit surface and a length ds. The following equation gives the decrease of energy due to the propagation of the wave inside this volume:

$$dI(r,s) = -N\left(\sigma_a + \sigma_d\right)Ids = N\sigma_e Ids \qquad (J.12)$$

where:

- σ_a and σ_d are the absorption and scattering cross sections respectively, whence one derives the amounts of power respectively absorbed $(\sigma_a I)$ and scattered $(\sigma_d I)$ by an individual particle,
- σ_e is the extinction cross section and is the sum of the absorption and scattering cross sections,
- N is the number of particles contained in an elementary volume.

Let us now determine the energy increase in this volume. This is achieved by considering the intensities arriving by multiple scattering from all directions into the elementary volume:

$$dI(r,s) = \frac{1}{4\pi} \int p\left(\overline{s},\overline{s}\,'\right)d\Omega \qquad (J.13)$$

where $p\left(\overline{s},\overline{s}\,'\right)$ is the phase function describing the spatial distribution of energy.

The fundamental equation describing the energy transfer is obtained by the addition of the two preceding terms:

$$\frac{dI(r,s)}{dr} = -k_e I(r,\overline{s}) - \int_{4\pi} p(\overline{s},\overline{s}') I(r,\overline{s}') d\Omega \qquad (J.14)$$

where k_e is the extinction coefficient defined by $k_e = N\,\sigma_e$.

This equation is a scalar equation, and is more particularly valid in the case of a non-polarised incident wave. A vector formulation of this equation is generally preferred in order to take into account the polarisation of the incident wave as well as the properties of the medium itself, which is characterised by global quantities.

Adopting an adequate system of coordinates, the equation for the energy transfer can be rewritten in the following form:

$$\frac{d\overline{I}(z,\theta,\varphi)}{dz}\cos\theta = -\overline{\overline{k}}_e(\theta,\varphi)\overline{I}(z,\theta,\varphi) + \int_0^{2\pi} d\varphi' \int_0^{\pi} \overline{\overline{P}}(\theta,\varphi;\theta',\varphi')\overline{I}(z,\theta',\varphi')d\theta' \qquad (J.15)$$

where:

- $\overline{I}(z,\theta,\varphi)$ is the intensity vector,

- $\overline{\overline{P}}(\theta,\varphi;\theta',\varphi')$ is the 4 x 4 phase matrix describing the scattering occurring from the θ,φ direction towards the θ',φ' direction,

- $\overline{\overline{k}}_e(\theta,\varphi)$ is the 4 x 4 matrix of the extinction coefficients.

The equation is complex and only in some specific cases, for instance isolated spherical scatterers, can it be solved. Different resolution methods have nevertheless been developed in order to solve this problem: these methods rely on some fundamental quantities, the simple and modified Stokes parameters, the scattering matrix, the Mueller matrix, the phase matrix and the extinction matrix, as well as on certain approximations, the physical optics approximation, valid in the case of high-frequency scattering, and the Rayleigh scattering approximation, valid in the case of low-frequency scattering.

Stokes Parameters

The polarisation state of a wave is commonly described in the form of a vector rotating in a two-dimensional space. In the general case, the extremity of this vector shall describe an ellipse. The norm of the vector is proportional to the total energy of the wave.

The polarisation state of a wave can also be represented by a stationary vector in a four-dimensional Euclidian space. The components of this vector have the same physical dimensions. The vector is written in the form:

$$\bar{S} = \begin{bmatrix} I_0 \\ Q \\ U \\ V \end{bmatrix} = \begin{bmatrix} E_v^2 + E_h^2 \\ E_v^2 - E_h^2 \\ 2\,\mathrm{Re}\left(E_v E_h^*\right) \\ 2\,\mathrm{Im}\left(E_v E_h^*\right) \end{bmatrix} \qquad (J.16)$$

where I_0 represents the total intensity of the wave, Q is the difference of intensity between the vertical and horizontal polarisations, while U and V represent the phase differences between the components in the vertical and horizontal polarisations respectively.

The relation between the components of the Stoke vector, three of which are independent from one another, is given by the following equation:

$$I_0^2 = Q^2 + U^2 + V^2 \qquad (J.17)$$

The h and v indices in the equation for the vector stand for the horizontal and vertical polarisation respectively. The components of the Stokes vector are proportional to the components of the intensity vector, the proportionality factor being the characteristic impedance of the scattering medium.

Modified Stokes Parameters

The resolution of problems associated with radiative transfer is made easier using the modified Stokes parameters. The vector for the modified Stokes parameters is written in the following way:

$$S_m = \begin{bmatrix} I_v \\ I_h \\ U \\ V \end{bmatrix} = \begin{bmatrix} < E_h^2 > \\ < E_v^2 > \\ < 2R_e\left(E_h E_v^*\right) > \\ < 2I_m\left(E_h E_v^*\right) > \end{bmatrix} \qquad (J.18)$$

where:

$$I_v = E_v^2 = \left(I_0 + Q\right)/2 \qquad (J.19)$$

and:

$$I_h = E_h^2 = (I_0 - Q)/2 \qquad (J.20)$$

Scattering Matrix

The introduction of the scattering matrix D allows determining the different components of the field scattered by a target as a function of the different components of the incident field. This leads to the following matrix equation:

$$\begin{bmatrix} E_h^d \\ E_v^d \end{bmatrix} = \frac{e^{-ikr}}{r}[D].\begin{bmatrix} E_h^i \\ E_h^i \end{bmatrix} = \frac{e^{-ikr}}{r}\begin{bmatrix} S_{hh} & S_{hv} \\ S_{vh} & S_{vv} \end{bmatrix}.\begin{bmatrix} E_h^i \\ E_h^i \end{bmatrix} \qquad (J.21)$$

where r is the distance between the scatterer (or the set of scatterers in the case where the target is of large size) and the reception antenna, while k is the wave number.

The elements S_{ij} in the matrix D are referred to as the complex scattering amplitudes. They are dependent on the frequency, on the observation angles and on the orientation of the scatterers.

Mueller Matrix

In the presence of a single scatterer, the relation existing between the incident and scattered fields can be demonstrated to be equivalent to the relation between the modified Stokes parameters of the scattered and incident fields. The modified Stokes parameters of the scattered fields can be mapped into the corresponding parameters of the incident fields using either the Mueller matrix M or the phase matrix. The Mueller matrix is one-to-one mapped onto the scattering matrix D (Ulaby 1990).

Phase Matrix

The Mueller matrix, averaged over all orientations and dimensions of the scatterer, is referred to as the phase matrix P. This matrix, used in the theory of radiative energy transfer, is expressed by the following equation:

$$I^d = \frac{1}{r^2}P.I^i \qquad (J.22)$$

where $P = <M>$, while $<M>$ represents the coefficients of the Mueller matrix averaged over all dimensions and orientations of the scatterer.

The phase matrix of a set of N particles with known shapes, randomly distributed in space, can be determined from the phase matrix of a single modelled element by averaging it over all dimensions, orientations and possible shapes of the scatterers.

Extinction Matrix

The extinction matrix k_e represents the attenuation due to absorption and scattering of the Stokes parameters. This matrix is deduced from the coefficients of the scattering matrix using approximations depending on the density of the scatterers and their relative size compared to the wavelength (Ulaby 1990).

Wave Scattering by a Particle

In order to determine the wave scattering in a forest medium, it is necessary to calculate first the wave scattering by the individual elements, the leaves, the branches, etc. Each such element will be represented by an elementary geometrical form, for instance a disk or a sphere, thus enabling to perform rigorous calculations. The different individual elements are regarded as point scatterers with respect to the remote region where the scattered field is determined. Only in the case of wave scattering by a spherical particle, also known as Mie scattering, can theoretical methods be used for the solution of this problem. In the general case, approximations, based on the relative ratio between the size of the scatterer and the wavelength, are used. The two most important such methods shall be reviewed hereafter: the physical optics approximation and the Rayleigh scattering approximation.

Physical Optics Approximation. This approximation can be used provided that the size of the diffusers is large with respect to the wavelength. At high frequencies, the surface of such diffusers can be considered as an infinite plane. The problem can therefore be reduced, like in geometrical Optics, to the problem of a wave plane striking or not an infinitely conducting plane surface. The Fresnel reflection coefficients are used in this case.

Rayleigh Scattering Approximation. This approximation holds as long as the size of the diffusers remains small compared to the wavelength, i.e. for $kd \ll 1$, where d is a characteristic dimension of the diffuser. This approximation relies on the assumptions that the electric field present inside the particle is uniform (electrostatic field) and that the radiated far-field has the same features than a dipolar field.

J.2.8 MIMICS Model (MIchigan MICrowave Scattering)

This model takes into account the scattering occurring in all regions of the vegetal cover, the crown (leaves, needles, primary or secondary branches), the trunks and the ground (Ulaby 1990). For each region, the input parameters consist of geometrical and radio parameters. These parameters are either given in the form of probability distribution functions, either measured or calculated by means of models applied to indirect measurements, as is the case for instance with dielectric parameters, which are determined from temperature and water content measurements.

The crown contains two types of dielectric scatterers with known densities:

- leaves are modelled in the form of flat rectangles with very limited thickness (i.e. the thickness is small with respect to the random dimensions of the rectangle),
- branches and needles are modelled in the form of circular cylinders with random orientations.

The trunk area contains vertical cylinders with finite length:

- the phase matrix of a basic leave is determined using the physical optics approximation,
- the phase matrix of the branches is determined from the field that would be diffused by a cylinder with infinite length compared to the wavelength,
- the phase matrix of the needles is determined by representing them in the form of spheroids taking into account their relative dimensions with respect to the wavelength.

The analysis of the problem proceeds in two steps:

- first, the intensity backscattered by vegetation is determined under the assumption that the ground constitutes a specular interface. Most of the energy arriving at the surface of the ground is redirected along the specular direction, provided that this surface is neither too rough nor too inclined with respect to the vertical plane. The total attenuation due to vegetation and its backscatter contribution can then be deduced. The interactions between the two layers of the vegetal cover remain relatively simple,
- the ground backscatter contribution is then determined under the assumption that this interface behaves like a rough surface.

The total scattered intensity is determined from the values obtained in the two steps of the calculation Further detail concerning this method can be found in the article by Ulaby et al. (Ulaby 1990).

J.2.9. Karam-Fung model

This model takes into account the multiple scattering which takes place inside the vegetation and more specifically inside the crown (Karam 1992). The vegetal cover is divided into two layers: depending on the tree species, the highest layer contains branches and either leaves or needles, while the lowest layer contains the trunks. In order to account for the complexity of the forest environment, it shall be assumed that within the same layer different categories of scatterers are represented: for instance, a distinction between primary and secondary branches will be drawn and similarly, different categories of trunks will be distinguished. Elements in the same category of scatterers have identical geometrical dimensions and are characterised by their diameter, length and density. The orientation and slope of each component of the cover are defined using probability distribution functions: unlike geometrical dimensions, which can be measured, probability distribution functions have to be introduced in order to statistically represent the spatial distribution of the branches and leaves.

The needles, branches and trunks are described as dielectric cylinders of finite length, while the leaves are modelled in the form of circular discs. The ground is described as a rough surface defined by its correlation length and its standard height deviation.

The backscatter contributions considered in this model are:

- the direct propagation towards the rough ground and the attenuation resulting from the forward and backward propagations through the vegetal cover (for example in monostatic remote detection),
- the direct scattering in the crown (backscattering contributions of the leaves, needles and branches),
- the interactions between the crown and the ground depending on whether the ground is smooth or rough,
- the direct scattering by trunks,
- the interactions between the trunks and the ground depending on whether the ground is smooth or rough,
- the multiple scattering in the crown.

J.3 Experimental Results

The empirical approach turns out to be the best suited for the planning of communication networks operating in rural environments, where only limited information concerning the vegetation is available. In order to achieve a better understanding of the effects of vegetation on propagation at the 900 MHz (GSM frequency band) and 2200 MHz frequency (DCS 1800 and UMTS frequency bands) frequencies, as well as the underlying physical phenomena, two different experiments were carried out:

- in the first experiment, a 2270 metre fixed link, with a 160 metre length inside the vegetation, and where the receiving antennas were located outside the vegetation area, was considered in order to investigate the daily and seasonal influence of vegetation depending on the meteorological parameters,

– in the second experiment, mobile links were set up in a wooded area in order to investigate the effects of the penetration distance inside the foliage, depending on the period of the day and the season (presence or absence of leaves). This experiment was carried out over a 52 kilometre long route in the Harth forest near Mulhouse in France, for vegetation depths ranging from a few metres and 6 kilometres.

After providing a brief description of each of these experiments, the main results concerning vegetation attenuation, its daily variations and the influence of leaves will be presented and compared to values predicted by theoretical models. An optimisation of the ITU-R model shall also be proposed in the course of this section.

J.3.1 Fixed Link Measurements

Description of the Experiment

The experiment was conducted near Belfort in France. As represented in Fig. J.2, the experimental arrangement consisted of four directive antennas: a group of two collocated transmitting antennas and a group of two collocated receiving antennas, separated from each other by vegetation. The transmitting and receiving antennas were located at a distance large enough to prevent coupling phenomena, and were placed upon a 20 metre high mast. The gains of the antennas were equal to 11 dB with horizontal and vertical apertures equal to 80 degrees and 36 degrees respectively at 3 dB. The experiment was conducted during a whole year in order to study seasonal variations.

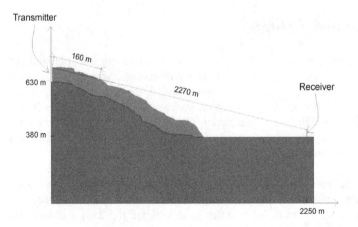

Fig. J.2. Cross-section of the experimental site

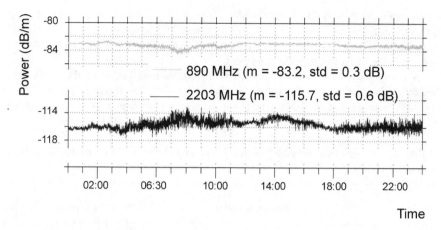

Fig. J.3. Received power in dB/m at the 890 and 2203 MHz frequencies on November 26, 1999

Results

The measured data were averaged over thirty samples using the sliding average method, in order to investigate the diurnal evolution of the daily, monthly and annual variations. Fig. J.3 represents a typical example of the variation of the received power in dB/m at the two considered frequencies on November 26, 1999.

The standard deviation (*std*) observed and represented in Fig. J.3 is generally small; the variations observed during a day present but a limited extent. The presence of scintillations, which are most of the time correlated at the two frequencies, is to be noted. The amplitudes of these scintillations are higher at the 2203 MHz frequency, where they are generally due to the movement of leaves caused by the wind (Pearce 1999). The maximum fading is equal to approximately 1 dB at the 890 MHz frequency and 4 dB at the 2203 MHz frequency.

The daily average received powers are about -83 dB/m and -116 dB/m at the 890 and 2203 MHz frequencies respectively. The attenuation due to vegetation was then determined from the received powers taking into account the equivalent isotropically radiated power (EIRP), the gains of the receivers, the cable losses, free-space attenuation, etc. The values 39.6 dB and 59.6dB were found at the 890 MHz and at the 2200 MHz frequencies respectively, corresponding to values for the specific attenuation equivalent to 0.25 dB/m and 0.37 dB/m. Due to the small standard deviations, it would be extremely difficult to draw a correlation between these variations in power and the meteorological parameters.

From yearly variations presented in Fig. J.4, the value of the equivalent specific attenuation was found to be equal to 0.25 dB/m and 0.4 dB/m at the 890 and 2203 MHz frequencies respectively. The equivalent specific attenuation at the 890 MHz frequency does not present any marked seasonal variation. A maximum value was observed during November 1998, which however does not appear to be correlated with any meteorological parameter (pressure, temperature, humidity or precipitation intensity). On the contrary, at the 2203 MHz frequency, a seasonal variation can be clearly observed, revealing a 4 dB additional attenuation due to the presence of leaves in deciduous trees.

A comparison between the experimentally measured values of the equivalent specific attenuation and the values predicted by different models (MED, RICE, ITU-R) is presented in Table J.1. These values can be seen to be in accordance with the values presented in Recommendation ITU-R P.833 (ITU-R 1990).

Table J.1. Comparison between predicted and experimental values of the equivalent specific attenuation

Model	890 MHz [dB/m]	2203 MHz [dB/m]
MED	0.16	0.21
RICE	0.38	0.48
ITU-R (receiver located outside the vegetation area)	0.2	0.4
Experimental measurements	0.25	0.4

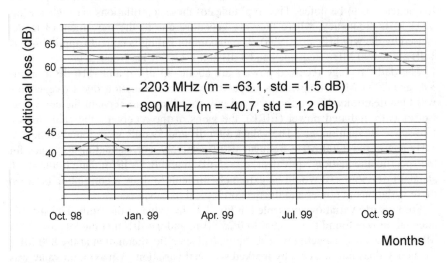

Fig. J.4. Annual variation of the attenuation due to vegetation

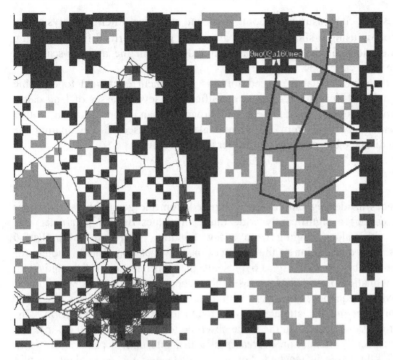

Fig. J.5. Route of the mobile measurements realised inside vegetation in the Harth forest, superimposed over geographical land data

3.2 Mobile Link Measurements

Description of the Experiment

Mobile links were set up in the Harth forest near Mulhouse, along a 52 kilometre long route, in order to realise measurements for distances inside vegetation ranging from a few metres to 6 kilometres. Fig. J.5 represents the geographical location of the transmitter and the measurement route in the forest area.

A transmitting omnidirectional $\lambda/2$ antenna was placed upon a 25 metre high mast, while a second, receiving omnidirectional antenna was mounted upon a vehicle moving at an altitude of 1.6 metre above the ground level. The equivalent isotropically radiated power was equal to 45 dB/m and 42 dB/m at the 936.9 MHz and 2203 MHz frequencies respectively. Measurements were carried out at two frequencies and at different times during the day, and repeated every 11 days from May to June and every 13 days from November to December. The amplitude of the field was recorded every $\lambda/4$ metre during the displacements of the vehicle. A numerical analysis was then performed in order to associate geographical coordinates to each measurement point. A second numerical filtering analysis over a 40 λ window using a sliding average algorithm was performed for eliminating the fast variations of the signal (Rayleigh fading) from the average-scale and large-scale slow variations.

The geographical databases used for this experiment contained topographic data with a 100x100 metre resolution classified into seven classes of land data : water, woods, open environment, suburban, urban, dense urban, other environment. Only data from the second class were used, for evaluating the vegetation distance between the transmitter and the receiver.

Results

The distances inside vegetation were computed from the geographical databases described earlier: a significantly wide range of values of this parameter was obtained, with values ranging from a few hundreds metres to more than 6 kilometres. This distribution of the distances inside vegetation is wide enough for defining a reliable model. Fig. J.6 presents the dependence of the equivalent specific attenuation on the distance inside the canopy as observed in May-June at the 2203 MHz frequency. For a path with a length equal to 160 metres, the value of the equivalent specific attenuation was found to be equal to 0.39 dB/m. This value is of the same order than the value observed for a fixed link (0.4 dB/m). The equivalent specific attenuation decreases at a fast rate until a penetration depth equal to 2 kilometres is reached. For longer distances, the equivalent specific attenuation decreases very slowly from about 0.013 dB/m at the distance of 4 kilometres to 0.009 dB/m at the distance of 6.4 kilometres.

The excess vegetation attenuation compared to free-space attenuation can be modelled in the form of a log(d) law defined by the following equation:

$$A_{ev} = a.\log(d) + b \qquad (J.23)$$

where d is the penetration depth in the canopy expressed in kilometres. Comparisons with values obtained with the Rizk model (Rizk 1996) were also drawn, although these measurements had been carried out in a microcell environment at the 900 MHz frequency. These comparisons are summarised in Table J.2. The parameter *std* expressed in dB represents the standard deviation of the differences between values predicted by the model and values actually measured.

For the Rizk model, the standard deviation was evaluated using values of attenuation measured in summer at the 936.6 MHz frequency. The average error is equal to 4.5 dB. Log(d) models lead to satisfactory results in terms of standard deviation.

The models thus defined bring to the fore the effects of seasonal variations: attenuations are more important in summer than in winter. This appears essentially through the decrease of the term b occurring between summer and winter: the values for this parameter decreases by 1.9 dB and 8.6 dB at the 939.6 MHz and

2203 MHz frequencies respectively. For distances ranging from 300 metres to 6 kilometres, the vegetation loss difference due to season goes from 1.5 to 2.5 dB at the 939.6 MHz frequency and from 8.8 to 8.3 dB at the 2203 MHz frequency. These figures are valid for distances smaller than 6 kilometres. Due to the difference between the values of the term a at the 939.6 MHz and 2203 MHz frequencies respectively, the slopes decrease from summer to winter in the first case, whereas they remain stable in the second one.

Table J.2. log(d) models

	A	B	Std (dB)
Rizk model at 939.6MHz	12	37.8	6.4
939.6 MHz (summer)	17.4	31.3	6.2
939.6 MHz (winter)	16.6	29.4	6
2203 MHz (summer)	14.2	41.5	6
2203 MHz (winter)	14.6	32.9	6.8

Table J.3. Fit of the ITU-R P.833-2 model

	Model	Mean error (dB)	Std (dB)
ITU-R model	$A_m = 32.45 \times f_{GHz}^{0.752}$	17.7	15.7
Fitted model	$A_m = 22.42 \times f_{GHz}^{0.43}$	0	8.7

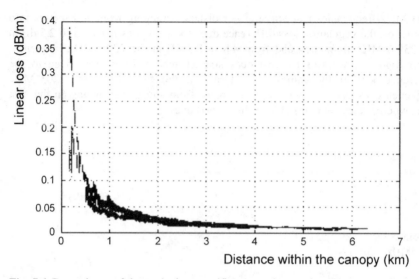

Fig. J.6. Dependence of the equivalent specific attenuation on the distance travelled inside the canopy at the 2203 MHz frequency

From the data thus obtained, a correction was introduced in the parameters of the model presented in Recommendation P833-2 ITU-R. Let us here recall that this model provides a frequency-dependent expression of the maximum attenuation for a receiver located inside a vegetation area with a given type and a characteristic depth A_m. For very short paths in a wooded region γ, the attenuation coefficients were evaluated on the basis of the measured data, leading to the values 0.25 and 0.4 dB/m at the 939.6 MHz and 2203 MHz frequencies respectively. The two parameters A_m were evaluated by considering their different combinations in the [0;100]x[0;1,5] interval and by minimising the sum of the squares of the mean error and of the standard deviation. Table J.3 summarises the results that were obtained.

These new parameters have been accepted by the ITU-R and are taken into account in Recommendation ITU-R P.833-3 (ITU-R 2000). Fig. J.7 presents the fitted ITU-R model; the strong dispersion of the measured data can also be seen in this figure. The maximum value of attenuation A_m, limited by the scattering of the surface wave, is reached very rapidly after a distance in vegetation of about 500 metres. This parameter depends on the tree species and on the density of vegetation as well as on the radiation pattern in vegetation of the receiving antenna and on the vertical distance between the antenna and the top of the vegetal cover (ITU-R P.833-2).

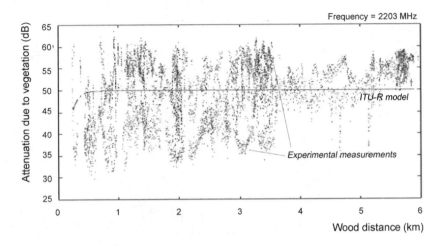

Fig. J.7. Optimised ITU-R model

References

Al-Nuami MO, Hammoudeh AM (1994) Measurements and predictions of attenuation and scatter of microwave signals by trees. IEE Proc. Microw. Antennas Propagation vol 114 2

Karam MA, Fung AK, Lang RH, Chauban NS (1992) A microwave scattering model for layered vegetation. IEEE Trans. On Geos. And Remote Sens. vol 30 4: 767-784

Karam MA, Fung AK, Mougin AK, Lopes E, Le Vine DM, Beaudoin A (1995) A microwave polarimetric scaterring mode for forest canopies based on vector radiative transfer theory. Remote Sens. Environ. 53: 16-30

Stephens RBL, Al-Nuami MO (1995) Attenuation measurements and modelling in vegetation media at 11.2 and 20 GHz. Electronics Letters vol. 31 20

Ulaby FT, Sarabandi K, McDonald K, Witt M, Dobson G (1990) Michigan microwave canopy scattering model. Int. J. Remote sensing vol 11: 1223-1253

Ulaby FT, Moore RK, Fung AK (1986) Microwave Remote Sensing Fundamentals and Radiometry. vol II Radar Remote Sensing and Surface Scattering and Emission Theory. vol III From Theory to Applications. Artech House Inc.

K Diffraction Models

1 Diffraction by a Single Knife Edge

The attenuation induced by the diffraction by a single knife-edge obstacle can be rigorously determined using Fresnel diffraction theory. Fig. 1 represents the diffraction by a single knife edge, first in the case where the line-of-sight path between the transmitter and the receiver is clear, and then in the case where the direct path is obstructed by the knife edge.

Let d be distance between the transmitter and the receiver, d_1 and d_2 be the respective distances from the edge to the transmitter and the receiver, h is the algebraic height of the edge above the straight line joining the transmitter to the receiver, α be the incidence angle and θ be the diffraction angle.

Let us now introduce a parameter v defined by the following equation:

$$v = h\cos\alpha\sqrt{\frac{2}{\lambda}\left(\frac{1}{d_1}+\frac{1}{d_2}\right)} = \theta\sqrt{\frac{2}{\lambda}\left(\frac{1}{d_1}+\frac{1}{d_2}\right)} \tag{K.1}$$

where θ is as defined by the equation:

$$\theta = h\frac{d_1+d_2}{d_1 d_2} \tag{K.2}$$

The field strength at the reception point is given by the following equation:

$$\left|\frac{E}{E_0}\right| = \left|\frac{1}{1+j}\int_v^\infty e^{j\pi\frac{t^2}{2}}dt\right| = \frac{\sqrt{2}}{2}\left[\left(\frac{1}{2}-\xi(v)\right)^2 + \left(\frac{1}{2}-\eta(v)\right)^2\right] \tag{K.3}$$

where E_0 is the field strength that would exist at the same distance in free space, i.e. in the absence of the edge, while $\xi(v)$ and $\mu(v)$ are defined by the Fresnel integrals:

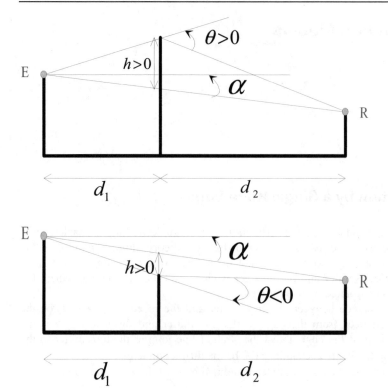

Fig. 1. Fresnel diffraction by a single knife edge

$$\xi(v) = \int_0^v \cos\frac{\pi t^2}{2}dt \qquad\qquad (K.4)$$

$$\mu(v) = \int_0^v \sin\frac{\pi t^2}{2}dt \qquad\qquad (K.5)$$

where v is the dimensionless parameter of the Fresnel-Kirchhoff diffraction formula, which expresses the obstruction by the obstacle of the direct path between the transmitter and the receiver.

The development in series of the Fresnel integrals leads to the following approximate relations:

– if $v \leq -0.71$:

$$A_{dB} = 0 \qquad\qquad (K.6)$$

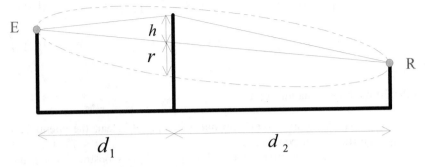

Fig. 2. Radius of the Fresnel ellipsoid

– if $-0.71 < v \le 2.3$:

$$A_{dB} = -6.9 + 20\log\left[\sqrt{(v-0.1)^2)+1} - v + 0.1\right]$$ (K.7)

– if $v > 2.3$:

$$A_{dB} = -12.95 - 20\log v$$ (K.8)

Some authors consider the $q = h/r$ parameter related to the Fresnel-Krichhoff parameter v, where r represents the radius at the level of the edge of the Fresnel ellipsoid with foci colocated at the emitter E and receiver R. The equation for this parameter is as follows:

$$q = \frac{h}{r} = \frac{v}{\sqrt{2}} = h\sqrt{\frac{1}{\lambda}\left(\frac{1}{d_1} + \frac{1}{d_2}\right)} = h\sqrt{\frac{(d_1 + d_2)}{\lambda d_1 d_2}}$$ (K.9)

The geometry of the Fresnel ellipsoid is represented here in Fig. 2.
  In practice the following expressions can be used (Deygout 1992):

$$A_{dB} = 0 \qquad \text{if } \frac{h}{r} < -0.5,$$ (K.10)

$$A_{dB} = 6 + 12\frac{h}{r} \qquad \text{if } -0.5 \le \frac{h}{r} < 0.5,$$ (K.11)

$$A_{dB} = 8 + 8\frac{h}{r} \qquad \text{if } 0.5 \le \frac{h}{r} < 1, \qquad\qquad \text{(K.11)}$$

$$A_{dB} = 16 + 20\log_{10}\left(\frac{h}{r}\right) \qquad \text{if } \frac{h}{r} > 1 \qquad\qquad \text{(K.12)}$$

It results from it these equations that:

-   if $h/r < -1$, the first Fresnel ellipsoid is completely clear: the attenuation due to diffraction is equal to zero,
-   if $-1 < h/r < 0.5$, the first Fresnel ellipsoid is slightly obstructed, and the direct path just grazes the obstacle: the attenuation due diffraction remains equal to zero,
-   if $-0.5 < h/r < 0$, the first Fresnel ellipsoid is obstructed, but the transmitter and the receiver are in optical line-of-sight, i.e. the direct path has some clearance: the attenuation due to diffraction increases from 0 to 6 dB in proportion to h/r. The value 6 dB is reached when the edge is tangent to the line-of-sight path between the emitter and the receiver. Since in this case the Fresnel-Kirchhoff parameter v is negative, the term negative height is used to describe this situation.
-   if $0 < h/r$, the first Fresnel ellipsoid is obstructed and the line-of-sight path is blocked: theoretically attenuation increases with h/r to infinity. In this case, the Fresnel-Kirchhoff parameter v is negative (positive height).

These four cases are successively represented in Fig. 3

2 Diffraction by Multiple Knife Edges

2.1 Theoretical Methods

Millington Method

This method expresses the excess attenuation caused by diffraction by two knife edges compared with attenuation in free space A_{dB} in the form of a surface Fresnel integral (Millington 1962). A schematic representation of this method is presented in Fig. 4. The attenuation due to equation is given by the following equation, which can be applied provided that $d_1, d_2\ d_3 > > h_1, h_2, \lambda$:

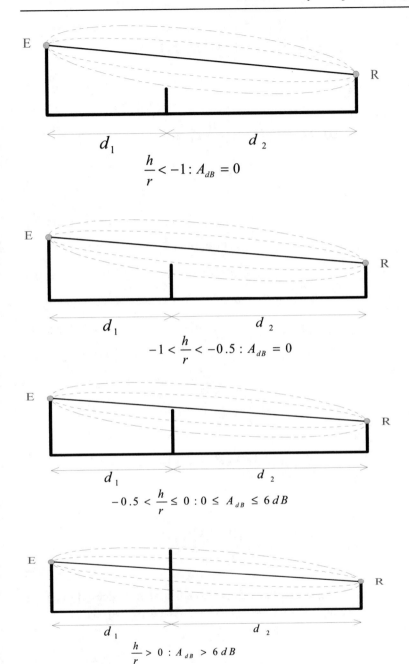

$$\frac{h}{r} < -1 : A_{dB} = 0$$

$$-1 < \frac{h}{r} < -0.5 : A_{dB} = 0$$

$$-0.5 < \frac{h}{r} \le 0 : 0 \le A_{dB} \le 6\,dB$$

$$\frac{h}{r} > 0 : A_{dB} > 6\,dB$$

Fig. 3. Radius of the Fresnel ellipsoid. The first ellipsoid is: *a.* clear *b.* slightly obstructed *c.*obstructed *d.* very obstructed.

$$A_{dB} = \frac{j}{2}\left[G(\rho_0,\gamma_1)+G(\rho_0,\gamma_2)\right] \tag{K.15}$$

with:

$$G(\rho_0,\gamma) = \int_S \exp\left(j\frac{\pi}{2}\rho_0^2\right)ds \tag{K.16}$$

where:

$$\gamma_1 = \tan^{-1}\left(\frac{p_2\sin(\alpha_{12})}{p_1-p_2\cos(\alpha_{12})}\right) \tag{K.17}$$

$$\gamma_2 = \tan^{-1}\left(\frac{p_1\sin(\alpha_{12})}{p_2-p_1\cos(\alpha_{12})}\right) \tag{K.18}$$

$$p_1 = \left[\frac{k(d_1+d_2+d_3)}{\pi d_1(d_2+d_3)}\right]^{\frac{1}{2}}h_1 \tag{K.19}$$

$$p_2 = \left[\frac{k(d_1+d_2+d_3)}{\pi d_3(d_1+d_2)}\right]^{\frac{1}{2}}h_2 \tag{K.20}$$

$$\alpha_{12} = \tan^{-1}\left(\left[\frac{d_2(d_1+d_2+d_3)}{d_1 d_3}\right]^{\frac{1}{2}}\right) \tag{K.21}$$

$$\rho_0 = \frac{\left(p_1^2+p_2^2-2p_1p_2\cos(\alpha_{12})\right)^{\frac{1}{2}}}{\sin(\alpha_{12})} \tag{K.22}$$

Parameters p_1 and p_2 represent the relative weights of the edges A_1 and A_2 respectively, while α_{12} accounts for the distance between these two edges.

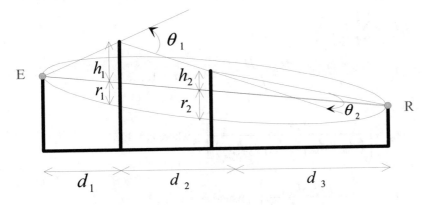

Fig. 4. Radius of the Fresnel ellipsoid (Millington method)

Vogler Method

This method expresses the excess attenuation due to the diffraction by several thin edges compared to atttenuation in free space (A_{dB}) in the form of a multiple integral developed in series in order to facilitate numerical applications (Vogler 1982):

$$A_{dB} = \frac{\exp(\sigma_N)}{2^N} C_N \left(\frac{2}{\sqrt{\pi}}\right)^N \int_{\beta_1}^{+\infty} \dots \int_{\beta_N}^{+\infty} \exp(2f) \exp\left(-\sum_{m=1}^{N} x_m^2\right) dx_1 \dots dx_r \quad \text{(K.23)}$$

where:

- N is the number of edges,

- $f = 0$ if $N = 1$ and $f = \sum_{m=1}^{N-1} \alpha_m (\alpha_m - \beta_m)(x_{m+1} - \beta_{m+1})$ if $N > 1$,

- $C_N = 1$ if $N = 1$ and $C_N = \left[\dfrac{\displaystyle\prod_{m=2}^{N} d_m \sum_{n=1}^{N+1} d_n}{\displaystyle\prod_{m=2}^{N}(d_m + d_{m+1})} \right]^{\frac{1}{2}}$ if $N > 1$,

- $\sigma_N = \sum_{m=1}^{N} \beta_m^2$,

$$- \quad d_T = \sum_{m=1}^{N+1} d_m,$$

$$- \quad \theta_m = \frac{H_m - H_{m-1}}{d_m} + \frac{H_m - H_{m+1}}{d_{m+1}} \quad \text{where } m = 1, ..., N,$$

$$- \quad \alpha_m = \left[\frac{d_m d_{m+2}}{(d_m + d_{m+1})(d_{m+1} + d_{m+2})} \right]^{\frac{1}{2}}, \text{ where } m = 1, ..., N-1,$$

$$- \quad \beta_m = \theta_m \left[\frac{jk d_m d_{m+1}}{2(d_m + d_{m+1})} \right]^{\frac{1}{2}}, \text{ where } m = 1, ..., N.$$

The parameters α_m account for the distances between the edges, while the parameters β_m account for the obstruction of the direct path by each edge. The solution is valid provided that $|\theta_m| < 12°$, $kd_m > 1000$ and $\beta_m > \beta_{min} = -0.89\,j^{1/2}$.

2.2 Approximate Methods

These methods rest on the principle of superposition which asserts that the diffraction by several knife edges can be treated like a succession of single knife edge diffractions. In the methods to be hereafter described (Epstein-Peterson method, Giovanelli method, Shibuya method, Deygout method), the diffraction by two knife edges will therefore be first considered, before extending the solution to the case of N successive edges.

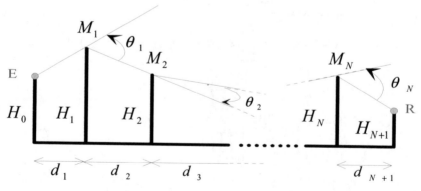

Fig. 13. Radius of the Fresnel ellipsoid (Vogler method)

Epstein-Peterson Method

This method is based on the assumption that each edge is illuminated by a transmitter or by the energy diffracted over the previous edge. The total attenuation is the sum of the attenuations induced by the diffraction by each edge (Epstein 1953)

Two edge case. The following equation yields the attenuation due to diffraction by two knifes edges M_1 and M_2:

$$A_{dB} = A_{dB}(M1) + A_{dB}(M2) \qquad \text{(K.24)}$$

The attenuations due to the diffraction by the edges M_1 and M_2 are given by the equations:

$$A_{dB}(M_1) = f(d_1, d_2, h_1) \qquad \text{(K.25)}$$

$$A_{dB}(M_2) = f(d_2, d_3, h_2) \qquad \text{(K.26)}$$

where h_1 and h_2 are defined as follows:

$$h_1 = (H_1 - H_0) - d_1 \frac{H_2 - H_0}{d_1 + d_2} \qquad \text{(K.27)}$$

$$h_2 = (H_2 - H_1) - d_2 \frac{H_3 - H_1}{d_2 + d_3} \qquad \text{(K.28)}$$

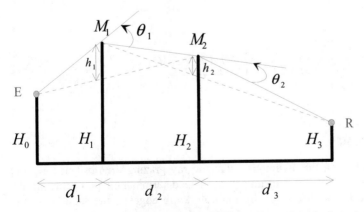

Fig. 5. Radius of the Fresnel ellipsoid: two edge case (Epstein-Peterson method)

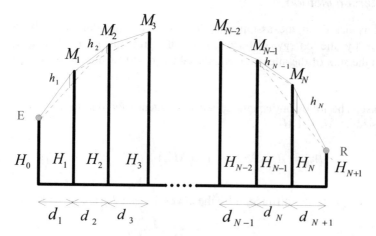

Fig. 6. Radius of the Fresnel ellipsoid: generalisation to N edges (Epstein-Peterson method)

Generalisation to N edges. The generalisation of the Epstein-Peterson method to the case of N edges leads to the following equation:

$$A_{dB} = \sum_{k=1}^{N} A_{dB}(M_k)$$ (K.29)

The attenuation due to each individual edge M_{Ik} is given by the equation:

$$A_{dB}(M_k) = f(d_k, d_{k+1}, h_k)$$ (K.30)

where:

$$h_k = (H_k - H_{k-1}) - \frac{d_k(H_{k+1} - H_{k-1})}{d_k + d_{k+1}}$$ (K.31)

Shibuya Method

The method relies on the assumption that the ray grazing the obstacles M_k and M_{k+1} generates a fictitious transmitter E_k. The determination procedure of the attenuation due to the diffraction by multiple knife edges is the same as in the Epstein-Peterson method with the difference however that the transmitter E is replaced here by a fictitious transmitter E_k (Shibuya 1983).

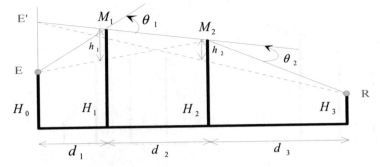

Fig. 8. Radius of the Fresnel ellipsoid: two edge case (Shibuya method)

Two edge case. The following equation yields the attenuation due to diffraction:

$$A_{dB} = A_{dB}(M1) + A_{dB}(M2) \tag{K.32}$$

The attenuations due to the diffraction by the edges M_1 and M_2 are given by the equations:

$$A_{dB}(M_1) = f(d_1, d_2, h_1) \tag{K.33}$$

$$A_{dB}(M_2) = f(d_1 + d_2, d_3, h_2) \tag{K.34}$$

where the heights h_1 and h_2 are as defined by the relations:

$$h_1 = (H_1 - H_0) - d_1 \frac{H_2 - H_0}{d_1 + d_2} \tag{K.35}$$

and:

$$h_2 = (H_2 - H_{E1}) - (d_1 + d_2) \frac{H_3 - H_{E1}}{d_1 + d_2 + d_3} \tag{K.36}$$

In these equations, H_{E1} represents the height of the fictitious transmitter, as defined by the relation:

$$H_{E1} = H_1 + d_1 \frac{H_1 - H_2}{d_2} \tag{K.37}$$

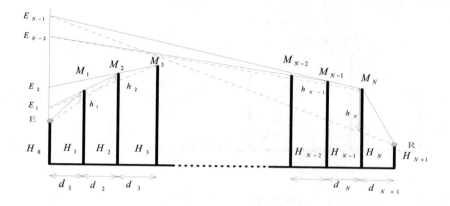

Fig. 8. Radius of the Fresnel ellipsoid: generalisation to N edges (Shibuya method)

Generalisation to N edges. The generalisation of the Shibuya method to N edges leads to the following equation:

$$A_{dB} = \sum_{k=1}^{N} A_{dB}\left(M_k\right) \tag{K.38}$$

where the attenuation due to each individual edge M_{lk} is given by the equation:

$$A_{dB}\left(M_k\right) = f\left(d_1 + ... + d_k, d_{k+1}, h_k\right) \tag{K.39}$$

where the heights h_k are given by the equations:

$$h_k = \left(H_k - H_{E_{k-1}}\right) - \frac{\left(d_1 + ... + d_k\right)\left(H_{k+1} - H_{E_{k-1}}\right)}{d_1 + ... + d_{k+1}} \tag{K.40}$$

while the heights of the fictitious transmitters are recursively defined as follows:

$$H_{E1} = H_1 + d_1 \frac{H_1 - H_2}{d_2} \tag{K.41}$$

$$H_{E_k} = H_k + \left(d_1 + ... + d_k\right)\frac{H_k - H_{k+1}}{d_{k+1}} \tag{K.42}$$

The Shibuya and Epstein-Peterson formulas are related to one another through the following relation:

$$A_{dB}\left(Shibuya\right) = A_{dB}\left(Epstein-Peterson\right) + 10\log_{10}\left[\frac{\displaystyle\prod_{k=1}^{N-1}(d_k + d_{k+1})}{\left(\displaystyle\prod_{k=2}^{N-1}d_k\right)\left(\displaystyle\prod_{k=1}^{N}d_k\right)}\right] \tag{K.43}$$

Deygout Method

This method is based on the assumption that one of the edges is dominant: this edge is primarily responsible for the attenuation due to diffraction, the other edges playing but a secondary role (Deygout 1966).

Two edge case. Let us here consider two knife edges M_1 and M_2. The determination of the main edge proceeds from the consideration of the parameter $q = h/r$ defined with respect to the radius of the first Fresnel ellipsoid. Fig. 9 represents the geometry associated with this method. The edge corresponding to the highest value of the parameter q is defined as the main edge, while the other edges are considered as secondary edges.

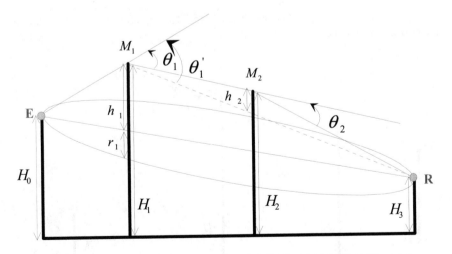

Fig. 9. Radius of the Fresnel ellipsoid: M_1 is the main edge (Deygout method)

1. Let us first assume that M_1 has been defined as the main edge. In this case, the following equation yields the attenuation due to diffraction:

$$A_{dB} = A_{dB}(M1) + A_{dB}(M2) \tag{K.44}$$

The attenuations due to the diffraction by the edges M_1 and M_2 are given by the equations:

$$A_{dB}(M_1) = f(d_1, d_2 + d_3, h_1) \tag{K.45}$$

$$A_{dB}(M_2) = f(d_2, d_3, h_2) \tag{K.46}$$

where the heights h_1 and h_2 are as defined by the relations:

$$h_1 = (H_1 - H_0) - d_1 \frac{H_R - H_0}{d_1 + d_2 + d_3} \tag{K.47}$$

and:

$$h_2 = (H_2 - H_1) - d_2 \frac{H_3 - H_1}{d_2 + d_3} \tag{K.48}$$

2. Let us now assume that M_2 is the main edge. The equation for the attenuation due to diffraction becomes in this case:

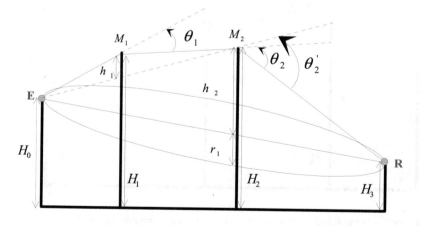

Fig. 10. Radius of the Fresnel ellipsoid: M_2 is the main edge (Deygout method)

$$A_{dB} = A_{dB}(M1) + A_{dB}(M2) \tag{K.49}$$

The attenuations due to the diffraction by the edges M_1 and M_2 are given by the two following equations:

$$A_{dB}(M_1) = f(d_1, d_2, h_1) \tag{K.50}$$

$$A_{dB}(M_2) = f(d_1 + d_2, d_3, h_2) \tag{K.51}$$

where the heights h_1 and h_2 are as defined by the relations:

$$h_1 = (H_1 - H_0) - d_1 \frac{H_2 - H_0}{d_1 + d_2} \tag{K.52}$$

$$h_2 = (H_2 - H_1) - (d_1 + d_2) \frac{H_3 - H_0}{d_1 + d_2 + d_3} \tag{K.53}$$

The attenuation values obtained with this method are higher than the values obtained with Millington rigorous method. This led to the introduction of a corrective factor T_c, defined as follows (Deygout 1991):

$$\tag{K.54}$$

$$T_c = \left[12 - 20 \log_{10} \left(\frac{2}{1 - \delta/\pi} \right) \right] \left[\frac{q}{p} \right]^{2p}, 0 < q \le p$$

$$T_c = 0, \ q \le 0 \text{ or } p = 0 \tag{K.55}$$

where:

$$\tan(\delta) = \sqrt{\frac{d_2(d_1 + d_2 + d_3)}{d_1 d_3}} \tag{K.56}$$

The equation for the loss path is thus:

$$A_{dB} = A_{dB}(M1) + A_{dB}(M2) - T_c \tag{K.57}$$

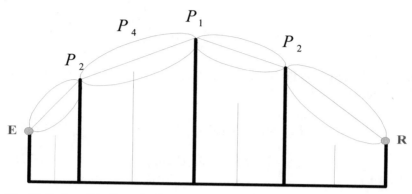

Fig. 11. Radius of the Fresnel ellipsoid: generalisation to N edges (Deygout method)

Generalisation to N edges. In order to extend this method to the case of N successive knife edges, the following procedure, schematically represented in Fig. 11, is applied:

1. one first determines the main edge P_1 between the transmitter E and the receiver R, i.e. the edge at which the ratio h/r assumes its highest value. The attenuation A_{ER} (P_1) associated with this edge is determined using Deygout approximate method for the diffraction by a single knife edge.
2. following the same procedure, one determines at the left and right of the edge P_1 a main edge P_2 between E and P_1, and a main edge P_3 between P_1 and R. The attenuations associated with these two edges are $A_{EP_1}\left(P_2\right)$ and $A_{P_1R}\left(P_2\right)$ respectively.
3. considering only the couples of edges consisting of a main edge and a secondary edge, i.e. in this case (P_1, P_2) and (P_1, P_3), two corrective terms $T_{c_{ER}}\left(P_1,P_2\right)$ and $T_{c_{ER}}\left(P_1,P_3\right)$ are calculated. The couples consisting of two secondary edges, i.e. in this case (P_2, P_3), are not considered, since the distance between these edges is large enough as to allow for neglecting the associated corrective term.
4. the resulting attenuations are then added.
5. steps 2, 3 and 4 are repeated at the left and right of the edges P_1 and P_2, until all the edges considered along the path between E and R are in line-of-sight, i.e. no edge is present in the first Fresnel ellipsoid between two such edges.

The equation for the total attenuation due to diffraction is therefore:

$$A_{dB} = A_{ER}\left(P_1\right) + A_{Ep_1}\left(P_2\right) - T_{c_{ER}}\left(P_1,P_2\right) + A_{P_1R}\left(P_3\right) - T_{c_{ER}}\left(P_1,P_3\right) + \dots \quad \text{(K.58)}$$

Giovanelli Method

Like the Deygout method, this method is based on the assumption that one of the edges is dominant and is primarily responsible for the attenuation due to diffraction, the other edges having but a secondary influence (Giovanelli 1984).

Two edge case. Let us here consider the case of two knife edges M_1 and M_2. The determination of the main edge proceeds from the consideration of the parameter q = h/r defined with respect to the radius of the first Fresnel ellipsoid. The edge corresponding to the highest value of the parameter q is defined as the main edge, while the other edges are considered as secondary edges.

1. We first consider the case, represented in Fig. 12, where M_1 is defined as the main edge.

The projection of the ray originating at the apex point of the edge M_1 and grazing the edge M_2 creates a fictitious receiver R' within the plane P_r. The Fresnel-Kirchhoff parameter ν is a function of the diffraction angle θ_1, and does not depend on the angle θ'_1. The effective height h_1 of the principal obstacle is determined from the obstruction by the edge M_1 of the Fresnel ellipsoid with foci at E and R':

$$h_1 = (H_1 - H_E) - d_1 \frac{H_{R'} - H_E}{d_1 + d_2 + d_3} \qquad (\text{K.59})$$

$$H_{R'} = H_1 + (d_2 + d_3) \frac{H_2 - H_1}{d_2} \qquad (\text{K.60})$$

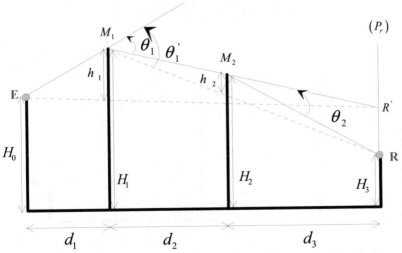

Fig. 12. Radius of the Fresnel ellipsoid: M_1 is the main edge (Giovanelli method)

The secondary obstacle M_2 is then considered along the propagation path M_1M_2R, with a diffraction angle θ_2 and an effective height h_2 given by the equation:

$$h_2 = (H_2 - H_1) - d_2 \frac{H_R - H_1}{d_2 + d_3} \qquad (K.61)$$

2. Let us now consider the case, represented in Fig. 13, where M_2 has been defined as the main edge.

The projection of the ray originating from the apex point of the edge M_1 and grazing the edge M_2 creates a fictitious transmitter E' within the plane E_p. The Fresnel-Kirchhoff parameter v is a function of the diffraction angle θ_2, and does not depend on the angle θ'_2. The effective height h_2 of the main obstacle is determined from the obstrcution by the edge M_2 of the Fresnel ellipsoid with loi located at E' and R:

$$h_2 = (H_2 - H_{E'}) - (d_1 + d_2) \frac{H_3 - H_{E'}}{d_1 + d_2 + d_3}, \qquad (K.62)$$

$$H_{E'} = H_1 + (d_1) \frac{H_1 - H_2}{d_2}. \qquad (K.63)$$

The secondary obstacle M_1 is then considered along the propagation path EM_2R, with a diffraction angle θ_1 and an effective height h_1 given by the equation:

$$h_1 = (H_1 - H_E) - d_1 \frac{H_2 - H_E}{d_1 + d_2}. \qquad (K.64)$$

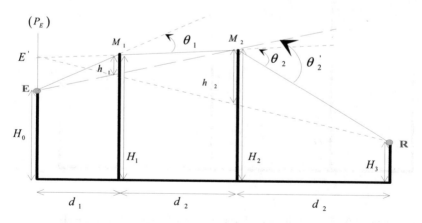

Fig. 13. Radius of the Fresnel ellipsoid: M_2 is the main edge (Giovanelli method)

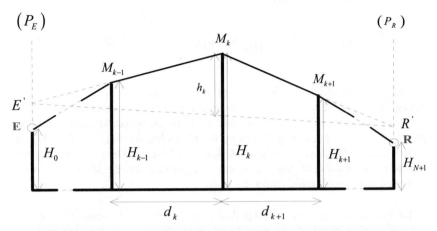

Fig. 14. Radius of the Fresnel ellipsoid : generalisation to N edges (Giovanelli method)

Generalisation to N edges. The determination of the attenuation due to the diffraction by N successive edges is based on the repetition of the two edge procedure. Different cases are to be considered, depending on which edge is the main obstacle between E and R:

-   if the main obstacle is an edge M_k located between the first edge M_1 and the last edge M_N, two observation planes P_E and P_R are introduced at the level of the transmitter E and receiver R. The projection of the light ray which grazes the edges M_k and M_{k-1} is unaffected by the obstacles M_{k-2} M_{k-3} M_2 M_1 and thus defines a fictitious transmitter E' within the plane P_E. Likewise the projection of the light ray which grazes the edges M_k and M_{k+1} is unaffected by the obstacles M_{k+2} M_{k+3} M_{N-1} M_N and thus defines a fictitious receiver R' within the plane P_R. The effective height h_k of the main obstacle M_k is determined from the obstruction by this edge of the Fresnel ellipsoid with foci located at E' and R'. This construction is then extended to the propagation paths located left and right of the main obstacle, by introducing a new observation plane at the level of M_k, with the same function as P_R for determining a new main edge at the left of M_k and as P_E for determining a new main edge at the right of M_k respectively. The procedure is then repeated. The equations for the heights of the fictitious transmitters and of the receivers are:

$$H_{E'} = H_k + \left(d_1 + \ldots + d_k\right)\tfrac{H_k - H_{k+1}}{d_{k+1}},$$ (K.65)

$$H_{R'} = H_k + (d_{k+1} + ... + d_{N+1})\frac{H_{k+1} - H_k}{d_{k+1}}, \tag{K.66}$$

$$h_k = (H_k - H_{E'}) - (d_1 + ... + d_k)\frac{H_{R'} - H_{E'}}{d_1 + ... + d_{k+1}}. \tag{K.67}$$

- if the main obstacle is the first edge M_1 located immediately after the transmitter, an observation plane P_R is introduced at the level of the receiver R: the projection of the light ray originating from the secondary source M_1 and grazing the edge M_2, is unaffected by the obstacles M_3, M_4 ..., M_{N-1}, M_N and thus defines a fictitious receiver R' within the plane P_R. The effective height h_1 of the main obstacle is determined from the obstruction by M_1 of the Fresnel ellipsoid with foci located R and R'.

- if the main obstacle is the last edge M_N located immediately before the receiver, an observation plane P_E is introduced at the level of the transmitter E: the projection of the light ray originating from the secondary source M_N and grazing the edge M_{N-1} is unaffected by the obstacles M_{N-2}, M_{N-3} ..., M_2, M_1 and thus defines a fictitious transmitter E' within the plane P_E. The effective height h_1 of the main obstacle is determined from the obstruction by M_N of the Fresnel ellipsoid with foci located at E' and R..

3 Diffraction by a Single Rounded Obstacle

In general, the diffraction caused by an actual obstacle located between the transmitter and the receiver is more adequately represented by a rounded obstacle than by a simple knife-edge obstacle. The vicinity of the obstacle will be represented here by a cylinder with a radius equal to the radius of curvature of the relief.

3.1 Wait Method

For a positive and small angle θ, and assuming that the transmitter and the receiver are both at a relatively large distance from the edge, the attenuation caused by the diffraction over a rounded obstacle can be expressed as the sum of three terms (Wait 1959):

$$A_{dB}(v, \rho) = A_{dB}(v, 0) + A_{dB}(0, \rho) + U(\chi), \tag{K.68}$$

where:

– v is the Fresnel-Kirchhoff diffraction parameter, as defined by the equation

$$v = \sqrt{\frac{2(d_1 + d_2)}{\lambda d_1 d_2}}, \qquad \text{(K.69)}$$

– ρ is a parameter accounting for the radius of curvature R of the relief:

$$\rho = \left(\frac{\lambda}{\pi}\right)^{\frac{1}{6}} R^{\frac{1}{3}} \left(\frac{d_1 + d_2}{d_1 d_2}\right)^{\frac{1}{2}}, \qquad \text{(K.70)}$$

– χ is a parameter proportional to the product of the first two parameters (v and ρ):

$$\chi = \left(\frac{\pi R}{\lambda}\right)^{\frac{1}{3}} \theta \cong \sqrt{\frac{\pi}{2}} v\rho, \qquad \text{(K.71)}$$

where:

   - $A(v, 0) = f(d_1, d_2, h)$ is the attenuation caused by the obstacle M considered as a sharp edge,
   - for $\rho < 3$, $A(0, \rho) \cong 7.2\rho - 2\rho^2 + 3.6\rho^3 - 0.8\rho^4$,
   - for $-\rho \leq \chi < 0$, $U(\chi) = A(0, \rho)\chi\rho^{-1}$,
   - for $0 \leq \chi < 4$, $U(\chi) = 12.5\chi$,
   - for $\chi \geq 4$, $U(\chi) = 17\chi - 6 - 20 \log(\chi)$.

As can be seen from Fig. 15, the radius of curvature radius can be determined using the following equation:

$$R = \frac{2D_s d_1 d_2}{\theta\left(d_1^2 + d_2^2\right)} \qquad \text{(K.72)}$$

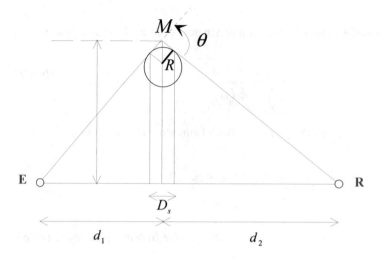

Fig. 15. Diffraction by a rounded obstacle (Wait method)

Note. If the radius of curvature R is equal to zero, then the parameters ρ and χ are cancelled: in this case, the total attenuation due to the diffraction caused by the rounded obstacle is reduced to the attenuation due to diffraction by a knife edge.

If the diffraction angle θ is equal to zero, then the parameter χ is cancelled: in this case, the total attenuation due to the diffraction caused by the rounded obstacle to the attenuation due to grazing angle knife-edge attenuation.

The previous restrictions can be reduced to the following conditions

$$\frac{2\pi}{\lambda}\sqrt{d_1 d_2} \geq 10 \qquad\qquad \text{(K.73)}$$

and:

$$\left(\frac{\pi R}{\lambda}\right)^{\frac{2}{3}} \frac{1}{R^2} d_1 d_2 \geq 0.1 \qquad\qquad \text{(K.74)}$$

3.2 ITU-R Method

The following equation yields the attenuation due to the diffraction by a rounded obstacle with radius R as represented in Fig. 16:

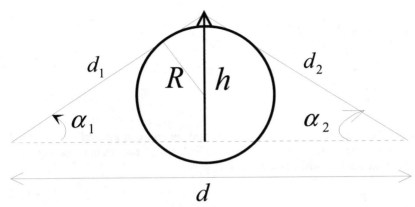

Fig. 16. Diffraction by a rounded obstacle (ITU-R method)

$$A_{dB} = J(v) + T(m,n)$$ (K.75)

where $J(v)$ is the Fresnel-Kirchoff attenuation induced by an equivalent knife edge whose apex point would be at the intersection point of the tangent rays. For $v > -0.7$:

$$J(v) = 6.9 + 20\log\left(\sqrt{(v-0.1)^2 + 1} + v - 0.1\right)$$ (K.76)

The following equation yields the dimensionless parameter v:

$$v = 0.0316h\left[\frac{2(d_1 + d_2)}{\lambda d_1 d_2}\right]^{\frac{1}{2}}$$ (K.77)

where h and λ are expressed in metres and d_1 and d_2 in kilometres.

$T(m,n)$ is the additional attenuation due to the curvature of the obstacle:

$$T(m,n) = km^b$$ (K.79)

where:

- $k = 8.2 + 12.0n$
- $b = 0.873 + 0.27\left[1 - \exp(-1.43n)\right]$

$$- \quad m = R \left[\frac{d_1 + d_2}{d_1 d_2} \right] / \left[\frac{\pi R}{\lambda} \right]^{\frac{1}{3}}$$

$$- \quad n = h \left[\frac{\pi R}{\lambda} \right] / R$$

As can be seen, when R tends towards zero, m and consequently $T(m,n)$ tend also to zero. Therefore, in the case of a cylinder with null radius, the equation reduces to the case of diffraction by a knife edge.

References

Deygout J (1966) Multiple knife-edge diffraction of microwaves. IEEE Transactions on Antennas and Propagation vol AP-14, 4: 480-489

Deygout J (1991) Correction factor for multiple knife-edge diffraction. IEEE Transactions on Antennas and Propagation vol AP-39, 8: 1256-1258

Deygout J (1992) Evaluation de l'affaiblissement de propagation des ondes décimétriques en zone rurale. L'onde radioélectrique vol 72, 3: 56-63

Epstein J Peterson W (1953) An experimental study of wave propagation at 850 MC. Proceedings of the I.R.E. vol 41: 595-611

Giovanelli CL (1984) An analysis of simplified solutions for multiple knife-edge diffraction. IEEE Transactions on Antennas and Propagation vol AP-32, 3: 297-301

ITU-R (2000) Propagation by diffraction. Rec. ITU-R P.527-7

Millington G Hewitt R Immerzi FS (1962) Double knife-edge diffraction in field-strength predictions. Proc. IEE Monogr., 507E: 419-429

Shibuya S (1983) A Basic Atlas of Radio-Wave Propagation. John Wiley & Sons, New York

Vogler LE (1982) An attenuation function for multiple knife-edge diffraction. Radio-science vol 17, 6:1541-1546

Wait JR Conda AM (1959) Diffraction of electromagnetic waves by smooth obstacles for grazing angles. Journal of Research of the N.B.S. vol 63D

L Mobile Radio Measurements

L.1 Measurement of the Field Strength

Prediction models of the field strength are used for determining the radio coverage of the emitters, i.e. for predict the local average value of the envelope of the electric field at a given point. Measurements conducted in different environments allow the development as well as the statistical optimisation and the validation of these models. These measurements allow to record both the instantaneous and the average value over a short time interval of the radio field strength along a path travelled by a mobile, or in this case by a vehicle.

The measurement principle, schematically represented in Fig. L.1, can be summarised as follows: a vehicle (or a truck in the case of measurements carried out indoors) equipped with a field intensity measuring device and a data acquisition receiver moves along a definite set of paths. An impulse generator bound to one of the wheels of the vehicle engages measurements of the instantaneous field. The measurements are taken at fixed distance intervals rather than at fixed time intervals, since this would force the vehicle to move at a constant speed and cannot be easily achieved in practice in urban environments. The distance interval Δx is called the spatial sampling step or the measurement step. The computer controlled receiver is coupled to a data acquisition system and magnetic storage system of the field strength measured in dBμV.

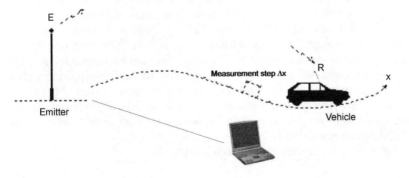

Fig. L.1. Measurement principle of the mobile radio field strength

A numerical analysis is then performed in order to associate geographical coordinates to each measurement point. This operation is realised using either a chart with a digitiser, or a geographical database with digitisation software. In general, real-time localisation systems such as the GPS system do not authorize the capture of the geographical coordinates with a sufficient degree of precision, especially in urban environments where it should be lower than one metre. The measurements thus performed are said to be as raw: they reproduce all the variations affecting the amplitude of the signal, in proportion to the envelope of the electric field. A data analysis associated with a numerical filtering is then performed in order to eliminate the fast variations due to Rayleigh fading from the slow medium-scale and large-scale variations which can be modelled and predicted.

The spatial sampling step Δx of the data acquisition system is selected so as to allow the reconstitution of the signal. The Nyquist theorem imposes a sampling no higher than $\lambda/2$, while the average length L of the interval must be selected in such a way as to minimise the local average error of the envelope of the radio field.

The mobile radio signal being assumed to be wide sense stationary over the average interval length, the latter must not be higher than the maximum distance at which the local average can be regarded as being constant. This maximum distance is a function of the local average distribution depending on the environment, which itself depends on the frequency. As an example, the effects induced by diffraction are more pronounced when the frequency increases. The sampling step is therefore empirically determined with respect to the environment (rural, mountainous, suburban, urban, interior of buildings, etc.) and to the frequency. In general an average length L equal to 40 λ is used. The local average value of the mobile radio signal is then estimated a more or less 1 dB in 90 percent of cases (Lee 1986).

L.2 Measurement of the Impulse Response

The methods used for measuring the impulse response of a system can be classified into two families: frequential methods and temporal methods.

Frequential methods measure the path loss in a given frequency band by performing a frequency scan. The intrinsic difficulty in these methods lies in obtaining the phase of the impulse response. The most commonly adopted methods, used for instance in network analysers, consists in synchronising with a cable the transmitter and the receiver. The impulse response is then obtained by Fourier transform. Frequential methods are more specifically adapted to the characterisation in broad frequency range of the propagation of millimetre waves inside buildings (Jimenez 1994).

Two different types of temporal methods exist:

- the simplest method is based on the transmission of a very short impulse as close as possible to a Dirac function. The instantaneous field measured at the reception point directly corresponds to the impulse response of the radio channel. Although this method is simple in theory, in practice its implementation turns out to be complicated: not only is it indeed difficult to emit high power impulses during a very short time, but the reception of such impulses is moreover delicate (Kauschke 1994).
- a second method consists in using a sounding signal in the form of a pseudo-random binary sequence with good autocorrelation properties and with a maximum length. This sequence modulates a carrier signal, which generally has a modulation of MDP2 type. The correlation of the received sequence with an image of the emitted sequence is then performed by the receiver: due to the autocorrelation property of sequences with maximum length sequences, a correlation peak can thus be obtained. Channel sounders with analog sliding correlators have also been used: however, these sounders were limited, most of all because they did not provide the phase information (Parsons 1992). It is indeed essential, especially in the case of hardware and software simulations, to determine the amplitude and phase of the impulse response.

With the advances in numerical analysis methods, recent sounders use in the presence of noise a digital processing related to inversion methods like the Wiener inversion. The use of such a digital processing also allows achieving both a higher resolution and a lesser sensitivity to the nonlinearities in the transmission chain. Several different systems have been developed according to this method (Kauschke 1994; Levy 1990; Zollinger 1998). The reader will find a more extensive bibliography in the COST 231 final report (COST 231 1999).

The main characteristics of a propagation channel sounder are the following ones:

- the analysis band, defined as the frequency band around the central frequency at which the measurement is performed. The analysis band must be higher of course than the frequency band planned for the useful signal. The wider the frequency band is, the better the temporal resolution of the sounder is, i.e. the possibility of separating close delays. Analysis bands typically range from approximately a few megahertz to several tens of megahertz.
- the maximum length of the impulse response: in mountainous environments, the maximum delay due to remote echoes may reach an order of a few tens of microseconds (1 μs corresponds roughly to a 300 metre path).
- the spatial or temporal sampling step which corresponds to the width of the analysable Doppler spectrum: this width is inversely proportional to the time interval separating two successive impulse responses.
- the measurement dynamics, which can be expressed through the relation between the power of the strongest path and the noise threshold. The dynamics generally obtained with present day propagation channel sounders are of the order of 30 dB.

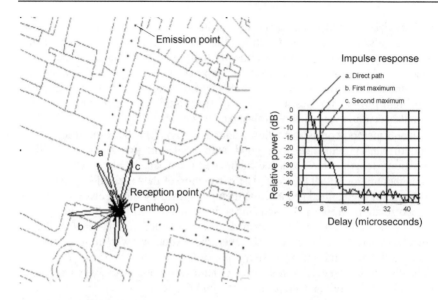

Fig. L.2. Measurement of the impulse response and of the received power depending on the angles of arrival at a point of a path in a microcell context (Paris, Pantheon district, 900 MHz, CNET propagation channel sounder). *a.* direct path, *b.* path with a reflection at a building, *c.* path with diffraction over roofs

L.3 Measurement of the Directions of Arrival

These measurements are intended at evaluating the angular distribution of the received energy, i.e. at determining the direction of arrival of the waves. The simplest idea consists in using a very directive rotating antenna. This however cannot be achieved at these frequencies. Although the use of a multi-sensor array antenna would be an ideal solution, its implementation would require N parallel measurement chains. Therefore, a simulated antenna array consisting of a rotating arm device coupled to a reference antenna is used in practice for the measurement of directions of arrival. The environment is assumed to be stationary during the time of rotation of the arm. This measurement can be demonstrated to be equivalent to a multi-sensor measurement along a circular antenna array.

In a narrow frequency range, the signal is subject to Rayleigh fading: therefore, no privileged direction of arrival of the signal can be identified at these frequencies, whereas in a broad frequency range, the instantaneous power is subject to greatly attenuated fast variations resulting from the frequency selectivity of the channel. This method is therefore generally adopted (Rossi 1997). The results, presented in the form of polar diagrams with an amplitude scale expressed in dB, allow the validation of prediction models of the field strength, using for instance the profile method or the ray launching method. A better understanding of the angular distribution of energy at the level of the base station enables the development of new spatial techniques based on the use of adaptive antennas, like the beam-forming method or the *spatial division multiple access* (SDMA) method (Guisnet 1996; Klein 1996; Thomson 1996; Ertel 1998; Guisnet 1999; Pajusco 1998). For more detail on the subject of directions of arrival, their experimental determination, their mathematical modelling as well as the different linear and high-resolution determination methods of the directions of arrival, the reader is referred to Appendix M.

L.4 AMERICC Propagation Channel Sounder

In order to conduct studies on propagation as well as set up the models integrated to the radio engineering devices, France Telecom R&D defined and developed in its laboratories in Belfort a propagation channel sounder called *AMERICC* (*Appareil de mesure de réponse impulsionnelle pour la caractérisation du canal radio,* impulse response measurement system for the characterisation of the radio channel). This device allows a more accurate characterisation of the radio propagation channel over a bandwidth ranging from 0 to 250 MHz around a carrier frequency ranging from 1.9 to 60 GHz (Conrat 2002).

This sounder can be used for the characterisation of the different multiple paths as regards their lengths, directions of arrival, amplitudes and phase terms. This very complete characterisation of the propagation channel allows defining with great precision the quality of future communication systems (UMTS, LMDS, HIPERLAN, etc.) and is absolutely essential for the development of realistic propagation models for the study of multiple-input multiple-output radio communication systems.

The technical features of this sounder include:

- a space resolution of the order of one metre owing to its data acquisition system at the 1 GHz frequency,
- a multi-sensor array system with ten antennas at the reception point,
- a carrier frequency adjustable from 1.9 GHz to 60 GHz,
- GPS positioning and remote piloting.

References

Conrat JM, Thiriet JY, Pajusco P (2002) AMERICC, l'outil de mesure du canal large bande radioélectrique développé par France Télécom R&D. 4ièmes journées d'études Propagation électromagnétique dans l'atmosphère du décamétrique à l'angström, Rennes

COST231 (1999) Evolution of land mobile radio (including personal) communications. Final report, Information, technologies and Sciences, European Commission

Ertel RB, Cardieri P, Sowerby KW, Rappaport TS, Reed JH (1988) Overview of spatial channels for antenna array communications systems. IEEE Personal Communications.

Guisnet B, Verolleman Y (1996) Evaluation of different methods of DOA using a circular array applied to indoor environments. Vehicular Technology Conference, Atlanta, Georgia

Guisnet B, Perreau X (1999) Validation of an accurate direction of arrival measurement set-up and experimental exploitation. 3rd European Personal Mobile Communications Conference, Paris

Jimenez J (1994) CODIT Propagation activities and simulation models. Proc. RACE MPLA Workshop, Amsterdam, pp 673-677

Kauschke U (1994) Wideband indoor channel measurements for DECT transmitter positioned inside and outside office building. Proc. PIMRC'94 and WIN, The Hague, vol IV pp 1070-1074

Klein A, Mohr W, Thomas R ,Weber P, Wirth B (1996) Direction of arrival of partial waves in wideband mobile radio channels for intelligent antenna concepts. Vehicular Technology Conference, Atlanta, Georgia

Lee WCY (1986) Mobile Communications Engineering. McGraw-Hill, New-York

Levy AJ, Rossi JP, Barbot JP, Martin J (1990) An improved sounding technique applied to wideband mobile 900 MHz propagation measurements. Proc. 40th IEEE Vehicular Technology Conference, Orlando, USA

Pajusco P (1998) Experimental characterisation of DOA at the station in rural and urban area. Vehicular Technology Conference, Ottawa, Ontario, Canada

Parsons JD (1992) The Mobile Radio Propagation Channel. Pentech Press Publishers

Rossi JP, Barbot JP, Levy AJ (1997) Theory and measurement of the angle of arrival and time delay of UHF radiowaves using a ring array. IEEE Transactions on Antennas an Propagation, vol 45 5: 876-884

Thompson J, Grant P, Mulgrew B (1996) Smart antenna arrays for CDMA systems. IEEE Personal Communications

Zollinger E (1988) Measured in house radio wave propagation characteristics for wideband communications systems. 8th European Conference on Electrotechnics, EUROCOM'88, Stockholm, Sweden, pp 314-317

M Directions of Arrival

M.1 Introduction

The actual knowledge of the directions of arrival of radio waves leads to a better understanding of propagation situations as well as to a better modelling of the propagation channel. The determination methods of the directions of arrival are part of the programmes developed for modelling propagation by ray tracing and ray launching.

The use of multisensor array antennas at the base stations of mobile radio communication networks may induce different advantages, including:

- the reduction of the number of base stations required for the coverage of low traffic areas,
- the improvement of the radio coverage in order to provide a greater penetration inside buildings,
- a capacity increase for a constant number of base stations in high traffic microcell environment.

The use of antenna arrays at base stations can be exploited in order to improve and reduce interferences. The *space division multiple access technique* (SDMA) leads to the optimisation of cellular networks: this technique consists, through the spatial discrimination of the radio paths, in simultaneously communicating with several mobiles using the same time interval and the same frequency band. This results in an increase of the bandwidth traffic within a given frequency band and for a fixed number of sites. This leads to the possibility of intelligent array antennas capable of modifying the orientation of the principal lobe in order to maximise the signal on noise ratio of the signals at the reception.

M.2 Experimental Determination of Directions of Arrival

The directions of arrival can be experimentally determined by simultaneously measuring the impulse response with several different sensors and by comparing the phase evolution for each sensor.

Several data acquisition systems rely on the concept of a linear antenna array. This concept has enabled the development of most methods used for the determination of directions of arrival. However these systems were found not to be very practical during measurement campaigns.

A recently developed method is based on the coupling at the reception point of a rotating antenna system to the channel sounder. For the sake of both convenience and cost-effectiveness, this system is not a real antenna array but an omnidirectional antenna describing a circle with radius r_0. Measurements are performed at constant angular step. In order to achieve a perfect synchronisation between the transmitter and the channel sounder, a second antenna used as a time reference is inserted in the measurement device. For each position of the antenna, two simultaneous measurements are therefore performed, along the circle and at the reference antenna. The value measured along the circle is then phase readjusted with respect to the value measured at the reference antenna. Assuming the propagation channel to be stationary during the time that the antenna moves along the circle, this phase adjustment can be demonstrated to be equivalent to a simultaneous measurement carried out on an antenna array.

M.3 Mathematical Modelling of the Signal

In broad frequency range, the received signals arrive at the antenna array with a delay τ. If $\Delta\tau$ is the time resolution of the impulse response, then the matrix containing the direction of arrival information is only weakly time-dependent, provided however that $\Delta\tau$ satisfies the following condition:

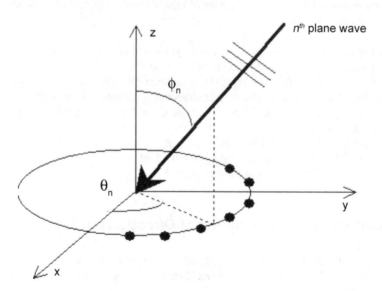

Fig. M.1. Circular antenna array

$$\frac{r_0}{c} \ll \Delta\tau \tag{M.1}$$

The time resolution $\Delta\tau$ of the impulse response is directly dependent on the frequency band B for which the impulse response was evaluated, as shown by the following equation:

$$\Delta\tau = \frac{1}{2B} \tag{M.2}$$

Let us here consider an array of M antennas distributed regularly along the circumference of a circle with radius r_0. Let us then assume that a given emission source has been chosen and that N multiple paths arrive at the antenna array. It shall be further assumed that the multiple paths correspond to vertically polarised plane waves arriving at the antenna array with an elevation φ_n and an azimuth θ_n.

The channel sounder performs measurements at a constant angular step along the circle described by the omni directional antenna at the angles $2\pi\, m\,/\,N$, where $m \in \{0, 1, ..., M-1\}$.

In far field assumption, the field received by the mth antenna is expressed in the following way:

$$y_m = E_z \exp\left(j\frac{2\pi r_0}{\lambda} \sin\varphi\cos\left(\theta - \frac{2\pi m}{M}\right)\right) \tag{M.3}$$

where $m \in \{0, 1, ..., M-1\}$ and E_z is the field received at the origin.

A white noise with variance σ_b^2 whose components are assumed to be independent from one sensor to another is superimposed over the signal. The output signal can therefore be written in the following way:

$$y_m = \sum_{n=1}^{N} a\left(\theta_n, \varphi_n\right)x_n + b_m = A\left(\Theta\right)x_m + b_m \tag{M.4}$$

where $m \in \{0, 1, ..., M-1\}$, while:

$$a\left(\theta_n, \varphi_n\right) = \left[e^{j\frac{2\pi r_0}{\lambda}\sin\varphi_n\cos\theta_n} \quad e^{j\frac{2\pi r_0}{\lambda}\sin\varphi_n\cos(\theta_n - \frac{2\pi}{M})} \quad ..... \quad e^{j\frac{2\pi r_0}{\lambda}\sin\varphi_n\cos(\theta_n - \frac{2\pi(M-1)}{M})} \right] \tag{M.5}$$

and:

$$x_n = \left[E_{z1} \quad ... \quad E_{zN}\right]^T. \tag{M.6}$$

The input signals $x_n(t)$ are assumed to be independent from the noise $b(t)$. Under this assumption, the covariance matrix can be written in the form:

$$R_y = A(\Theta) R_x A(\Theta)^T + \sigma_b^2 I \tag{M.7}$$

where I is the unit matrix.

In all the methods of determination of the directions of arrival that will be discussed hereafter, the following assumptions are made:

– the number of sensors is assumed to be larger than the number of sources,
– the antenna is non-ambiguous, i.e. for a set of N angles of arrival the matrix $A(\Theta)$ of the directional vectors is of full row, i.e. is of row N,
– the signals reaching the antenna network are coherent: all these signals result from plane waves in a narrow frequency band,
– the additive noise of thermal origin is a Gaussian white noise.

M.4 Determination Methods of the Directions of Arrival

Methods developed for the determination of directions of arrival fall into two categories: linear methods, based for instance on a Fourier analysis associated with a Wiener inversion or on a phase reconstruction, and nonlinear methods, like the MUSIC (MUltiple Signal Classification) method, the method based on the maximum probability estimate, the Esprit method, etc. These methods are characterised through the robustness of their respective algorithms.

M.4.1 Linear Methods

Two different types of linear methods have been developed for the determination of the directions of arrival: Fourier analysis associated with a Wiener inversion and phase reconstruction.

Fourier Analysis with Wiener Inversion

For a system in broad frequency range, we assume that the electromagnetic field received at the antenna array with a delay t can be expressed as a continuous sum of plane waves. The field received at the mth antenna is thus written in the form (Barbot 1993):

$$y(m,t) = \int_{-\pi/2}^{\pi/2} \int_{0}^{2\pi} E(\theta,\varphi,\tau).e^{j\frac{2\pi r_0}{\lambda}\sin\varphi\cos(\theta-\frac{2\pi n}{M})} \partial\varphi\partial\theta \tag{M.8}$$

where $m \in \{0, 1, ..., M$ -$1\}$. If the plane waves arrive at the circular antenna array with a uniform azimuthal distribution and a constant elevation φ, this equation becomes:

$$y(m,t) = \sum_{n=0}^{M-1} E(n,t).e^{j\frac{2\pi r_0}{\lambda}\sin\varphi\cos(\frac{2\pi(m-n)}{M})} \tag{M.9}$$

where $(m,n) \in [0, M$ - $1]^2$. The complex amplitude of a source is given by $E(n,t)$, while the vector x contains the complex amplitude of each source. The aim of the Fourier analysis method consists in determining the energy distribution of these sources. Eq. M.9 can be rewritten in the form of a convolution product given by the following equation:

$$y(n,t) = Y * E(t) \tag{M.10}$$

where:

$$Y = e^{j\frac{2\pi r_0}{\lambda}\sin\varphi\cos\left(\frac{2\pi n}{M}\right)} \tag{M.11}$$

The amplitude E of a source is determined by inversion from Eq. M.10. By performing a discrete Fourier transform, we arrive to the following equation:

$$\tilde{y} = \tilde{Y}.\tilde{E} \tag{M.12}$$

whence an expression for E can be deduced:

$$\tilde{E} = \tilde{Y}^{-1}.\tilde{y} \tag{M.13}$$

Unfortunately, as the values for Y tend very rapidly towards 0, the inversion of Y cannot be directly performed. A Wiener inversion associated with a window is therefore used: a pseudo inversion is evaluated as follows:

$$Y_n^{-1} = W\frac{Y_n^*}{Y_n^*Y_n + b_s}. \tag{M.14}$$

where W is the window, while the positive constant b_s is a regularisation term.

The main disadvantage of the method lies in the strong dependence of the choice of both the regularisation term and the window on the results of the direction of arrival. As an example, we give in Table M.1, some characteristics of the commonly used windows.

Table M.1. Window values

Window	Level of the first secondary lobe	Peak width (3 dB)	Peak width (6 dB)
Rectangle	-13	1[a]	1[a]
Hanning ($\alpha = 0.2$)	-32	1.61	1.65
Hamming	-43	1.46	1.49
Blackman	-58	1.88	1.94

[a] reference value

Phase Reconstruction

The synthesis of an antenna through the linear recombination of the excitations leads to the following response z (Balanis 1990):

$$z = \sum_{k=1}^{M} s_k y_k \tag{M.15}$$

In the case of a plane wave (θ, φ), the equation for z assumes the following form:

$$z = \sum_{k=1}^{M} s_k e^{j \frac{2\pi r_0}{\lambda} \sin \varphi \cos(\theta - \theta_k)} \tag{M.16}$$

where s_k is the excitation coefficient (amplitude and phase) of the kth antenna and θ_k is the angular position of the kth element in the x-y plane as defined by the equation:

$$\theta_k = \frac{2\pi k}{M} . \tag{M.17}$$

The excitation coefficient can ordinarily be written in the form:

$$s_k = I_k e^{j\alpha_k} \tag{M.18}$$

with:

$$\alpha_k = -\frac{2\pi r_0}{\lambda} \sin \varphi_0 \cos(\theta_0 - \theta_k) . \tag{M.19}$$

where (θ_0, φ_0) is the direction along which the main peak can be located. The equation for the field can be therefore rewritten in the following form:

$$z = \sum_{k=1}^{M} I_k e^{j\frac{2\pi r_0}{\lambda}\left[\sin\varphi\cos(\theta-\theta_k)-\sin\varphi_0\cos(\theta_0-\theta_k)\right]} . \qquad (M.20)$$

After transformation, this equation becomes:

$$z = \sum_{k=1}^{M} I_k e^{j\frac{2\pi r_0}{\lambda}\rho_0\cos(\theta_k-\zeta)} \qquad (M.21)$$

where:

$$\zeta = \tan^{-1}\left[\frac{\sin\varphi\sin\theta - \sin\varphi_0\sin\theta_0}{\sin\varphi\cos\theta - \sin\varphi_0\cos\theta_0}\right] \qquad (M.22)$$

and:

$$\rho_0 = r_0\left(\left(\sin\varphi\cos\theta - \sin\varphi_0\cos\theta_0\right)^2 + \left(\sin\varphi\sin\theta - \sin\varphi_0\sin\theta_0\right)^2\right)^{1/2} \qquad (M.23)$$

Eq. M.21 being periodic, it can be rewritten in the form of a complex Fourier series:

$$f(k) = e^{j\frac{2\pi}{\lambda}\rho_0\cos(\theta_k-\zeta)} = \sum_{m=-\infty}^{\infty} C_m e^{-jm\zeta} \qquad (M.24)$$

where:

$$C_m = \frac{1}{2\pi}\int_{-\pi}^{\pi} e^{j\frac{2\pi}{\lambda}\cos(\theta_k-\zeta)} e^{jm\zeta}\,d\zeta . \qquad (M.25)$$

The field received at the antenna array can therefore be written as:

$$z = \sum_{k=1}^{M}\sum_{m=-\infty}^{\infty} I_k J_m\left(\frac{2\pi\rho_0}{\lambda}\right) e^{jm\left(\frac{\pi}{2}+\theta_k-\zeta\right)} . \qquad (M.26)$$

Thus, assuming that $m = M + r$, we are presented with the following two equations:

$$z = \sum_{m=-\infty}^{\infty} J_m\left(\frac{2\pi}{\lambda}\rho_0\right)p(l,r) \tag{M.27}$$

and:

$$p(l,r) = \left\{\sum_{k=1}^{M} I_k e^{j(lM+r)\left(\frac{\pi}{2}+\theta_k-\zeta\right)}\right\}. \tag{M.28}$$

Noting that if $r \neq 0$, then $p(l,r) = 0$, we are thus presented with the following equation:

$$z = M\left(\sum_{k=1}^{M} I_k\right)\sum_{l=-\infty}^{\infty} J_{LM}\left(\frac{2\pi\rho_0}{\lambda}\right)e^{jLM\left(\frac{\pi}{2}-\zeta\right)}. \tag{M.29}$$

The implementation of this method is relatively simple, and can be achieved in two different ways. The first approach consists in programming it with respect to the equation of the field received on the antenna array under the assumption that J_0 (.) is the main term. The second method consists in the convolution of the matrix of the received signals with the directive matrix with opposite phase and with azimuth and elevation equal to zero.

Since the level of the secondary lobes is high, a filtering window is applied to the reception samples.

M.4.2 Non-Linear or High Resolution Methods

High-resolution methods allow to a more accurate estimation of the parameters associated with the different multiple paths of a radio propagation channel (amplitude, delay, direction of arrival and Doppler frequency) than the traditional linear methods. We shall hereafter discuss below two of these methods: the MUSIC (MUltiple SIgnal Classification) method and the method based on the maximum probability estimate.

MUSIC Method

The high-resolution algorithms which have been developed for the study of linear antenna arrays can be applied to the study of circular antenna arrays. Indeed, it can be demonstrated that a uniform circular antenna array may be regarded as quasi-equivalent to a linear antenna array. The differences lie in the geometry of the antenna array and in the form of the signals arriving at the array, i.e. their phase. A linear antenna array can therefore be simulated from data collected by the circular antenna array under consideration. Two different methods can be used for achieving this spatial transformation: the sample average method (Tewfik 1990) and the beamspace transformation method (Zoltowski 1992).

Before discussing these two methods, the problem can be summarised as follows (Tewfik 1990). Let us here consider a plane wave arriving in the absence of noise at the antenna array with the angular position (θ, φ). Using a complex notation, the amplitude of the field received at the kth antenna can be expressed by the following equation:

$$y(k,t) = E(t)\exp\left(j\frac{2\pi r_0}{\lambda}\sin\varphi\cos\left(\theta - \frac{2\pi k}{M}\right)\right) \tag{M.30}$$

Relying on the properties of Bessel functions, i.e. their periodic character and their decomposition in Fourier series, the previous equation can be rewritten in the form:

$$y(k,t) = E(t)\sum_{m=-\infty}^{\infty} j^m J_m(\xi)\exp\left(jm\left(\frac{2\pi k}{M} - \theta\right)\right) \tag{M.31}$$

where $J_m(\xi)$ is the Bessel function of the first kind and order m defined by the equation:

$$J_m(\xi) = \frac{j^{-m}}{\pi}\int_0^\pi \exp(j\xi\cos\theta)\cos(m\theta)d\theta \tag{M.32}$$

with:

$$\xi = \frac{2\pi r_0}{\lambda}\sin\varphi. \tag{M.33}$$

$y(k,t)$ may be regarded as a sample at frequencies equal to $2\pi k/M$ of the Fourier transform $y(m)$ defined by the equation:

$$y(m) = A(t)j^m J_m(\xi)\exp(-jm\theta) \tag{M.34}$$

The discrete Fourier transform $y(n,t)$ of $y(k,t)$ can therefore be expressed by the relation:

$$y(n,t) = E(t)\sum_{m=-\infty}^{\infty} j^m J_{n+mM}(\xi)\exp(-j(n+mM)\theta) \tag{M.35}$$

where $0 \leq n \leq M - 1$. Since the number M of antennas is higher than ξ, it turns out that for $1 \leq m$ and $m \leq -2$, $J_{n+mN}(.) = 0$. The discrete transform $y(n,t)$ can therefore be written in the form:

$$y(n,t) = E(t)j^m \left(J_n(\xi)\exp(-jn\theta) + (-1)^{M-n} J_{M-n}(\xi)\exp(-j(n-M)\theta) \right) \quad \text{(M.36)}$$

If $M - n$ is large enough, then $J_{M-n}(.) = 0$. Hence,

$$y(n,t) = E(t)j^m J_n(\xi)\exp(-jn\theta). \quad \text{(M.37)}$$

Under these assumptions, we must consider that the number M of antennas is larger than the integer B defined by the equation:

$$B = \frac{2\pi r_0}{\lambda} \sin\varphi \quad \text{(M.38)}$$

Sample Average. Averaging the samples allows isolating the complex amplitude of the plane waves arriving at the circular antenna array. We introduce a new series $z(n,t)$ by averaging $x(n,t)$ and $x^*(M-n,t)$, where M is a natural integer assumed to be even. The procedure can be described as follows:

$$\text{if } n = 0, \text{ then } z(n,t) = x(0,t) \quad \text{(M.39)}$$

$$\text{if } n \text{ is odd, then } z(n,t) = \left[x(n,t) - x^*(M-n,t) \right]/2 \quad \text{(M.40)}$$

$$\text{if } n \text{ is even, then } z(n,t) = \left[x(n,t) + x^*(M-n,t) \right]/2 \quad \text{(M.41)}$$

where:

$$z(n,t) \approx E(t)j^n J_n(\xi)\exp(-jn\theta) \quad \text{(M.42)}$$

The amplitude of the sources is determined from the following set of equations:

$$\text{if } n = 0, \; z'(n,t) = x(0,t)/J_0(\xi) \quad \text{(M.43)}$$

$$\text{if } n \neq 0, \; z'(n,t) = j^{-n} \left[x(,t) + (-1)^n x^*(M-n,t) \right]/2J_n(\xi) \quad \text{(M.44)}$$

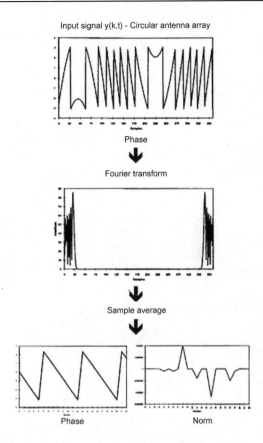

Fig. M.2. Space transformation of a circular antenna array into a linear antenna array, applied to 23 sensors (Tewfig & Hong method)

The space transformation for N plane waves is obtained by computing the expression:

$$z'(n,t) = \sum_{m=1}^{N} E_m(t) \exp(-jn\theta_n). \tag{M.45}$$

As can be seen in Fig. M.2, the outputs of the modified antenna network are similar to the outputs of a linear network. The direction of arrival can therefore be estimated using high-resolution algorithms, like the MUSIC algorithm.

Beamspace Transformation (Zoltowski 1992). Let us consider the already introduced equation for the discrete Fourier transform:

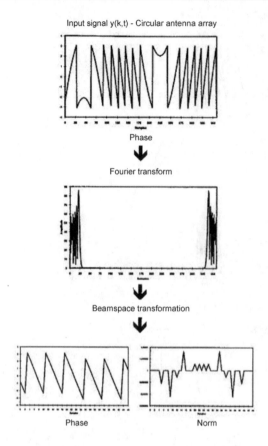

Input signal y(k,t) - Circular antenna array

Phase

⬇

Fourier transform

⬇

Beamspace transformation

⬇

Phase   Norm

Fig. M.3. Space transformation of a circular antenna array into a linear antenna array, applied to 47 sensors (Matthews & Zoltowski method)

$$y(n,t) = E(t)j^m \left(J_n(\xi)\exp(-jn\theta) + (-1)^{M-n} J_{M-n}(\xi)\exp(-j(n-M)\theta) \right) \quad \text{(M.46)}$$

The samples resulting from the discrete Fourier transform can be directly calculated without having to perform an averaging operation. The space transformation consists in decomposing this equation into two expressions and in analysing separately the different samples of the spectrum. The first $B + 1$ elements of the spectrum are associated with the first expression, while the $M - B$ remaining elements are associated with the second expression. This leads to a matrix containing the $2B + 1$ values of the space transformation realised following the beamspace transformation method. The procedure is schematised in Fig. M.3: as can be seen, the output signals are equivalent to the signals arriving at a circular antenna array.

Estimation of the Directions of Arrival of Coherent Sources. Signals received in urban environments present a high correlation. In order to compensate for this correlation, we proceed to a separation of the received signals through a redefinition of the structure of the covariance matrix. The algorithms used for determining the angles of arrival are indeed extremely sensitive with respect to the detection of multipath sources.

We shall use here the *forward only spatial smoothing* (FOSS) method which has a high performance as regards the estimation of correlated sources (Fuhl 1993). This preprocessing method disrupts the coherence properties of the signals and allows obtaining a covariance matrix of full row which can be used with the MUSIC algorithm.

The aim of the FOSS method is to divide a table consisting of the signals received at the M sensors received signals $\{1, 2, ..., M\}$ into a panel consisting of $(L + 1)$ elements. For this purpose, a $(M - L - 1) \times (L + 1)$ dimension smoothing window is therefore used.

Let y^f be the transformation sub-matrix, defined as follows:

$$.y^f = \begin{bmatrix} y_1 & y_2 & \cdots & y_{L+1} \\ y_2 & y_3 & \cdots & y_{L+2} \\ \cdots & \cdots & \cdots & \cdots \\ y_{M-L} & y_{M-L+1} & \cdots & y_M \end{bmatrix} \tag{M.47}$$

The covariance matrix is therefore given by the equation:

$$R^f = y^f y^{f^H} . \tag{M.48}$$

MUltiple SIgnal Classification (Music) (Schmidt 1986). Depending on the analysis of the samples provided by the sensors, two different types of MUSIC algorithms, respectively based on two different types of space transformation, can be implemented. These algorithms include the FOSS method for the analysis of coherent signals.

Sample Average

The averaging operation is performed over a number B of values. Therefore, whatever the number of sensors is, we use only B samples of the discrete Fourier transform for programming the MUSIC algorithm.

In this case, the FOSS sub-matrix is of $(B - L + 1) * (L+1)$ dimension. Since the MUSIC method is based on the decomposition of the covariance matrix into eigenvalues, the implementation of this method is carried out on $(L + 1)$ data.

Beamspace Transformation

The same operation is performed over $2B + 1$ values of the Fourier transform. The dimension of the FOSS sub-matrix in this case is therefore $(2B - L)*(L + 1)$. The number L of samples used here is larger than with the previous method: the beamspace transformation algorithm thus presents the best characteristics in terms of precision.

MUSIC Algorithm

The main characteristic of this method is the decomposition of the covariance matrix into eigenvalues, leading to the definition of two subspaces: the noise subspace and the signal subspace. In order to develop the MUSIC algorithm, we shall first consider all the $(L + 1)*(L + 1)$ dimension averaged covariance matrix R^f for a set of N correlated signals. The covariance matrix can therefore be expressed by the equation:

$$R^f = AR_x A^H + \sigma^2 I .$$

(M.49)

Let $\lambda_1 \geq \lambda_2 \geq .... \geq \lambda_{L+1}$ be the eigenvalues of the covariance matrix and $v_1 \geq v_2 \geq$ .... $\geq v_{L+1}$ be the eigenvectors of $AR_x A^H$. Assuming that the directional matrix A is K full row, it can be shown that the $(L + 1 - K)$ smallest values of the matrix $AR_x A^H$ are null. The eigenvalues of the covariance matrix constitute an orthonormal basis in C^{L+1} and define two subspaces:

- the signal subspace, composed of eigenvectors associated to the K highest eigenvalues of R^f:

$$.U_S = \begin{bmatrix} v_1 & \cdots & v_K \end{bmatrix}$$

(M.50)

– the noise subspace, defined as the orthogonal complement of the signal subspace. This subspace is composed of eigenvectors associated to the $(L + 1 - K)$ smallest values of R_f:

$$U_S = \begin{bmatrix} v_{K+1} & \cdots & v_{L+1} \end{bmatrix}.$$

(M.51)

The values of the signal subspace and the values of the noise subspace are by definition orthogonal. The following equation is therefore valid:

$$U_N^H \cdot U_S = 0$$

(M.52)

The next step is to search for signal vectors as orthogonal as possible to the noise subspace. Following the MUSIC algorithm, the direction of arrival is then estimated by looking for the main peaks of the spectrum defined by the following equation:

$$MUSIC(\theta, \varphi) = \frac{1}{a^H(\theta, \varphi) U_N U_N^H a(\theta, \varphi)} \qquad (M.53)$$

The signal vector $a(\theta, \varphi)x$ is defined by the equation:

$$a^T(\theta, \varphi) = \left[e^{-j\xi\cos\theta} \quad e^{-j\xi\cos(\theta - 2\pi/(L+1))} \quad \cdots \quad e^{-j\xi\cos(\theta - 2\pi L/(L+1))} \right]. \qquad (M.53)$$

Method based on the Maximum Probability Estimate

Under the previously developed assumptions, the signals received by antenna network follow a Gaussian law whose mean and variance are $A(\theta, \varphi)x$ and σ^2 respectively. The following probability function can therefore be defined (Lähteenmäki 1993):

$$p(y) = \frac{1}{\pi^N \sigma^{2N}} .e^{-\sigma^2 (y - A(\theta, \varphi)x)^H (y - A(\theta, \varphi)x)}. \qquad (M.54)$$

where $(y - A(\theta, \varphi)x)^H$ represents the cross-conjugated of the vector $(y - A(\theta, \varphi)x)$.

The maximum probability estimate of the angular position of the sources can be expressed as the least square between the measured values and the signals of the parametric model $\hat{y} = A\hat{x}$. The purpose of this method is to minimise the quadratic error between the received and estimated signals. Accordingly, the following performance index can be defined:

$$E(\theta, \varphi) = \|y - A\hat{x}\|^2 \qquad (M.55)$$

For an azimuth estimate at constant elevation, the solution of this equation is as follows:

$$\hat{x}_{MLE} = Z_{MLE}x \qquad (M.56)$$

$$Z_{MLE} = \left(A^H A\right)^{-1} A^H = A^\circ.$$

where A^H and A° are the cross-conjugated and pseudo-inversion of the A matrix respectively.

Since the maximum probability estimate is a non-biased minimum variance estimator, the solution is obtained by searching on azimuth for a maximum of the following local function:

$$f(\theta) = \frac{1}{E(\theta, \varphi_0)} \qquad (M.57)$$

Since the number of analysed sources is limited, i.e. no higher than 2, this method is not well-suited to the determination of the directions of arrival, and was presented here only as rough guide.

References

Barbot JP, Levy AJ (1993) Indoor wideband measurements at 2.2 GHz in a shopping center. PIMRC'93 Conference, Yokohama

Balanis CA (1990) Antenna Theory : Analysis and Design. Arizona State University

Fuhl J, Molisch AF, Bonek E (1993) A new single snapshot algorithm for direction of arrival (DAO) estimation of coherent signals. Special issue on signal separation and interference cancellation for PIRMRC

Guisnet B, Verolleman Y (1996) Evaluation of different methods of DOA using a circular array applied to indoor environment. Vehicular Technology Conference, Atlanta, Georgia, USA

Guisnet B, Perreau X (1999) Validation of an accurate direction of arrival measurement setup and experimental exploitation. 3th European Personal Mobile Communications Conference, Paris

Lähteenmäki J (1993) Determination of dominant signal paths for indoor radio channel at 1.7 GHz. PIMRC'93 Conference, Yokohama

Schmidt RO (1986) Multiple signal location and signal parameter estimation. IEEE Transactions an Antenna and Propagation, vol 34, 3

Tewfik AH, Hong W (1990) On the equivalence of uniform circular arrays and uniform linear arrays. Proc. of the Fifth ASSP Workshop on Spect. Est. and Modeling, pp 139-143

Zoltowski MD, Mathews (1992) Direction finding with uniform circular arrays via phase mode excitation and beamspace root – Music. ICASSP

N Geographical Databases

N.1 Introduction

The description of propagation environments is achieved through the use of geographical databases. These databases contain data relating to the topography and relief, to the vegetation and the use (wood, forest, road, buildings, etc.), to the streets axes and to buildings. They result from a complex process based on the use of satellite or air photographs combined with complex digitisation processes. The computation methods used in radio prediction models are applied to the analysis of the propagation environment (urban, suburban, rural, and indoor). Outdoor data are processed either in raster mode or in vector mode. In the first case, data are presented in the form of a uniform grid of numerical values. Depending on the nature and on the source of the data, the values in the raster grid may represent a point or a complete mesh. In the vector mode, data are digitised in vector form.

The following types of geographical databases will be considered in this appendix: digital terrain model databases, global land use databases, local databases (building contours, communications lines, 1:25 000 raster data or 1:100 000 local land use data) and indoor databases.

N.2 Digital Terrain Model

The Digital Terrain Model (DTM) is a dataset representing the elevation of the ground surface with a 100 metre resolution. It contains data extracted from the databases of local geographical institutes providing a description of the territory in the form of level lines. A grid with a step size equal to 100 metres is defined over these level lines and the altitude of each point of the grid is interpolated from the level lines. This model can therefore be accurately described as a raster data model: the reported altitude for each point in the grid is not an average value calculated from a sample of altitudes observed inside the mesh. The precision claimed by the *Institut National Geographique* in France for the altitude of the calculated point directly depends on the equidistance of the level lines, i.e. on the vertical resolution. The vertical resolution is not uniform over the whole territory and assumes values in the following set: 5, 10, 20 and 40 metres. The standard deviation of the altitude error is equal to 3 metres over areas where the equidistance is equal to 5 or 10 metres, and 6 to 15 metres over areas where the equidistance is 20 or 40 metres.

Fig. N.1 presents two examples of digital terrain model maps with a 100 metre and 25 metre resolution respectively in Belfort.

N.3 Global Land Use Data

Global land use data provide a statistical description of the land use using data acquired from photographs taken by the land satellite LANDSAT from 1987 to 1989. The maps produced from these data are generated using a mesh with step size equal to 400 metres and applying a classification into seven classes. The following classes are considered:

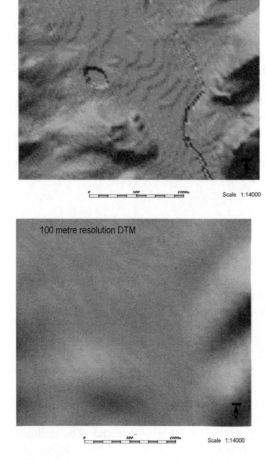

Fig. N.1. Examples of digital terrain model maps

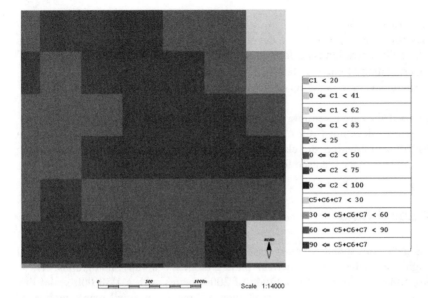

| C1 < 20 |
| 0 <= C1 < 41 |
| 0 <= C1 < 62 |
| 0 <= C1 < 83 |
| C2 < 25 |
| 0 <= C2 < 50 |
| 0 <= C2 < 75 |
| 0 <= C2 < 100 |
| C5+C6+C7 < 30 |
| 30 <= C5+C6+C7 < 60 |
| 60 <= C5+C6+C7 < 90 |
| 90 <= C5+C6+C7 |

Scale 1:14000

Fig. N.2. Example of a global land use data file in Belfort

- C_1: percentage of water,
- C_2: percentage of forest areas,
- C_3: percentage of other types of surfaces,
- C_4: percentage of bare mineral surfaces,
- C_5: percentage of dispersed settlement,
- C_6: percentage of mixed settlement,
- C_7: percentage of dense settlement.

Fig. N.2 presents an example of a global land use data file with a 400 metre resolution in Belfort.

In practice the definition of satellite images acquired with LANDSAT is a pixel representing a 60 x 80 square metre land surface at temperate latitudes. The pixels are then classified into the following classes: water, forest areas, bare mineral surfaces (rock), dispersed settlement, mixed settlement, dense settlement and other types of surfaces (meadows, arable soil, etc).

When the classification has been performed, the data are averaged through the application of a mesh with step size equal to 400 metres in order to determine the percentage of surface for each class in the mesh.

The classification process rests on the analysis of the signature of the LANDSAT pixel. For each class, the operator defines ten non-ambiguous reference areas which are to be used as reference signature for initialising the process; an automatic spectral classification is then performed. This process can be decomposed into a sequence consisting of four stages:

- classification into classes,
- classification into classes of developed sites: for each town or village, limits where this operation is performed are defined,
- superposition of the two classifications and possible adjustments,
- meshing and determination of the number of pixels.

Developed sites are taken into account only if they are visible on satellite images: accordingly, developed sites extending over too small a surface will not be considered. In practice, the classification process proceeds from the examination of a complementary file used for the localisation of developed sites.

N.4 Local Data

Local data correspond in most cases to large cities. These data are acquired through the analysis by numerical photogrammetry of air photographs. These photographs are taken at the altitude of 2000 metres with a 1:30 000 scale: this high degree of coverage allows producing a three-dimensional representation of the selected area. The measuring campaigns conducted in order to obtain these photographs are therefore localised in time: the older the photographs, the higher the probability for the data contained in the digitised file to diverge from the reality on the field.

The digitisation process is conducted using ground control points whose exact x,y,z coordinates are known with precision. These points provide a ground support for the aerial triangulation before proceeding to the reconstruction of adjoining photographs. The planimetric and altimetric reconstruction is realised by an operator using a stereo plotter. The digitisation is then performed in vector form on the following objects:

- building contours,
- communication lines.

The objects digitised during the restitution process are then analysed by readjusting the ground control points with respect to the IGN height database (contour line). The standard deviation of the error claimed by the IGN as regards the vector digitisation of building contours and major trunk roads varies from:

- 3 to 5 metres for the planimetric error,
- 2 to 3 metres for the altimetric error.

Among local data, only building contours and communication lines require an operator. The 1:25 000 raster data files and the 1:100 000 statistical land use data files are constructed from data collected during the reconstruction process. The actual heights of land use (buildings or forests) are evaluated through a cross-checking with the height database. The surface percentages during the classification into classes are also evaluated from data digitised by space division.

N.4.1 Building Contours

The tracing of the building contours during the restitution process is conducted over the ridges of roofs and over parklands. Chimneys, lift shafts and other objects which may be on top of roofs are not taken into account. In the case of high domes, the height taken into account is the height at the top of the dome.

Elements with a ground surface smaller than 25 square metres or with a height lower than 3 metres are generally not taken into account. Each contour is regarded as being at a constant altitude compared to the digital terrain model database. Fig. N.3 presents an example of base contour map in Belfort.

Interior courtyards are considered as being on the ground.

Forests are treated in the same way as buildings and are divided into several parts if their surface or the differences in height require it. Single isolated trees and tree alleys are not taken into account.

Bridges are treated in the same way as buildings: the reported height of buildings is the height measured above the mean sea level. Similarly, water surfaces are regarded as being at a constant altitude compared to digital terrain model data.

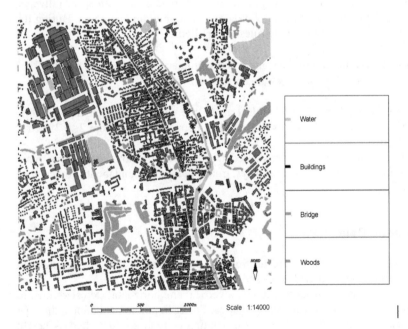

Scale 1:14000

Fig. N.3. Building contour map in Belfort

The limit between two contiguous buildings is digitised if the difference in height is higher or equal to two floors: in this case, each building is digitised with its exact height. In the opposite they are digitised as a single building with a height equal to the height of the highest building. The same rules apply to blocks of buildings.

Houses are subdivided into for classes depending on their ground surface: houses with a ground surface smaller than 65 square metres, houses with a ground surface ranging from 65 to 100 square metres, houses with a ground surface ranging from 100 to 150 square metres. These houses are represented in the form of points located at the centre of the ground surface. Houses extending over a surface larger than 150 square metres are represented like buildings with their exact contour. The following classification is therefore used:

- objects represented as a contour: developed sites with the exception of houses extending over a surface no larger than 150 square metres, forests, pedestrian malls and sidewalks, bridges, water surfaces.
- objects represented as single points: houses with surface ranging from 40 to 150 square metres.

N.4.2 Communication Lines

Lines of communication, for instance motorways, roads, main streets or railways, are digitised from the same aerial photographs than building contours. During the digitisation process, the following rules are applied:

- in rural areas, only motorways, main and secondary roads are considered,
- in urban areas, other road lines are considered.

The precision claimed by the IGN is of the same order than for building contours, i.e. from 3 to 5 metres for the planimetric error and from 2 to 3 metres for the altimetric error.

The following classes are considered for the classification: motorways, main and secondary roads, streets, railways, roadsides and roundabouts, other road lines.

N.4.3 Raster Data

The 1: 25 000 scale raster data result from the processing of data obtained through the reconstruction of wooded areas and building contours.

A raster digitisation using a grid with a vertical and horizontal step size equal to 25 metre is performed over the selected area. For each point thus defined, the type of land use and the altitude of the corresponding contour are identified. A height database enriched with the ground control points defined during the reconstruction of adjoining photographs is used for determining the height associated with the selected point. Raster data files provide information concerning:

- the altitude of a point at the ground,
- the type of land use at this point (bare ground, developed site, bridge, vegetation, water)
- the height of the land use (calculated by cross-checking)

The following classes are considered:

- C_1: ground altitude (in metres),
- C_2: height above ground of buildings (in metres),
- C_3: height above ground of vegetation (in metres),
- C_4: percentage of water,
- C_5: percentage of bare ground.

These classes are mutually exclusive, i.e. if information concerning one of these classes is provided for a given mesh, then the others classes of this mesh have null values.

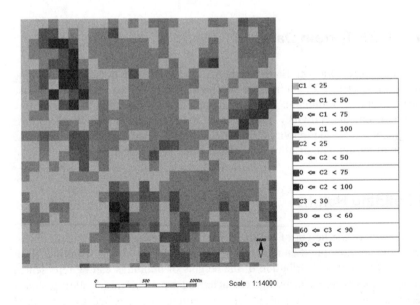

| C1 < 25 |
| 0 <= C1 < 50 |
| 0 <= C1 < 75 |
| 0 <= C1 < 100 |
| C2 < 25 |
| 0 <= C2 < 50 |
| 0 <= C2 < 75 |
| 0 <= C2 < 100 |
| C3 < 30 |
| 30 <= C3 < 60 |
| 60 <= C3 < 90 |
| 90 <= C3 |

Scale 1:14000

Fig. N.4. Local land use map with a 100 metre resolution in Belfort

N.4.4 Local Land Use Data

Local land use data with a 100 metre resolution are also produced during the processing of the vector data. A mesh with step size equal to 100 metres is applied to the area according to a division of this area into regions defined by integer Lambert coordinates. The different elements present inside each mesh (ground, developed sites, wood, water) are automatically indexed. Areas where no information is provided are regarded as being bare ground areas.

For each mesh, the following classes are considered:

- C_1: percentage of developed sites,
- C_2: percentage of wooded or forested rural areas,
- C_3: percentage of water (rivers, lakes, seas),
- C_4: percentage of land use with height larger than 10 metres,
- C_5: percentage of land use with height larger than 20 metres,
- C_6: percentage of land use with height larger than 40 metres,
- C_7: maximum height of land use inside the mesh.

Fig. N.4 presents an example of a local land use map with a 100 metre mesh near Belfort in France.

N.5 GLOBE Terrain Data

The Global Land One-km Base Elevation (*GLOBE*) is a database containing digital elevation data with a thirty- arc-second nominal grid spacing. This database includes average, minimum and maximum digital elevation values. More information on the GLOBE project can be obtained at the web address www.ngdc.noaa.gov/seg/topo/topo.shtml A Fortran source code enabling the extraction of terrain path profiles is available at the ITU Bureau.

N.6 Building Height

Measurements of the height of buildings have been conducted in different towns, for example in London (Saunders 1991; Parsons 1992) and in Guilford (Butt 1992). The probability distribution of the heights of buildings can be represented either by a lognormal distribution characterised by an average value μ and a standard deviation σ, or by a Rayleigh distribution characterised by a standard deviation σ.

The lognormal probability density function can be defined by the following equation:

$$p(h) = \frac{1}{h\sqrt{2\pi\sigma}} \exp\left(-\left(\frac{1}{2\sigma^2}\right)Log_n^2\left(\frac{h}{\mu}\right)\right).$$ (N.1)

where the values for μ and σ values are as reported in the following table (COST 255 2002):

Table N.1. Values for the mean and for the standard deviation

City	Mean μ [m]	Standard deviation σ [m]
Westminster	20.6	0.44
Guilford	7.1	0.27
Soho (London)	17.6	0.25

The Rayleigh probability density function is defined by the equation:

$$p(h) = \frac{h}{\sigma^2} \exp\left(-\frac{h^2}{2\sigma^2}\right).$$ (N.2)

where the values for σ are as reported Table N.2 (COST 255 2002):

N.7 Street Width

The widths of streets are extremely variable, especially in cities. The probability density $p(w)$ can be represented through a lognormal distribution. The values for the parameters μ and σ, estimated from experimental measurements, are equal to 14.9 and 0.65 metres respectively (COST 255 2002)

Table N.2. Values for the standard deviation

City	Standard deviation σ [m]
Westminster	17.6
Guilford	6.4

N.8 Indoor Data

Data of this type are significantly more accurate, with a precision below one metre. These data are produced from digitised maps of the building under consideration using software such as for instance Autocad, and provide a representation of the constitutive elements of the building, like the walls, the partitions, the doors or the windows. These data are primarily used in indoor propagation models and in penetration models.

N.9 Future Developments

In the last recent years, major advances have been made in the acquisition and processing of geographical data, due in particular to advances in the fields of aerial and spatial remote detection (high resolution images) and geomatics. Spatial analysis and the possible integration of relevant data into a Geographical Information System (GIS) database should lead to a better insight into the influence of geographical data on the development of propagation models as far as the land use is concerned (scale, quality, typology and nature of the elements), and to identify algorithms used for the determination of the radio coverage which are the most suited to the optimisation of numerical analyses (Turck 2002).

References

Butt G (1992) Narrowband characterization of high elevation angle land mobile satellite channel. Ph. D. Thesis, University of Surrey

COST Action 255 (2002) Radiowave propagation modeling for SatCom services at Ku-band and above; Cost 255 Final Worshop, Bech, Luxemburg

Parsons D (1992) The Mobile Radio Propagation Channel. Pentech Press, London

Saunders SR (1991) Diffraction modelling of mobile radio wave propagation in built-up areas. Ph. D. Thesis, Brunel University

Turck C, Thome D, Weber C (2002) Prédiction de couverture de champ radioélectrique pour les réseaux radiomobiles: L'apport de l'analyse spatiale et des systèmes d'information géographique; application en milieu urbain. $4^{\text{ièmes}}$ journées d'études. Propagation électromagnétique dans l'atmosphère terrestre du décamétrique à l'angström, Rennes, France

O Determination of the Electromagnetic Field after Interaction with Structures

O.1 Introduction

This appendix will be devoted to the determination of the evaluation of the electromagnetic field following its interaction with a metallic or dielectric structure, for instance a half-plane or a dihedron.

The parameters characterising the electromagnetic field (E, D, H, B) are described by Maxwell's equations. However, the application of these equations to complex environments, as well as their resolution, may turn out to be complicated and to require computation times incompatible with operational constraints in the field deployment of communications networks. Simpler methods and models had therefore to be developed in order to account for the interactions between an electromagnetic wave and a structure present along the propagation path of this wave, which generates different reflection, transmission and diffraction phenomena. The following methods will be successively described in this appendix: the geometrical optics method, the geometrical theory of diffraction, the uniform theory of diffraction, the finite difference in time domain method (FDTD), the moment method and the parabolic equation method.

O.2 Geometrical Optics Method

In this model, waves are assumed to propagate in the form of rectilinear rays within a homogeneous medium. The wave phase regularly varies along the rays. At the interface between two media, the ray follows Snell-Descartes law on reflection and transmission. This approximation can be derived from Maxwell's equations by considering only the $1/\omega$ first-order terms.

This method can be used for determining the interactions between a wave and a structure as well as for predicting, with a relatively high degree of accuracy, the propagation of this wave provided however that the observation is conducted in the region directly illuminated by the incident wave. In this case, this method leads to reasonable approximations, even though the total field cannot be in full rigour considered as being the sum of the geometrical optics fields, i.e. of the incident and reflected fields only.

This method is no longer applicable when considering the field received in the shadow region. Indeed, since no incident or reflected rays may reach the receiver, the field predicted in this region by geometrical Optics is null. This theory leads therefore to a discontinuity at the boundary between shadow and light which from a physical point of view is impossible within a homogeneous medium. This discontinuity limits the applicability of geometrical Optics, and led Sommerfeld to investigate the interactions between a plane wave and the straight edge of a half-plane and to determine the amplitude of the field in the shadow produced by this half-plane (Sommerfeld 1896).

Depending on the region of space considered around the half-plane, the total electromagnetic field is composed of the following fields:

$$U_{total} = U_{incident} + U_{refleted} + U_{diffracted} \quad \text{if } 0 \leq \varphi < LOR \tag{O.1}$$

$$U_{total} = U_{incident} \quad \text{if } LOR < \varphi < LOI \tag{O.2}$$

$$U_{total} = 0 \quad \text{if } LOI < \varphi \leq 2\pi \tag{O.3}$$

where *LOI* is the shadow boundary associated with the incident field, beyond which this field vanishes and *LOR* is the shadow boundary associated with the reflected field.

The total field can then be written in the following form:

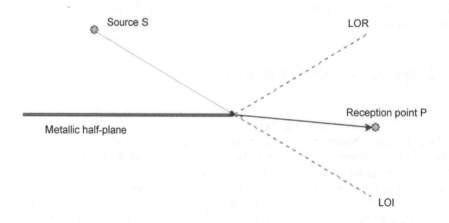

Fig. O.1. Interaction between an electromagnetic wave and a metallic half-plane

$$U = \frac{U(s,\varphi - \phi)}{2} \cdot + \tag{O.4}$$

$$U(s,\varphi - \phi)\frac{e^{-j\frac{\pi}{4}}}{\sqrt{2}} \rho_{-} \int_0^{\frac{\pi x^2}{2}} e^{\frac{\pi x^2}{2}} dx \mp \frac{U(s,\varphi + \phi)}{2} \mp U(s,\varphi + \phi)\frac{e^{-j\frac{\pi}{4}}}{\sqrt{2}} \rho_{+} \int_0^{\frac{\pi x^2}{2}} e^{\frac{\pi x^2}{2}} dx$$

where U_{total} is the total field, $U_{incident}$ is the incident field and:

$$\rho\pm = 2\sqrt{\frac{k_0 s}{\pi}} \cos\left(\frac{\varphi \pm \varphi'}{2}\right) \tag{O.6}$$

O.3 Geometrical Theory of Diffraction

The geometrical theory of diffraction (GTD) developed by Keller represents a generalisation of geometrical Optics accounting for diffraction phenomena. This generalisation is achieved through the introduction of a new class of rays, diffracted rays, along which the diffracted field can be determined. These rays follow similar reflection and refraction laws.

Keller demonstrated that when an incident wave propagates along a normal direction at the edge of a metallic half-plane, it generates a diffracted wave with cylindrical wave surfaces normal to the edge leaving in all directions.

The value of the diffracted field can therefore be determined with respect to a reference value. This reference value is set to the initial value of the diffracted field at the level of the diffracting structure.

Depending on the region of space considered around the half-plane, the total field is composed of the following fields:

$$U_{total} = U_{incident} + U_{refleted} + U_{diffracted} \qquad \text{if } 0 \le \varphi < LOR \tag{O.7}$$

$$U_{total} = U_{incident} + U_{diffracted} \qquad \text{if } LOR < \varphi < LOI \tag{O.8}$$

$$U_{total} = U_{diffracted} \qquad \text{if } LOI < \varphi \le 2\pi \tag{O.9}$$

The diffracted field is written as a product of four terms:

$$U^d = D.A_i \exp(j\varphi_i).\sqrt{\frac{\rho}{s(\rho + s)}}.\exp(-jk_0 s) \tag{O.10}$$

where:

− $A_i \exp(j\varphi_i)$ is the incident field with respect to the diffraction edge,

− $\sqrt{\dfrac{\rho}{s(\rho+s)}}$ is an amplitude term, and is a function of the observation distance

$s = QP$ and of caustic distance ρ. The caustic distance depends itself on the radius of curvature of the edge and on the radius of curvature of the wavefront. The following equation yields this parameter:

$$\frac{1}{\rho} = \frac{1}{\rho_e^i} - \frac{\bar{n}.(\vec{s}' - \vec{s})}{\rho_c \sin^2(\beta_0)} \qquad (O.11)$$

where:

− ρ_c is the radius of curvature of the edge, and is infinite in the case of a straight edge,
− ρ_e^i is the radius of curvature of the incident wavefront in the $(\vec{s}'; \vec{e})$ plane, where:
1. $\vec{e}$ is the direction vector tangent to the edge,
2. $\vec{s}$ is the direction vector of the direction of diffraction,
3. $\vec{s}'$ is the direction vector of the direction of incidence,
4. $\vec{n}$ is the direction vector normal to the diffraction edge,
− β_0 is the angle formed by the direction of incidence $\vec{s}'$ and the normal to the edge $\vec{n}$: β_0 is equal to $\pi/2$ in the normal incidence case.
− $\exp(-jk_0 s)$ is the phase term associated with distance $s = QP$,
− D is the diffraction coefficient. D is a function of the reflection coefficients R_p and $R_\perp$ and is written in the following form :

$$D_{//,\perp} = \frac{-\exp\left[\dfrac{-j\pi}{4}\right]\sin\left(\dfrac{\pi}{n}\right)}{n\sqrt{2\pi k_0}.\sin(\beta_0)}\left[\frac{1}{\cos\left(\dfrac{\pi}{n}\right) - \cos\left(\dfrac{\varphi - \varphi'}{2}\right)} + R_{//,\perp}\frac{1}{\cos\left(\dfrac{\pi}{n}\right) - \cos\left(\dfrac{\varphi + \varphi'}{2}\right)}\right] \qquad (O.12)$$

The details of the calculations are developed by Combes (Combes 1978). It should be noted that this approximation is valid only for large values of ρ, defined by the equation:

$$\rho\pm = 2\sqrt{\frac{k_0 s}{\pi}}\cos\left(\frac{\varphi\pm\varphi'}{2}\right) \tag{O.13}$$

The following consequences can therefore be drawn:

- the distance s between the diffraction and the observation points must be large enough,

- $\cos\left(\dfrac{\varphi\mp\varphi'}{2}\right)\neq 0$ or $\varphi\pm\varphi'\neq(2k+1)\pi$: the observation point should not be located neither at the shadow boundaries nor in their vicinity.

It might be pointed out that due to the wave number k_0 the values for $D_{//}$ and $D\perp$ are inversely proportional to the square root of the frequency. In the case of a metallic structure, the reflection coefficients $R_{//}$ and $R_\perp$ are equal to - 1 and + 1 respectively.

O.4 Uniform Theory of Diffraction

The geometrical theory of diffraction (GTD) improves significantly on geometrical Optics in so far as it allows the determination of the electromagnetic field in the shadow region. Unfortunately this theory displays a discontinuity at the *LOI* and *LOR* shadow boundaries, where the total electromagnetic field is infinite.

A theory allowing the determination of a spatially uniform electromagnetic field was therefore developed on the basis of geometrical theory of diffraction by Kouyoumjian and Pathak (Kouyoumjian 1974): this theory is known as the uniform theory of diffraction (UTD). In order to compensate for the divergence of the diffraction coefficient $D_{//}$ and $D\perp$ at and in the vicinity of the *LOI* and *LOR* shadow boundaries, this theory introduces in the diffraction coefficient a function referred to as the transition function *F(x)*.

The diffraction coefficients used in the uniform theory of diffraction are defined by the following equation:

$$D_{//,\perp} = \frac{-\exp\left[\dfrac{-j\pi}{4}\right]}{2n\sqrt{2\pi k_0}\cdot\sin(\beta_0)}\left[\frac{F(k_0 L^i a(\varphi-\varphi'))}{\cos\left(\dfrac{(\varphi-\varphi')}{2}\right)}+R_{//,\perp}\frac{F(k_0 L^i a(\varphi+\varphi'))}{\cos\left(\dfrac{(\varphi+\varphi')}{2}\right)}\right] \tag{O.14}$$

The transition function $F(x)$ is equal to unit in regions at a great distance from the *LOI* and *LOR* shadow boundaries, and tends to zero as the distance to the shadow boundaries decreases. This function is defined by the equation:

$$F(x) = 2j\sqrt{x}e^{jx}\int_{\sqrt{x}}^{\infty}e^{-j\tau^2}d\tau \qquad (O.15)$$

The argument x of the transition function is a function of the bending parameter L^i defined by the equation:

$$L^i = \frac{s(\rho_e^i + s)\rho_1^i\rho_2^i}{\rho_e^i(\rho_1^i + s)(\rho_2^i + s)}\sin^2(\beta_0) \qquad (O.16)$$

where:

- ρ_e^i is the radius of curvature of the incident wavefront in the $(\vec{s'}, \vec{e})$ plane, where the vectors $\vec{s'}$ and $\vec{e}$ are defined as follows:
1. $\vec{s'}$ is the direction vector of the incident ray between the source point S and the diffraction point Q,
2. $\vec{e}$ is the tangent vector with respect to the diffraction edge at the diffraction point.
- ρ_1^i and ρ_2^i are the principal radii of curvature of the incident wavefront in the incidence plane and in the transversal plane with respect to the incidence plane respectively. They intrinsically depend on the form of the incident wave.

By noting that $s = QP$ and that $s' = SQ$, the bending parameter in normal incidence can be expressed in the form:

$$L^i = s \text{ in the case of a plane wave,} \qquad (O.17)$$

$$L^i = \frac{ss'}{s + s'} \text{ in the case of a spherical wave.} \qquad (O.18)$$

- $a^{\pm}(\varphi \pm \varphi')$ is a function of the incidence and observation angles φ' and φ defined by the equation:

$$a^{\pm}(\varphi \pm \varphi') = 2\cos^2\left(\frac{2n\pi N^{\pm} - (\varphi \pm \varphi')}{2}\right) \qquad (O.19)$$

- N^+ is the nearest integer to the non-integer solution of the equation:

$$2\pi n N^+ - (\varphi \pm \varphi') = \pi \qquad (\text{O.20})$$

- N^- is the nearest integer to the non-integer solution of the equation:

$$2\pi n N^- - (\varphi \pm \varphi') = -\pi \qquad (\text{O.21})$$

For further detail on the uniform theory of diffraction, the reader is referred to the articles by Kouyoumjian and Mc Namara (Kouyoumjian 1974; McNamara 1990).

O.5 Finite Difference in Time Domain Method

This method is a time-domain numerical method. An extensive literature can be found on this subject: Taflove has presented a valuable synthesis which is very useful for understanding this method (Taflove 1994). The robustness of this method has been demonstrated in the study of diffraction by different structures with very short electromagnetic impulses (Jaureguy 1995).

This method is based on the discretisation of Maxwell's equations representing the evolution of the electromagnetic field in a computer domain through the superposition of two space-time grids shifted by $dx/2$, $dy/2$ and $dt/2$. The algorithm used here is the Yee algorithm (Yee 1969). The domain consists of a regular rectangular grid composed of unit cells with dimensions (dx, dy). The component E_y of the electric field on the cell with coordinates (x, y) at time t is determined from the components H_z of the magnetic field on the cells $(x - dx/2, y)$ and $(x + dx/2, y)$ previously determined at time $t - dt/2$.

In a continuous, isotropic and linear medium, the electromagnetic field is described by the two vector equations:

$$\frac{\partial \vec{H}}{\partial t} = -\frac{1}{\mu} \nabla \times \vec{E} \qquad (\text{O.22})$$

$$\frac{\partial \vec{E}}{\partial t} = -\frac{1}{\varepsilon} \nabla \times \vec{H} - \frac{\sigma}{\varepsilon} \vec{E} \qquad (\text{O.23})$$

In order to derive dual Maxwell's equations, Taflove introduced a real number ρ' meant to account for the magnetic losses (Taflove 1994):

$$\frac{\partial \vec{H}}{\partial t} = -\frac{1}{\mu}\nabla \times \vec{E} - \frac{\rho'}{\mu}\vec{H} \qquad (O.24)$$

$$\frac{\partial \vec{E}}{\partial t} = -\frac{1}{\varepsilon}\nabla \times \vec{H} - \frac{\sigma}{\varepsilon}\vec{E} \qquad (O.25)$$

where:

- E is the electric field (V/m),
- H is the magnetic field (A/m),
- ε is the electric permittivity (F/m),
- σ is the electric conductivity (S/m),
- μ is the magnetic permeability (H/m),
- ρ' represents the magnetic losses (Ω/ m)

In the case of an isotropic medium, the above system can be written in Cartesian co-ordinates (x, y, z):

$$\frac{\partial H_x}{\partial t} = \frac{1}{\mu}\left(\frac{\partial E_y}{\partial z} - \frac{\partial E_z}{\partial y}\right) - \frac{\rho'}{\mu}H_x \qquad (O.26)$$

$$\frac{\partial H_y}{\partial t} = \frac{1}{\mu}\left(\frac{\partial E_z}{\partial x} - \frac{\partial E_x}{\partial z}\right) - \frac{\rho'}{\mu}H_y \qquad (O.27)$$

$$\frac{\partial H_z}{\partial t} = \frac{1}{\mu}\left(\frac{\partial E_x}{\partial y} - \frac{\partial E_y}{\partial x}\right) - \frac{\rho'}{\mu}H_z \qquad (O.28)$$

$$\frac{\partial E_x}{\partial t} = \frac{1}{\varepsilon}\left(\frac{\partial H_z}{\partial y} - \frac{\partial H_y}{\partial z}\right) - \frac{\rho}{\varepsilon}E_x \qquad (O.29)$$

$$\frac{\partial E_y}{\partial t} = \frac{1}{\varepsilon}\left(\frac{\partial H_x}{\partial z} - \frac{\partial H_z}{\partial x}\right) - \frac{\rho}{\varepsilon}E_y \qquad (O.30)$$

$$\frac{\partial E_z}{\partial t} = \frac{1}{\varepsilon}\left(\frac{\partial H_y}{\partial x} - \frac{\partial H_x}{\partial y}\right) - \frac{\rho}{\varepsilon}E_z \qquad (O.31)$$

The FDTD method is based on the discretisation of the partial differential equation system. However, if the electromagnetic fields and the structure under consideration are assumed to be invariant with respect to a given direction, for instance along the z-direction, the previously described equation system can be decoupled into two systems consisting each of three partial differential equations.

In parallel polarisation, the system is reduced to:

$$\frac{\partial H_x}{\partial t} = -\frac{1}{\mu}\left(\frac{\partial E_z}{\partial y} + \rho' H_x\right) \qquad (O.32)$$

$$\frac{\partial H_y}{\partial t} = \frac{1}{\mu}\left(\frac{\partial E_z}{\partial x} - \rho' H_y\right) \qquad (O.33)$$

$$\frac{\partial E_z}{\partial t} = \frac{1}{\varepsilon}\left(\frac{\partial H_y}{\partial x} - \frac{\partial H_x}{\partial y} - \sigma E_z\right) \qquad (O.34)$$

while in perpendicular polarisation the system becomes:

$$\frac{\partial E_x}{\partial t} = \frac{1}{\varepsilon}\left(\frac{\partial H_z}{\partial y} - \sigma E_x\right) \qquad (O.35)$$

$$\frac{\partial E_y}{\partial t} = -\frac{1}{\varepsilon}\left(\frac{\partial H_z}{\partial x} + \sigma E_y\right) \qquad (O.36)$$

$$\frac{\partial H_z}{\partial t} = \frac{1}{\mu}\left(\frac{\partial E_x}{\partial y} - \frac{\partial E_y}{\partial x} - \rho' H_z\right) \qquad (O.37)$$

The spatial discretisation steps dx and dy are functions of the permittivity of the modelled structural material and on its form. The permittivity of the material is equal to:

- $L/400$ for metallic structures,
- from $\dfrac{L}{64\sqrt{\varepsilon_r}}$ to $\dfrac{L}{32\sqrt{\varepsilon_r}}$ for dielectric structures,
- $L/40$ in free space

where $L = c / f_c$ is the temporal length of the impulse, while f_c is the frequency at which the spectrum of the impulse is maximum. The calculation algorithm is designed to function with an ultra broad impulsive excitation.

O.6 Moment Method

This method consists in the numerical resolution of integro-differential equations expressing the diffracted field in terms of the vector and scalar potentials resulting from the currents induced on the surface of the diffracting structure.

The electric field diffracted by a structure at a point in space denoted by the vector $\vec{r}$ is given by the equation:

$$\vec{E}_{dif}(\vec{r}) = -j\omega\vec{A}(\vec{J}_s,\vec{r}) - \overline{grad}(\Phi(\rho_s\vec{r}))$$ (O.38)

where $\vec{A}$ is the vector potential and Φ is the electric scalar potential.

$\vec{A}$ and Φ can be defined from the current density $\vec{J}_s$ and from the charge density ρ_s on the surface at a point denoted by vector $\vec{r}'$ according to the following equations:

$$\vec{A}(\vec{J}_s,\vec{r}) = \mu_0 \oiint_S \vec{J}_s(\vec{r}').G(\|\vec{r}-\vec{r}'\|)dS$$ (O.39)

$$\Phi(\rho_s,\vec{r}) = \frac{1}{\varepsilon_0} \oiint_S \rho_s(\vec{r}').G(\|\vec{r}-\vec{r}'\|)dS$$ (O.40)

where $G(x) = \dfrac{e^{-jk_0x}}{4\pi x}$ is the free space Green function.

The induced currents and charges are connected to one another through the following continuity equation:

$$\rho_s(\vec{r}') = -\frac{1}{j\omega}div\vec{J}_s(\vec{r}')$$ (O.41)

The boundary condition on the surface S with local normal $\vec{n}_s$ connecting the incident and diffracted electric fields can then be expressed in the form:

$$\vec{n}_s \wedge (\vec{E}_{inc}(\vec{r}) + \vec{E}_{dif}(\vec{r})) = \vec{0}$$ (O.42)

From the different equations described above, the integral equation of the electric field at the surface S can be rewritten as follows:

$$[\vec{E}_{inc}(\vec{r})] = [\vec{E}_{dif}(\vec{r})] = j\omega\mu_0 \oiint_S \vec{J}_s(\vec{r}').G(\|\vec{r}-\vec{r}'\|)dS - \frac{1}{j\omega\varepsilon_0}\overline{grad}\left(\oiint_S div\vec{J}_s(\vec{r}').G(\|\vec{r}-\vec{r}'\|)dS\right)$$ (O.43)

The current density on the diffracting surface S can be determined from the incident field and from this integro-differential equation. The determination of the current density allows then the calculation of the field radiated in vacuum by this surface. This equation can be solved using the moment method through the description of the surface S in terms of elementary triangular shaped surface elements. This decomposition is generally performed using a surface mesher which provides a geometrical description of any surface. The currents are then defined for each such element and approximated at their edges. The diffracted field can then be calculated by integration of the currents induced on each element of the surface S according to the method outlined above.

Software tools, known as SWJ3D and SR3D respectively, have been developed in the group of Professor Mittra and at France Telecom R&D (Veihl 1993; SR3D 1995).

O.7 Parabolic Equation Method

The parabolic equation represents an approximation of the wave equation and allows a modelling of the propagation of a wave inside a cone centred along a privileged direction. This equation was introduced by Leontovich and Fock during the 1940s, with the aim in mind of treating the problem of the diffraction of electromagnetic waves around the Earth (Leontovich 1946; Fock 1965).

In the case of a homogeneous propagation medium with a refractive index N, the components of an electromagnetic wave satisfy the following scalar equation:

$$\frac{\partial^2 \psi}{\partial x^2} + \frac{\partial^2 \psi}{\partial z^2} + k^2 n^2 \psi = 0 \qquad (\text{O.44})$$

where $\psi(x,z) = E_y(x,z)$ in horizontal polarisation, $\psi(x,z) = H_y(x,z)$ in vertical polarisation and k is the wave number.

The introduction of a reduced function $u(x,z) = e^{-ikx} \psi(x,z)$ leads to the following equations:

$$\frac{\partial^2 u}{\partial x^2} + 2ik \frac{\partial u}{\partial u} + \frac{\partial^2 u}{\partial z^2} + k^2 \left(n^2 - 1\right) u = 0 \qquad (\text{O.45})$$

$$\left(\frac{\partial}{\partial x} + ik(1-Q)\right)\left(\frac{\partial}{\partial x} + ik(1+Q)\right) u = 0 \qquad (\text{O.46})$$

where:

$$Q = \left(\frac{1}{k^2}\frac{\partial^2}{\partial z^2} + n^2(x,z)\right)^{1/2} \qquad (\text{O.47})$$

The system is reduced to the propagation of two waves, a forward propagating wave u_+ and a backward propagating wave u_-. These waves satisfy the following system of equations:

$$u = u_+ + u_-$$

(O.48)

$$\frac{\partial u_+}{\partial x} = -ik\left(1-Q\right)u_+$$

(O.49)

$$\frac{\partial u_-}{\partial x} = -ik\left(1+Q\right)u_-$$

(O.50)

The numerical resolution of this equations system leads to the determination of the electromagnetic field in the *(x,z)* propagation plane. For more detail concerning this method the reader is referred to the book published by M. Levy (Levy 2000).

This method has several different applications for the study of the propagation of ultrasounds in marine environments, of light waves, of seismic waves or of atmospheric waves.

References

Combes PF (1978) Introduction à la théorie géométrique de la diffraction et aux coefficients de diffraction. Revue du CETHEDEC, 55

Fock VA (1965) Electromagnetic diffraction and propagation problems. Pergamon Press

Jaureguy M (1995) Etude de la diffraction par impulsions électromagnétiques très courtes d'objets en espace libre ou enfouis : modélisation numérique et extraction des paramètres caractéristiques. Ph. D. Thesis, Ecole Nationale Supérieure de l'Aéronautique et de l'Espace, spécialité électronique, Toulouse

Kouyoumjian RG, Pathak PH (1974) A uniform geometrical theory of diffraction for an edge in a perfectly conducting surface. Procedings of the IEEE vol 62 11 pp 1448-1561

Leontovich MA, Fock VA (1946) Solution of propagation of electromagnetic waves along the Earth's surface by the method of parabolic equations. J. Phys. USSR vol 10: 13-23

Levy M (2000) Parabolic equation methods for electromagnetic wave propagation. IEE Electromagnetic Waves Series 45, published by The Institution of Electrical Engineers, London

McNamara DA, Pistorius CWI, Malherbe JAG (1990) Introduction to the Uniform Geometrical Theory of Diffraction. Artech House Microwave Library, Boston, London

SR3D (1995) Centre National d'Etudes des Télécommunications, CNET/PAB/RSH/ANT

Taflove A, Umashankar KR (1994) The finite difference time domain method for numerical modeling of electromagnetic wave interaction with arbitrary structure. PIER 2 Finite element and finite difference methods in electromagnetic scattering, Chap. 8

Veihl JC (1993) Antenna analysis in complex environments. Ph. D. Thesis, Illinois University, Urbana-Champaign

Yee KS (1969) Numerical solution of initial boundary value problems involving Maxwell's equations in isotropic media. IEEE Transaction Antennas Propagation AP-14: 302-307